THE CAMBRIDGE ILLUSTRATED HISTORY OF
Astronomy

THE CAMBRIDGE ILLUSTRATED HISTORY OF

Astronomy

Edited By

MICHAEL HOSKIN

CAMBRIDGE
UNIVERSITY PRESS

PUBLISHED BY THE PRESS SYNDICATE OF THE UNIVERSITY OF CAMBRIDGE
The Pitt Building, Trumpington Street, Cambridge, United Kingdom

CAMBRIDGE UNIVERSITY PRESS
The Edinburgh Building, Cambridge CB2 2RU, UK   http://www.cup.cam.ac.uk
40 West 20th Street, New York, NY 10011–4211, USA   http://www.cup.org
10 Stamford Road, Oakleigh, Melbourne 3166, Australia
Ruiz de Alarcón 13, 28014 Madrid, Spain

© Cambridge University Press 1997

First published 1997
Reprinted 2000

Printed in the United Kingdom at the University Press, Cambridge

Typeset in Berkeley 10.25/14 pt

Project management, line text editing, typesetting, page make-up and design
by Stuart McCready, 23 Southdale Road, Oxford OX2 7SE

Picture research: Callie Kendall, Cambridge CB3 7HR
Artwork: European Map Graphics Limited, Finchamstead RG11 4RF

Imagesetting by Hilo Offset, Colchester CO4 4PQ

A catalogue record for this book is available from the British Library

Library of Congress Cataloguing in Publication data
The Cambridge illustrated history of astronomy / edited by Michael Hoskin.
        p.   cm. – (Cambridge illustrated histories)
        Includes bibliographical references and index.
                ISBN 0-521-41158-0 (hc)
                1. Astronomy–History.
        I. Hoskin, Michael A.   II. Series: Cambridge illustrated history.
                QB15.C36   1997
                520′.9–dc20   95-40923   CIP

ISBN 0 521 41158 0 hardback

Page i: Regiomontanus observing
with the astrolabe.

Pages ii–iii: Emission nebulae (clouds
of gas lit up by nearby stars) and dark
nebulae (clouds of dust) in the
constellation of Scorpius.

Page iii: Johannes Hevelius and his
wife Elisabetha observing
with a sextant.

Contents

The Contributors

Michael Hoskin is a Fellow of Churchill College, Cambridge. He recently took early retirement after three decades of teaching history of astronomy to Cambridge undergraduates, part of the time as Head of the Department of History and Philosophy of Science. In 1970 he founded *Journal for the History of Astronomy*, which he has edited ever since. He is the General Editor of the multi-volume *General History of Astronomy*, being published by Cambridge University Press under the auspices of the International Astronomical Union and the International Union for History and Philosophy of Science.

J.A. Bennett is Keeper of the Museum of the History of Science, University of Oxford, and a Fellow of Linacre College. He was previously Curator of the Whipple Museum of the History of Science at Cambridge. His books include *The Mathematical Science of Christopher Wren* and *The Divided Circle*.

Christopher Cullen is Senior Lecturer in the History of Chinese Science and Medicine in the Department of History at the School of Oriental and African Studies, University of London, and Chairman of the Publications Board of the Needham Research Institute, Cambridge. He is the author of *Astronomy and Mathematics in Early China: The Zhou Bi Suan Jing*.

David Dewhirst is a Fellow of Corpus Christi College, Cambridge. He recently retired as Astronomer and Librarian of the Institute of Astronomy, Cambridge University.

Owen Gingerich is Professor of Astronomy and History of Science at the Harvard-Smithsonian Center for Astrophysics in Cambridge, Massachusetts. He is Chairman of the editorial board for Cambridge University Press's *General History of Astronomy*. His interests cover a wide historical span, but he has concentrated particularly on the Copernican revolution. He has published two anthologies of articles: *The Great Copernicus Chase* and *The Eye of Heaven: Ptolemy, Copernicus, Kepler*.

Clive Ruggles is Senior Lecturer in Archaeological Studies at Leicester University, and edits the *Archaeoastronomy* supplement to *Journal for the History of Astronomy*. He has written and edited/co-edited several books, including *Astronomy and Society in Britain during the period 4000–1500 BC*, *Megalithic Astronomy*, *Records in Stone*, *Archaeoastronomy in the 1990s*, *Astronomies and Cultures*, and *Astronomy in Prehistoric Britain and Ireland*.

Preface

University teachers believe that in order to learn a subject, you have first to teach it. The thread running through this book is the history of astronomy as I learned it in three decades of lecturing to Cambridge undergraduates.

All teachers eventually convince themselves that they have seen the wood for the trees. I am no exception, and so I have elected to discuss at length, sometimes at considerable length, those few issues I believe to be of fundamental importance. To make room, questions that other historians might consider important, as well as innumerable lesser topics, are mentioned in passing, if at all.

We concentrate on the development, in the Near East and Europe, of the science of astronomy as the whole world knows it today. Other traditions, such as astronomy in China, and the sophisticated astronomies developed in the New World before the arrival of the *conquistadores*, occupy the attentions of respected historians of astronomy; but here they are described only briefly.

Readers sometimes come to the history of astronomy expecting the discussion to focus on 'who first got it right'. In the present work these expectations will be fulfilled very imperfectly, and this for two reasons.

First, 'getting it right' assumes that science is an onward and uninterrupted accumulation of truth, with theory approximating ever closer to reality. At the factual level, there is something in this. It is difficult to imagine that the claim, dating from Antiquity, that the Earth is roughly spherical will ever be abandoned, or that we shall one day discover that Venus is in fact closer to the Sun than is Mercury. But at a deeper, theoretical level, the development of science is immensely more complex. What has been termed 'normal science' often consists in the gradual clarification and elaboration of what is at first confused, with contributions from many hands. But there are sometimes dramatic and disturbing developments. A century after Isaac Newton's death it was generally believed that he alone of the whole human race had been privileged to announce the fundamental truths of the physical universe – that this announcement had been made once, in 1687, and that the feat could never be repeated. But this complacent view was destroyed by Einstein's root-and-branch reform of the most fundamental Newtonian concepts of space, time, gravitation, and so forth. Yet it would be a poor historian who declared Newton simply to have been 'wrong' and his work therefore unworthy of attention.

Second, today's historians of astronomy see it their duty not to award medals to past astronomers whose opinions coincided with those of their modern counterparts, but to take their readers on an exciting journey. This journey introduces them to lands that are conceptually foreign – to past cultures, that sought as we do to make sense of the heavens, but did so by asking questions often very different from those that we take for granted, and who looked for answers strange to our way of thinking. Historians invite their readers to venture with them into these alien ideas, leaving behind modern assumptions as to the nature and purpose of astronomy, and putting much of our modern knowledge of the heavens onto 'hold'.

For example, Plato's contemporaries observed that the heavens were rotating night after night with constant speed. They saw that there were myriads of 'fixed' stars which, while sharing in this rotation, preserved their positions relative to each other without change; but they also saw, moving among the fixed stars in puzzling fashion, seven 'wanderers' or 'planets': the Sun, the Moon, Mercury, and so forth. If, therefore, we are to understand astronomy in the nineteen centuries between Plato and Copernicus, we must put on one side the modern concept of 'planet', and accept the Sun and the Moon as planets. More important still, we must put on one side what we nowadays think of as the job of astronomers, for we are studying cultures in which their job was to contrive, for each of the seven wanderers, a geometrical model from which accurate tables of its future positions could be calculated.

This means that for nearly two millennia, astronomy was applied geometry. Fortunately, the non-mathematical reader need not despair, for the concepts involved were elementary – little more than circles rotating, singly or in pairs, with constant speed. The models themselves, with their sophisticated combinations of circles, were far from elementary; but we shall content ourselves with offering samples, and even these will be independent of the discussion in the main text.

No one could describe as elementary the mathematical techniques used by the successors of Newton. In the *Principia*, Newton claimed that the law of gravitational attraction was the key to understanding the physical universe. The test of this claim was whether or not the law, when applied to the dauntingly complex solar system, could account for the observed motions of the planets and their moons, and of the comets. For a century and more this question occupied the attentions of a tiny band of mathematicians of outstanding genius, and how to deal with their work is a problem for the historian of astronomy. But while their conclusions were of the keenest interest to astronomers, they were not themselves astronomers but mathematicians working in the service of astronomy, and so we can disregard the details of their calculations with a clear conscience.

These 'celestial mechanicians', like their ancient and medieval precursors, were preoccupied with the solar system. The stars were still little more than an unchanging – and therefore uninteresting – backdrop to the movements of the planets, and there was little to be done about them beyond the cataloguing of their positions and brightnesses. Even as late as 1833, the leading authority on the stars and nebulae, John Herschel, published *A Treatise on Astronomy* in which he dealt with these bodies in a single chapter. With rare exceptions, his contemporaries, professionals and amateurs alike, were preoccupied with just one star – namely the Sun – and its satellites.

But since then the balance has tilted sharply in the opposite direction, and today's historian sees that the pioneering eighteenth- and nineteenth-century investigations into stars, nebulae, and 'the construction of the heavens' were to have a profound influence on future astronomical thinking. This book therefore gives more space to early explorations beyond the confines of the solar system than would have seemed proper to astronomers alive at the time.

One issue recurs throughout our account of astronomy in recent centuries: distances. The observer sees the celestial bodies as spread out on the surface of the heavenly sphere; the evidence, that is, is two-dimensional. To theorize about the three-dimensional universe, observers must investigate the third co-ordinate, that of distance.

The story of this investigation is an exciting one, for the successful measurement of the distances of unimaginably remote objects is one of the astonishing achievements of astronomy – even the nearest stars are so far away that their light takes years to reach us. But this remoteness of celestial bodies brings an unexpected bonus, for we see them, not as they are now, but as they were when their light set out on its journey through space. This enables the astronomer to do the seemingly impossible, and look back in time. The more distant the object, the further back in time its light takes us; and today the distances studied are sometimes so great that the objects involved are cited in evidence, for and against cosmological theories of how the universe appeared in its infancy.

When does history end and science begin? Historians are themselves too close to contemporary astronomy to be able to offer a considered perspective. But although it is too soon to see 'Astronomy Today' with historical eyes, astronomy has clearly been transformed in recent decades, and the changes are too dramatic and too exciting for the historian simply to ignore. We therefore end our historical journey by looking around us, at how things stand today in the quest we share with our ancestors both ancient and modern: to understand the universe in which we find ourselves.

Michael Hoskin
Cambridge

The Editor thanks four colleagues who generously acted as Advisors to *The Cambridge Illustrated History of Astronomy*: J. A. Bennett (Museum of the History of Science, University of Oxford), Owen Gingerich (Harvard-Smithsonian Center for Astrophysics), Simon Mitton (St Edmund's College, Cambridge), and Curtis Wilson (St John's College, Annapolis, Maryland). He is also grateful to David A. King (Johann Wolfgang Goethe University, Frankfurt) and Noel M. Swerdlow (University of Chicago) for their help. For assistance with the illustrations, he thanks Owen Gingerich; the library of the Institute of Astronomy, Cambridge University (including the archives of the Cambridge Observatories); the library of the Royal Astronomical Society (librarian: Mr P. D. Hingley); the Museum of the History of Science, University of Oxford; and the Whipple Museum of the History of Science, Cambridge University.

CHAPTER 1   *Astronomy Before History*

Most historians of astronomy spend their days reading documents and books in libraries and archives. A few devote themselves to the study of the hardware – astrolabes, telescopes, and so forth – to be found in museums and the older observatories. But long before the invention of writing or the construction of observing instruments, the sky was a cultural resource among peoples throughout the world. Seafarers navigated by the stars; agricultural communities used the stars to help determine when to plant their crops; ideological systems linked the celestial bodies to objects, events, and cycles of activity in both the terrestrial and the divine worlds; and we cannot exclude the possibility that some prehistoric and proto-historic peoples possessed a genuinely predictive science of astronomy that might have allowed them, for example, to forecast eclipses.

This *History* will concentrate on the emergence of the science of astronomy as we know it today. The historical record shows this development to have taken place in the Near East and, more particularly, in Europe. We therefore begin by asking if anything is known of how prehistoric Europeans viewed the sky, and whether there is any evidence of predictive astronomy. Because it is all too easy for us to fall into the trap of imposing our Western thought-patterns and preconceptions onto the archaeological remains, we also look, by way of comparison, at members of two other groups who viewed or view the sky with minds untouched by Western ideas: the peoples who lived in America before the Spanish conquest, and peoples living today who pursue their traditional ways of life in relative isolation from the rest of mankind.

The celestial phenomena in the two regions most intensively investigated by students of prehistoric and protohistoric astronomy – northwest Europe and the American tropics – are very different. In the tropics the Sun and the other celestial bodies rise and set almost vertically, and for people living there the two times in the year when the Sun passes directly overhead often have special significance. At the higher European latitudes the celestial bodies rise and set along a slanting path and culminate in the south. Around midsummer the days are long, but thereafter the Sun's rising and setting points move steadily further south and the days get shorter and colder: a pattern that threatened disaster, unless the Sun could be persuaded to turn back. Although modern 'Druids' gather at Stonehenge at the midsummer sunrise, the monument's orientation in the opposite direction, towards the midwinter sunset, may well have held powerful symbolism for its builders.

THE SKY AS A CULTURAL RESOURCE IN PREHISTORIC EUROPE

Europeans living today enjoy at best the flimsiest of links with the prehistoric peoples who occupied the region. Some links may nevertheless exist. Legends associated with the huge passage tomb at Newgrange in County Meath, Ireland,

Opposite: The Inuit Moon-spirit. The board around the face of this mask represents air, the hoops the levels of the cosmos, and the feathers stars. Legends of the Arctic Inuit offer an interesting variation on the usual dominance of the Sun in world views, for the Inuit are a people who live so far north that for them the Moon is in many ways more important. In the brief Arctic summer there is little or no darkness, while during the long winter there is little or no daylight. Planets and stars are of minor interest, as they are often invisible because of sunlight or moonlight or bad weather; the Sun is mostly absent; and so it is the Moon, which provides light during the winter months, that is most prominent in their cosmology.

make the all-powerful god Dagda (or his son) dwell in the monument; the Dagda's cauldron was the vault of the sky, and the modern discovery (see page 9) that the winter sunrise penetrates the furthest recesses of the tomb might indicate a long-forgotten expression of the legend in stone. Again, it has been maintained that in Bronze Age Britain a calendar was in use whereby the year was divided into four by the solstices and equinoxes, and each of these four into two and then into two again, giving in all sixteen 'months' of from twenty-two to twenty-four days each; and it may be that vestiges of an eight-fold division of the year survived into Celtic times and hence into the Middle Ages where they were represented by the feasts of Martinmas, Candlemas, May Day, and Lammas in addition to the four Christianized solstices and equinoxes.

But was there in prehistoric Europe a quasiscientific astronomy of precise observation, perhaps even leading to the prediction of astronomical events? In Britain the suggestion that megalithic monuments, now known to have been built in the third and early second millennia BC, incorporated alignments chosen for astronomical reasons goes back to the eighteenth century. And at the beginning of the twentieth century an astronomer of the calibre of Sir Norman Lockyer could write: 'For my own part I consider that the view that our ancient monuments were built to observe and to mark the rising and setting places of the heavenly bodies is now fully established.'

The subject came to popular attention in the 1960s with the publication of a book on Stonehenge in which the author – himself an astronomer – claimed that in addition to the well-known phenomenon of the midsummer Sun rising over the Heel Stone, a great many other astronomical alignments were built into the configuration of the monument. He showed that, given regular observations

Stonehenge: An ancient 'observatory'?

The alignment of Stonehenge on the midsummer sunrise is perhaps the best known manifestation of ancient astronomy in northwestern Europe. Yet despite strong popular belief, it is unlikely that Stonehenge incorporated precise astronomical alignments or functioned as an ancient 'observatory' in any sense that would be meaningful to a modern astronomer. Nonetheless, astronomical symbolism was of great importance at Stonehenge, as at other ritual sites erected in the British Isles during the late Neolithic and Early Bronze Age. The reason had more to do with religion, ideology, and politics than with what we might regard as practical ends, such as telling the time of year for agricultural purposes.

The least contentious statement that can be made about Stonehenge is that the general orientation of the axis of the monument at various stages in its development was towards sunrise at the summer solstice in one direction, and towards sunset at the winter solstice in the other, and that this may well have been deliberate. A precision equivalent to, at best, two or three solar diameters is involved: the popular notion that the Heel Stone defined the direction of solstitial sunrise more precisely is quite unsupportable, because the supposed observing position (the centre of the monument) cannot be defined precisely enough, while the Heel Stone is too near to provide an accurate foresight and the horizon behind it is featureless.

Most students of Stonehenge have identified certain features at the site and tried to invent a theory to 'explain' them. Even when this is done impartially there are grave dangers in imposing astronomical (and geometrical) frameworks onto what is a very limited sample of the features at this much-altered site — those that today are superficially obvious, those that happen to have been excavated (while large areas of the site are still unexplored), and so on. For example, the Heel Stone is now known to have had a companion, long since destroyed, whose existence was discovered during rescue operations in 1979.

Some of the most famous astronomical theories regarding Stonehenge depend upon statistical arguments that the number of astronomical alignments between pairs of points selected are of possible significance. These arguments fall down on many different grounds: lack of prior justification for the points chosen, and archaeological doubts about some of those that were chosen; numerical flaws in the probability calculation; and, perhaps most importantly, the non-independence of data (for example, except in hilly regions, a line that roughly points towards midsummer sunrise in one direction will automatically point towards midwinter sunset in the other). When these errors are taken into account, no evidence whatsoever remains for preferred astronomical orientations of this sort.

It has been proposed that the fifty-six Aubrey Holes (named after their seventeenth-century discoverer, John Aubrey) could have been used as an eclipse predictor, if markers were moved around from hole to hole. The problem here is that while this undoubtedly represents a way in which a modern astronomer could use a structure at Stonehenge to predict eclipses, there is ample archaeological evidence to suggest that the prehistoric users of Stonehenge did no such thing. There are in fact dozens of so-called 'henge' monuments (monuments that resemble the first phases of Stonehenge before it acquired its distinctive structures of Bluestones and Sarsens) where rings of post-holes inside a ditch have been found, and in these the holes vary in number from under twenty to over 100.

extending over many years, it was technically possible to use elements of Stonehenge to keep track of the solar calendar, to study the more complex cycles of the Moon, and even to predict eclipses. And this, the author insisted, had indeed been one of its purposes.

If Stonehenge had been one among many similar monuments, these other monuments could have been examined to see if they displayed the same features. Unfortunately, Stonehenge is without parallel anywhere in the world – it was an object of wonder even in Antiquity. Its explanation is further complicated by the fact that it was constructed, modified, and reconstructed, over a period of some two millennia. Moreover, the stones we see today may not be exactly in the position they occupied when first erected; and when erected, they may not have been exactly in

On the other hand, in the region around Stonehenge there appears to have been a shift from lunar to solar symbolism as development progressed from the Neolithic into the Bronze Age. This is reflected in the directions in which the burial cairns from each period are aligned, and also in the apparent shift in the axis of Stonehenge from lunar alignment in the earlier phases to solar alignment in the later. A group of post-holes situated in the northeastern 'entrance' – a gap in the ditch between the Aubrey circle and the Heel Stone – may represent evidence that the original construction of the axis was oriented on an extreme rising position of the Moon, though this interpretation remains controversial.

In short, there is good reason to think that the construction of Stonehenge and related monuments embodied astronomical symbolism, but we have as yet no convincing evidence that what we might think of as scientific astronomy was practised there.

Stonehenge from the air. The axis of the monument is roughly aligned with the Heel Stone, seen on the lower left edge of the picture. The Aubrey Holes, which a modern astronomer could use for eclipse prediction, are seen in the outer ring of white dots.

the position the builders intended. As we cannot interrogate the builders, and as they left no written records of their intentions, we are forced to fall back on probability: we must ask ourselves how likely it is that an arrangement of the stones, that to our eyes is of astronomical significance, has occurred by design rather than by chance. That is, the study of Stonehenge involves us in statistics – and for statistical investigation a unique monument is unsatisfactory.

While Stonehenge was thus attracting popular attention (and controversy) in the 1960s, Alexander Thom (1894–1985), a retired Oxford professor of engineering, was quietly continuing the mammoth task he had set himself, of surveying to professional standards the many hundreds of stone rings and other megalithic monuments that survive in Britain, Ireland, and northern France. Thom was a collector of facts, and most collectors of facts shy away from speculation. Not so Thom. He maintained, not only that these megalithic monuments were constructed according to complex geometrical designs and laid out using carefully determined units of measurement, but that they were precisely located in order to facilitate astronomical observations of great accuracy.

In 1632, in his *Dialogue on the Two Great World Systems*, Galileo has one of his characters relate how he found himself making an accurate determination of the summer solstice, with an instrument provided by Nature free of charge:

> From a country home of mine near Florence I plainly observed the Sun's arrival at, and departure from, the summer solstice, while one evening at the time of its setting it vanished behind the top of a rock on the mountains of Pietrapana, about 60 miles away, leaving uncovered a small streak of filament of itself towards the north, whose breadth was not the hundredth part of its diameter. And the following evening, at the similar setting, it showed another such part of it, but noticeably smaller, a necessary argument that it had begun to recede from the tropic.

Thom believed that the constructors of the megalithic monuments he was studying had anticipated Galileo by three millennia or more. Some standing stones, he maintained, were astronomical backsights; their locations had been carefully selected so that, for example, the Sun at a solstice, or the Moon at one of its extremes, might be glimpsed setting behind a distant mountain, very much as Galileo describes. Priests with knowledge of the dates of these significant solar and lunar events, he suggested, might even have been able to predict eclipses and thus reinforce their privileged status in the community.

Not surprisingly, Thom became the centre of controversy: such prehistoric sophistication, especially among the inhabitants of regions remote from the supposed cradle of civilization in the eastern Mediterranean, appeared incredible to many archaeologists. To assess the plausibility of Thom's claims it was necessary to decide whether Thom had focused attention on a particular feature of the skyline as seen from the given site because he already knew it lay in a direction

The taula sanctuaries of Menorca

In the Neolithic and Bronze Ages (typically 3500 to 600 BC), many hundreds of tombs and a smaller number of sanctuaries were built in stone by the various cultures and subcultures that occupied the central and western Mediterranean basin. The great majority of the monuments have an entrance and an axis of symmetry, and so have a well-defined orientation – the direction in which they look. Fieldwork shows beyond doubt that builders nearly always observed a local custom when selecting the orientation of a new tomb or sanctuary, although these customs varied from place to place.

It is usually hazardous to speculate about the motivations underlying these customs, which may, or may not, have had something to do with the heavens. But we have a persuasive example of an astronomical custom at work in the *taula* sanctuaries on the Spanish island of Menorca, where a Bronze Age culture was at its height around 1000 BC. Such a sanctuary consisted of a walled precinct in the centre of which was the taula, a flat vertical slab of stone set into the ground, with a horizontal stone on top. The front face of the taula looked out through the entrance, nearly always in a southerly direction. Significantly, taulas were invariably located so that worshippers within had a perfect view of the horizon. Why was this important, when today there is nothing of interest to be seen away to the south?

We can find the likely answer by calculating backwards the effect of the wobble ('precession') of the Earth's axis caused by the pull of Sun and Moon on the non-spherical Earth, which over the centuries alters the stars to be seen from any given location. We discover that in Menorca in 1000 BC the Southern Cross was visible: it rose well to the south, being followed shortly by the bright star Beta Centauri, and then by Alpha Centauri, the second brightest star visible from the island. This prominent star grouping (see pages 18–19) has been of great importance in many cultures. If, as seems probable, it was associated with the rituals in the taula sanctuaries, we learn something of the religion of the prehistoric people of Menorca; and it may well be that they had links with Egypt, where constellations were routinely identified with deities.

The taula of Trepucó, in what is now a suburb of the port of Maó, Menorca. In the foreground is the precinct entrance, through which the taula faces, and to left and right are the remains of the precinct wall and of the vertical stones ('pilasters') that were erected at intervals around the inside of the precinct wall. Behind the taula is a tower or *talayot*; the purpose of these is uncertain, but they may have been status symbols for the villages that erected them.

of astronomical interest. Objectors argued that if the skyline contained numerous mountain peaks, one of which was in the direction of (say) the winter solstice, then the alignment of this particular peak with the solstice may well have been accidental.

Thom's sites have since been re-examined under procedures carefully designed to ensure objectivity. The controversy continues, but the re-examination has greatly reduced the plausibility of his claims to have demonstrated the existence in prehistoric Britain of a science of predictive astronomy. Yet the true value of Thom's work endures: today, prehistorians everywhere work with an increased awareness of the sky as a cultural resource for the people they are investigating.

It is important to avoid a false dichotomy between ritual on the one hand and high-level predictive astronomy on the other. Hesiod's description of an early Greek farmer's use of the reappearance of a constellation at dawn after some weeks of absence lost in the glare of the Sun (the constellation's 'heliacal rising') to tell the season favourable to planting, is an example of a low-level prediction. And since Galilean-type precision observations could have been recorded adequately by backsights consisting simply of poles inserted in the ground, then if stone

The entrance to the massive passage tomb at Newgrange, County Meath, Ireland, and above it the 'roof-box' through which the Sun at its midwinter rising shone down the length of the passage. In the fore-ground is the magnificently carved entrance stone, whose decorations may relate to the solstice. Further decorations hidden from sight on the interior of the roofbox include a circle with short radiating lines, which may have symbolized the rising sun.

Newgrange, diagrammed from above (top) and in cross-section (below), showing the path of the Sun's rays at midwinter sunrise. The tumulus covering the tomb is some 250 feet across and over 30 feet high.

monuments were indeed erected as backsights, they must also have served another and presumably ritualistic purpose.

How does the debate now stand? In northern Europe, the most convincing evidence of a prehistoric concern for astronomy has already been mentioned: it involves the Newgrange passage tomb in Ireland, built around 3000 BC. From an entrance on the southeastern side, a 62-foot passage leads to a central chamber 20 feet high, from which three side chambers open out. Some time after construction, when the bones of many bodies had been placed in the tomb, the entrance was blocked by a large stone. Yet although the living were excluded, the light of the midwinter Sun continued to enter via an otherwise-inexplicable 'roof-box', a slit constructed above the entrance. For some two weeks either side of the winter solstice, the Sun, on rising, shone down the length of the entrance passage and illuminated the central chamber – as it still does. That this should happen by chance, and that the 'roof-box' has some other explanation, is so unlikely that there is little doubt that Newgrange was deliberately constructed to face sunrise at the winter solstice.

But the sunlight was intended to fall upon the bones of the dead, not to be seen by the living: even a living occupant of the central chamber would have learned only a very approximate date for the solstice. Here we clearly have no predictive science of astronomy such as Thom claimed to be the *raison d'être* of other sites.

A particularly interesting example of Thom's sites is Ballochroy in the Kintyre peninsula in Scotland. Here there is a row of three standing stones, two of which are thin slabs oriented across the alignment of the row. A few yards away is a rectangular burial cist; this is aligned with the stones, and its longer sides are oriented in the same direction.

Around the solstices – midwinter's or midsummer's day – the Sun's rising and settings positions are changing almost imperceptibly: thus in the week before or the week after a solstice, the Sun's rising and setting positions at this latitude alter by only one-third of its diameter. This makes determination of the actual solstices difficult, and the solstices are basic to a knowledge of the annual cycle of the Sun. Thom, however, believed that at Ballochroy the prehistoric erectors of the stones had overcome this problem by the location they had contrived for the stones – one from which the Sun was to be seen at the winter solstice setting behind Cara Island which is on the horizon 7 miles away, and at the summer behind Corra Bheinn, a mountain more than 19 miles distant. Even though the Sun is then altering its

The three standing stones at Ballochroy in the Kintyre peninsula, Scotland. The stone at the centre is a slab seen edge-on, believed by Alexander Thom to 'indicate' the distant mountain, Corra Bheinn, where the Sun was to be seen setting at the summer solstice. However, the stone to the right is a slab that has no significant 'indication', while that to the left is not a slab. In the direction to the left, the line of stones pointed to midwinter sunset, but the view was interrupted by the cairn covering a nearby burial cist.

setting position from one night to the next by only a few arc minutes (an arc minute is one-sixtieth of a degree), this change becomes apparent to the observer within a very few days of the actual solstice, because of the sensitivity of the vast measuring instrument that Nature has provided. According to Thom, the direction of midwinter sunset was indicated by the alignment of the stones, and that of midsummer sunset by the flat faces of the central stone.

One problem with testing such a theory arises from our ignorance of when, to within several centuries, the stones were erected. Although the northern and southern limits of the Sun's annual path alter only slightly from one millennium to the next, this is enough to make an important difference when we are observing with instruments tens of miles in length. At a site with distant mountains in roughly the right direction, it may well be possible to find a date for the site when it would

The 150-foot image of a spider seen in this aerial view is one of the famous stylistic depictions of animals and birds that occupy a small corner of the Nazca pampa. Much more extensive, however, are trapezoids, rectangles, spirals, and long straight lines, both narrow and wide, some of which can also be seen in the picture. The various features, some obliterating earlier ones, may well span several centuries. Some lines are astronomically aligned, but astronomy is only a small part of the symbolism that brought ritualistic order to an other-wise empty and infertile part of the landscape.

have had the exceptional characteristics that Thom's theory requires. As to the 'indications' supposedly built into the stones themselves, these are of the kind that tend to be identified by the investigator after he has already convinced himself of the astronomical purpose of the site. It is then that he is likely (in this example) to focus attention on the middle slab (which points roughly in the 'right' direction) rather than on the northernmost (which does not), and to specify the 'intended' alignment of the stones themselves, to a precision quite unjustified for a despoiled (and originally longer) row of three closely-placed, large and irregular stones, two of which are slabs set across the axis. At Ballochroy there is the additional difficulty that the nearby burial cist would have been covered by a cairn in prehistoric times, and this cairn would have obscured the view towards the midwinter sunset; indeed, the cairn is still to be seen in a seventeenth-century sketch of the site.

All in all, then, while there is no doubt that what we may term the Galilean method was feasible in prehistoric Europe (as elsewhere), the claims of this Scottish engineer to have discovered a prehistoric science of predictive astronomy at present merit the peculiarly Scottish verdict of *not proven*.

EARLY ASTRONOMY IN THE AMERICAS

The student of prehistoric Europe has virtually no written or oral evidence to guide him, and the monuments he studies are usually modest structures. The complex societies that developed in the American tropics have left a much richer legacy. Many of the buildings that have survived are of great sophistication; investigators have the opportunity to question living descendants; and we possess written records of various kinds – stone inscriptions and other meaningful carvings, documents such as the handful of Mayan bark books known as codices, and detailed accounts from the first Spaniards to come into contact with these cultures.

A strange aspect of Inca society that flourished in Peru at the time of the conquest (in the middle decades of the sixteenth century) has been revealed largely through the study of accounts written by Spanish settlers shortly thereafter. This is the system of *ceques*, conceptual straight lines radiating out from the *Coricancha* or Temple of the Sun, the central religious monument in the Inca capital of Cuzco. There were forty-one ceque lines, along which sacred monuments were located and which served to divide society into different groups. Some ceque lines were oriented astronomically, for example on the rising position of the Sun on the day when it could be seen directly overhead (that is, in the zenith), while others were oriented upon sacred mountains on the horizon; still others were related to water flow and irrigation. Thus we see that such systems of radial lines related spatial divisions on the ground to the divisions in society, to geographical features, and to astronomical events. Astronomy was merely one component of a highly complex system covering many different aspects of society.

Such systems were also present, albeit in less complex forms, in other Inca cities. Indeed, the concept of radiality seems have existed in the Andes even earlier, in pre-Inca times, when systems of straight lines, radiating out from features such as hill tops and cairns, were given physical expression on desert pampas. The most famous such pampa is that at Nazca, in the coastal region of Peru, where there are several dozen 'line centres' with radiating lines constructed by brushing aside the thin layer of brown surface stones to reveal the bright yellow sandy soil beneath. Weather conditions in this region are so stable that these lines have survived to the present day.

Many of the radiating lines join one line centre to another; many are perfectly straight and run for several miles. It appears that they were sacred pathways, and that many factors may have influenced their orientation, just as was the case in the later ceques. Astronomy was one such factor; the direction of water flow was another. While the Nazca lines do not, as has been suggested, represent 'the largest astronomy book in the world', there is little doubt that astronomical symbolism, including alignments on the rising and setting Sun on significant days such as those of its passage through the zenith, features in the construction and use of the lines.

The significance of the Sun's zenith passage is easy to understand. At any given place in the tropics, the Sun spends part of the year to the north, and the rest of the year to the south; but at noon on the day when the Sun passes from north to south, or vice versa, the Sun stands directly overhead. Not surprisingly, zenith passages were also a focus of interest in Mesoamerica, many hundreds of miles to the north. The two days when this occurred could be pinpointed in a simple but very spectacular fashion by the use of so-called 'zenith tubes': when the Sun was directly overhead, its light shone down the tube onto the floor below.

An outside view of the 'zenith tube' at Xochicalco, Mexico. In a simple but dramatic fashion, the tube pinpointed for the ancient Mesoamericans the two days when the Sun passed directly overhead: at the moment of the zenith passage, light shone down the tube onto the floor of the chamber below.

Opposite: Two pages of the solar eclipse table in the Mayan codex now in Dresden. Dots represent the number 1 and bars the number 5, and the places, which are to be read vertically downwards, are in powers of 20 (where we count in powers of 10). For example, at the foot of the upper half of the right-hand page, on the left and right of the enthroned figure, five of the six vertical pairs represent $20 \times (5+3) + (5+5+5+2)$, that is, 177. The vertical pair immediately to the left of the enthroned figure represents $20 \times (5+2) + (5+3)$, that is, 148.

One of the most extraordinary civilizations known to history was the classic Mayan, which flourished in parts of what are now southern Mexico, Guatemala, and Belize. The Maya wrote in hieroglyphics, and although nearly all of their bark books were destroyed in the mid-sixteenth century by the invading Spaniards, a handful survived, including two that appear to be detailed astronomical (or rather, astrological) almanacs.

The Maya were obsessed with the passage of time, records of which were inscribed on every form of structure. They used three separate counts of days. The first was a year of 365 days, formed of eighteen months of twenty days each plus an (unlucky) additional period of five days. The second was a year of 360 days or *tun*, used in the calculation of very long periods of time. The third and most significant was a sacred almanac of 260 days. Each day of the almanac had a compound name, formed of one of the numbers one to thirteen and one of twenty names.

The day 1 Ahau belonged to Venus, and it was on this day that the cycle of revolutions of Venus had to begin and end. As viewed from Earth, Venus appears to have a cycle (its 'synodical revolution') of fractionally under 584 days, and sixty-five times 584 days equals 146 sacred almanacs of 260 days each. Accordingly, the Venus table in the bark book known as the Dresden Codex covers sixty-five synodical revolutions. However, since the period of 584 days is roughly two hours too long, an adjustment was needed, and this had to be one that somehow preserved 1 Ahau as the beginning date of the Venus cycle. After the sixty-first revolution, the error was between four and five days, the revolution ending on the day 5 Kan which, as luck would have it, was four days after 1 Ahau. The Maya therefore took the opportunity to subtract four days at this point, so that next cycle began again at 1 Ahau. Even this correction left a residual error, and the codex reveals how further corrections of a similar form were made when the opportunity offered. The final sequence was accurate to some two hours in 481 years.

The obsession with Venus was motivated by astrology: its dawn reappearance after the period of invisibility while it passed between the Earth and the Sun was a time of great peril, and the tables would give forewarning of this and so allow ceremonies to take place that might succeed in warding off the threatened evil. There is no evidence that the Maya took the least interest in other Venus events that, to our geometrical outlook, are of equal if not greater importance.

Immediately following the Venus table in the Dresden Codex is a table that occupies eight pages. It covers some 11,960 days, which equals forty-six sacred almanacs of 260 days. Early in the present century it was noticed that the numbers of days separating the pictures on the pages are familiar to astronomers as intervals between solar eclipses, and it emerged that one function of the table was to give forewarning of these perilous events. The Maya did not possess the knowledge needed to determine whether a solar eclipse would actually be visible from their territory, but no doubt those that were not seen had been averted by the ceremonies prompted by the table.

The table consists principally of totals of 177 (or 178) days, which is the period occupied by six lunar-phase months, with occasional 148-day (five-month) periods. Eclipses take place only when Earth, Sun, and Moon are (roughly) in line; seen from Earth, the paths of the Sun and Moon cross every 173.31 days, and a solar eclipse can occur only within a few days of this event. Mayan records must have shown that eclipses took place only during 'danger periods' that occurred every six months (177 days), but the four-day error required the occasional substitution of

The so-called Governor's Palace at Uxmal, Yucatan, Mexico. The alignment from the palace to a huge pyramid to the southeast points towards the southernmost rising point of Venus, while the alignment in the opposite direction points towards the planet's northernmost setting point. There are several hundred Venus glyphs on the palace façade.

an interval of five months. The table makes 405 months equal to 11,958 days, which would imply a length of the month only eight minutes short of the modern value.

The table also provides the required calendar pattern involving the 260-day almanac, which in this case had to commence with the day 12 Lamat. The pattern depends upon a remarkable coincidence: three of these natural intervals of 173.31 days equal two of the Mayan 260-day almanacs to within a couple of hours. Occasional adjustments were made for this error, in a way similar to the treatment of Venus intervals.

It must be emphasized that we have been able to give no more than a taste of the tortuous complexities of Mayan calendrics. This unparalleled obsession with the interrelationship of time-intervals, some of them man-made and others supplied by Nature, would be well-nigh inconceivable if we did not possess the written record.

But given this obsession, it would not be surprising to find that their buildings incorporated astronomical alignments. The investigator, however, once again faces the problem that each of the complex structures is unique in form, and it is difficult in any given instance to prove that the alignment is intentional rather than accidental. A good example is the Governor's Palace at the great Mayan city of Uxmal in Yucatan (Mexico). This vast building has a different orientation from other buildings on the site, and faces towards what seemed a bump on the horizon but proved to be a huge pyramid some 3 miles to the southeast. Measurements showed that the alignment pointed to the southernmost rising point of Venus, and the suggestion that this was the motivation for the orientation of the Palace found support in the Venus glyphs that are carved on the building. Many similar astronomical alignments have been proposed, and – given what we know of the Maya mentality – it would be surprising if all of them have come about by chance.

THE SKY AS A CULTURAL RESOURCE TODAY

The Maya are an example of a highly sophisticated civilization whose interests in astronomy were alien to our own: we shall go astray if we impose our own thought patterns on those outside the European tradition.

The same lesson is taught us with equal force by living societies that are close to nature. In Africa, an example of a people whose calendric preoccupations are very different from ours is provided by the present-day Mursi of southwest Ethiopia. They depend for their subsistence on rain cultivation, flood cultivation, and herding, and the timing of their annual migrations from one region to another is crucial. Yet they have no 'scientific' calendar such as we might expect. Their year consists of thirteen months, and so is eighteen days longer than the solar year. But their calendar keeps in step with the seasonal year, not by the occasional omission of a month, but by a process of institutionalized disagreement with continuous adjustment. The balance between divergent opinions as to what stage of the year has been reached is influenced by discussion of seasonal markers such as the appearance of birds, the flowering of plants, and horizon observations of the Sun, *all* of which are seen as inexact. However, one crucial event – the annual flooding of the River Omo – is monitored outside the calendar by the heliacal setting (the last sunset appearance for some weeks) of four stars in Centaurus and the Southern Cross. Thus, to our eyes it would seem that in this one instance where precision in timing is vital, the Mursi do in fact use a precise rather than a haphazard method of determining the time of year. But the Mursi do not see it in this way. To them, there is a direct association between these stars and the River Omo and certain flowers and plants: the successive disappearance of the stars in the morning sky is correlated with terrestrial events such as the flooding of the river and the flowering of the plants.

Such direct associations between the celestial and the terrestrial are common among native societies. The Barasana people of the Columbian Amazon, for example, perceive a 'Caterpillar Jaguar' constellation which is the father of caterpillars on Earth: as the constellation rises higher and higher in the sky at dusk, so terrestrial caterpillars increase in numbers. To us this is the result of the coincidence whereby the constellation happens to be in the eastern sky at dusk at

The Southern Cross (in the right of the photograph) standing almost upright; when this is the case, its direction is due south. Following the Cross, to the left of the picture, are the bright stars Beta and (far left) Alpha in the constellation of the Centaur. In prehistoric times, this star group was visible from the Mediterranean, and may have been an object of cult in the taula sanctuaries of Menorca (see page 7). Today the stars, which are represented on the flags of Australia and New Zealand, are considered significant in many parts of the globe. For example, the Mursi of southern Ethiopia use them to time a critical event in their calendar (see page 17). The 'Coal Sack', a dark patch clearly visible close to the Southern Cross, forms the head of an emu in Aboriginal folklore, and of a llama for the Andean village of Misminay.

the time of year when the caterpillars pupate and come down from the trees on which they feed; but to the Barasana this is cause and effect.

To the people of the remote Andean village of Misminay, the association between earth and sky is stronger still. The Milky Way is regarded as a celestial river which

Navigating by the stars

Navigation out of sight of land must often have been based on observation of the stars, but such techniques leave few clues for the archaeologist. In some regions, however, remnants of such skills survive to the present day.

Of all the world's peoples, few have been more dependent on such navigational skills than the inhabitants of Polynesia and Micronesia in the Pacific. The navigators were specialists, the possessors of a secret lore. Daytime navigation was by the Sun, while at night the succession of stars that rose or set on the horizon in the same direction made up a star path. Navigators of the Caroline Islands refined star paths by use of what was in effect a star compass, with the Pole Star in the north, the upright Southern Cross in the south, and in between them, fifteen rising points and fifteen setting points of stars or star groups. They used the thirty-two directions they identified at these points much as we use our compass points south, south by southeast, southeast, and so on. The five most southerly directions in the Caroline Islanders' system – all indicated by the Southern Cross and its follower Alpha Centauri as they climb and descend their star path – are shown below.

Five directions to steer by at night

A Hopi horizon calendar, sketched by the anthropologist Alexander M. Stephen, who lived among the Hopi in the 1890s. It shows a Sun priest's observations for the purpose of timing midwinter ceremonies. The Sun's setting in the notch that it reached on 10 December (between peaks of the San Francisco range near Flagstaff, Arizona) was the signal to begin in four days the nine-day celebration of the winter solstice. The solstice occurs on or about 22 December.

is a reflection of the terrestrial Vilcanota river, and the two are perceived as parts of an integrated system that serves to circulate water through the terrestrial and celestial spheres. The Milky Way is directly overhead the village twice in every twenty-four hours, and it chances that the directions it then makes are at right angles to each other. This results in a conceptual quartering of the sky, which is reflected on the ground in the very layout of the village itself: radiating out from the central building (now a Catholic church) are four paths, together with irrigation channels, that divide the village into four quarters. For the villagers of Misminay, observations of the various celestial bodies are an integral part of their agricultural and pastoral activities and festivals. Some aspects of the system of practice and beliefs found here and in neighbouring villages can be traced back at least as far as Inca times.

Past traditions of the Native Americans can sometimes be recovered, for their descendants are with us and can be questioned, and aspects of their practices still survive. A particular study has been made of the Hopi of Arizona. Their cardinal directions are not our north/south/east/west, but the directions to the points on the skyline where the Sun rises and sets at the solstices. The beginning of their winter solstice ceremony (Soyal) is decided by the Sun Chief and the Soyal Chief, who together observe the Sun from their village as it sets in a distant notch in the San Francisco peaks 80 miles away. Soyal lasts for nine days, and begins four days after the chiefs have made satisfactory observations. Calculations show that the Sun sets in the notch between one and two weeks before the solstice, so that the solstice occurs after the ceremonies are well under way. It is interesting that even when the notch in the skyline is so very distant, it has proved desirable to observe the Sun some time prior to the actual solstice, when it is still moving perceptibly from one day to the next.

The remaining chapters of this book will be devoted to the developments in the Near East and Europe that led to the science of astronomy as the whole world knows it today. They will rely mainly on written evidence, supplemented by what can be learned from the astronomical instruments that are preserved in museums and observatories.

Written evidence survives in quantity only from the last centuries before Christ. In the present chapter we have seen something of cultures that antedated those of Babylon, Egypt, and Greece: cultures that flourished in Europe in the second, third, and even fourth millennia BC. Our attempt to infer what was in the minds of such prehistoric builders has been based mainly on what is left of the stones they used in their monuments. By looking at modern peoples whose cultures have been little affected by Western ideas – and by seeing something of American cultures that preceded the Spanish conquest – it has become evident that our preoccupations in studying the sky are by no means the only ones, and that our attempts to interpret these silent stones are fraught with danger.

As we now turn to the historical records, and read what our predecessors actually wrote, we are on safer ground. But the temptation to impose onto these writings our attitudes, our interests, and our factual knowledge of astronomy, is all the more insidious. It must be remembered that history of astronomy is a journey back in time to cultures alien to our modern thinking, and that, like good anthropologists, we must try to see the world through the eyes and minds of those cultures. What gives the history of astronomy its special interest, is the fact that its object of investigation – the sky that prehistoric, ancient, and medieval cultures sought to understand – is the same sky that modern astronomers explore.

CHAPTER 2

Astronomy in Antiquity

From late Antiquity to the seventeenth century, astronomy had two related goals: to show that the movements of the planets were not haphazard but regular and therefore predictable, and to predict them with accuracy. All else that concerned astronomers was peripheral.

The first of the two goals was Greek, a product of the intense philosophical activity in Athens in the fourth century BC. The commitment to accuracy in prediction came from the quite different Babylonian tradition, which merged with Greek astronomy in the aftermath of the conquests of Alexander the Great in the third quarter of the same century.

THE SOURCES

The written materials, Babylonian and Greek, that have survived for the historian of astronomy to study are very limited. From Babylon we have clay tablets – mostly smaller than a man's hand – that have been recovered by excavation, legal or otherwise. The great majority are from the last seven centuries before Christ. Indeed, most were written in the 'Hellenistic era', after the death of Alexander in 323 BC, and in some cases even after the death in about 120 BC of Hipparchus, arguably the greatest of the Greek astronomers.

These Babylonian tablets, with their columns of numbers, have survived for over 2,000 years. The Greeks wrote on less durable materials, and most of their scientific writings have succumbed to the ravages of time and the pressure of economic forces. Some vanished because no copy was lucky enough to outlast the turmoil of later ages; but others were simply superseded, and thereafter the cost of copying them was no longer justified. Euclid's famous *Elements*, for example, incorporated – and therefore superseded – many of the mathematical works of his predecessors; these no longer survive, and the historian of mathematics must reconstruct pre-Euclidean history, by dismembering the *Elements* and relocating each section in its original context.

The situation is especially difficult for the historian of Greek astronomy, because the great work of synthesis – Ptolemy's *Almagest* – was composed late in Antiquity, in the second century *after* Christ, and half a millennium after Aristotle was teaching in Athens. Ptolemy's monumental achievement would have been awesome even if all the earlier writings on which he drew – most notably those of

A Mesopotamian boundary stone from about 1100 BC, now in the British Museum. In the centre are a scorpion and a lion, both still among the zodiacal constellations, and at the top are representations of Venus, the Moon, and the Sun.

102 485

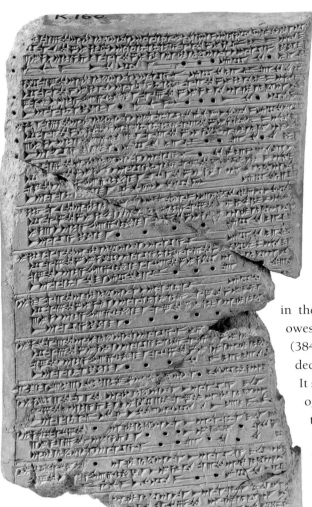

Hipparchus – had survived. But after the *Almagest* became available, Hipparchus's works lost their value, and most of them vanished. In their absence, the *Almagest* was without peer, a daunting legacy to the Arabic and Latin civilizations of the Middle Ages, whose astronomers were in awe of their towering predecessor.

When we come to study the earliest Greek attempts to make sense of the universe, we find that our sources are fragmentary in the extreme. What little we have owes much to the custom of Aristotle (384–322 BC), of citing his predecessors before demolishing them. It seems that alongside the mythologies inherited from earlier times, there gradually emerged a speculative interest in the natural world, which led to efforts to make sense of nature – rather than to attempts to answer quantitative questions, such as led to the columns of numbers on the Babylonian tablets. But what we would recognize as a mature, predictive science of astronomy was to develop only in the Hellenistic era, when these two approaches, the Greek and the Babylonian, merged.

A record of the earliest astronomical observations in Mesopotamia that have come down to us: the appearances and disappearances of Venus in the reign of King Ammisaduqa, of the Hammurabi dynasty. This tablet lists these events according to the lunar calendar then in use. Modern computations of the behaviour of Venus in the second quarter of the second millennium BC have been compared with the data on this and similar tablets in order to determine the years 'before Christ' of Ammisaduqa's reign, and hence of the Hammurabi dynasty.

ASTRONOMY IN BABYLON

The city of Babylon lay on the left bank of the Euphrates river, some 70 miles south of modern Baghdad. During what is known as the Old Babylonian period (possibly 1830–1531 BC), the city was ruled by the dynasty of Hammurabi. Babylon then fell to the Hittites, but was soon incorporated into the Cassite empire, after which followed a long period of Assyrian domination. This ended with the destruction of Niniveh and its great library in 612 BC. After a period of independence, Babylon came under Persian rule; but in 331 BC the city was conquered by the army of Alexander the Great, and thereafter the cultures of Babylon and Greece were in direct contact.

The tablets that survive from the Old Babylonian period are more important for the history of mathematics than for that of astronomy. They do, however, display a technique that was to be crucial for the later development of astronomy: the employment of an efficient numerical notation. To write the number 1, the Babylonian scribe pressed his stylus into the clay edgeways on; to mark a 10, flatways.

The calendars of ancient Egypt

The richness of the mythology with which the ancient Egyptians invested the heavens contrasts with the poverty of their grasp of the movements of the celestial bodies. In part this reflects the limitations of their geometry, and their failure to match the Babylonians in developing a place-value system of numerical notation (that is, one in which the value of a digit depends upon its place in the number). Instead, they had symbols for 1, 10, 100..., and they simply repeated these as often as was necessary. For numbers less than 1 they used fractions that (with the exception of $2/3$) always had unity in the numerator: $1/2$, $1/3$, and so on; other fractions had to be expressed in sums of these (for example, $2/5 = 1/3 + 1/15$), while multiplication was achieved by successive doubling. Clearly, such a primitive arithmetic could permit only the most elementary grasp of celestial movements.

One astronomical question, however, was of such importance to the elaborately structured society of Egypt that it could not be ignored: the calendar. Egyptian life revolved around an annual event, when the river Nile mysteriously rose and covered much of the land in the valley; then, as the waters subsided, planting could be carried out; and finally there followed a period of growth and harvest. From the middle of the third millennium BC, we have records that show that the custom was already established of dividing the year into three seasons — the flooding, the subsidence of the river, and the harvest — each normally of four lunar months. However, since the lunar month lasts about twenty-nine and a half days, twelve such months fell well short of the average interval between one rising of the river and the next. From time to time, therefore, one of the three seasons would have to be given a fifth, additional ('intercalary') month. But how was this to be regulated?

The brightest star in the sky is Sirius. For part of the year the Sun is close to Sirius, and Sirius is then invisible, hidden from sight in the glare of the Sun. Then, one morning, in the eastern sky before dawn, Sirius briefly reappears, in what is termed its 'heliacal rising'. By chance, the heliacal rising of Sirius took place around the time when the waters of the Nile began to flood, and some unknown genius saw that it was

possible to control the calendar year by requiring that Sirius should always rise in the twelfth month, and by devising a rule to ensure this.

The seasonal year exceeds twelve lunar months by some eleven days. This meant that if Sirius rose in the current year in the last eleven days of the twelfth month, then — unless action was taken — a year later Sirius would rise in the first (and therefore wrong) month. To avoid this, whenever Sirius rose late in the twelfth month, an intercalary thirteenth month was announced for that same year.

This calendar was satisfactory for religious festivals. But Egypt had developed into a highly organized society, and to have months that were sometimes of twenty-nine and sometimes of thirty days, and years that were sometimes of twelve and sometimes of thirteen months, must have been very inconvenient. It was realized that the length of the seasonal year was close to 365 days, and another unknown genius proposed a system whereby every year was to contain exactly twelve months, and every month exactly thirty days divided into three 'weeks' of ten days each. At the end of the twelve calendar months there were to be five extra days, to make 365 days in all, a total that never varied.

This administrative calendar was probably introduced soon after 3000 BC, and it existed alongside the earlier, religious calendar. Because the administrative calendar allowed of no exceptions whatever to the number of days in the year, the interval between any two dates in this calendar was easy to calculate; so convenient was this that astronomers used the calendar until early modern times. But the natural year is in fact a few hours more than 365 days (which is why we have our leap years). This meant that before long the new administrative calendar began to get seriously out of step with the seasonal year; but instead of modifying the administrative calendar so that it fitted nature, the bureaucratic Egyptians persisted with it (and even invented a third calendar, a lunar calendar tied to the administrative one). Only as late as 239 BC was an attempt made to introduce a system of leap years so that the administrative calendar would keep in step with the seasonal year.

The astronomical ceiling of a chapel in the temple at Dendera, engraved by scholars of the Napoleonic expedition to Egypt (1798–1801). The zodiacal signs form an inner ring whose centre is somewhat above the midpoint of the ceiling, and the planets are depicted among them, as gods holding staffs. Around the outside (adjacent to the hands of the supporting goddesses) are the thirty-six decans, which may have been deities participating in rituals as well as indicators of the night hours.

Night-time by the stars

Temple records from Egypt show that some festivals were observed at a particular night hour. But how did they measure time at night?

The division of the administrative year into twelve months, each of which was divided into three 'weeks' of ten days, created a total of thirty-six such 'weeks'. The heliacal rising of Sirius (its first reappearance in the dawn sky after a lengthy absence hidden in the glare of the Sun) had long been used to control the religious calendar, and to Sirius were added a further thirty-five star groups or constellations whose heliacal risings were separated by roughly ten-day intervals (for which reason we term these star groups 'decans').

Suppose that around dawn on a given night one such decan rose heliacally. Throughout that same night the immediately preceding decans had been rising at regular intervals, and so these risings had marked the passing of time, rather like the striking of a modern church clock.

There were thirty-six decans in all, but the existence of the Earth's atmosphere leads to the persistence of twilight in the period following sunset, and to the advent of dawn some time before sunrise, and the effect of this is to shorten the period of darkness. As a result, the number of decans that came into play in an typical night was smaller than the theoretical eighteen, and there would be fewer still in summer. It therefore proved convenient to take just twelve decans into the reckoning each night, the winter nights being assigned longer first and last 'hours' to accommodate the greater length of night-time. The Egyptians may have divided daytime into twelve parts by analogy.

Unfortunately, our knowledge of decans comes mainly from a handful of coffin-lids on which they are represented, and one does not expect to find a scientifically-accurate document on a coffin-lid. As a result, for us the identification of most decans is very uncertain.

A Babylonian tablet from about 1700 BC (right) and a diagrammed equivalent. The scribe has pressed his stylus edgeways on to indicate the number 1, and sideways on to indicate 10, and the numbers are written, not in powers of 10 as with us, but in powers of 60. Along the side to the upper left the scribe has written 30, and below the horizontal diagonal he has written 42,25,35, or $42 + 25/60 + 35/60^2$. If the value of this latter number is divided by 30, we obtain $1 + 24/60 + 51/60^2 + 10/60^3$, which is written along the horizontal diagonal. This converts into 1.414213 in our decimal notation, which is the square root of 2 (correct to 0.000001). We see that the scribe knew that the diagonal of a square is the side multiplied by the square root of 2, a special case of what we term Pythagoras's Theorem; and that as early as 1700 BC, scribes possessed the arithmetical techniques to calculate the square root of 2 with great accuracy.

Combinations of these two marks were used to write numbers up to 59. For 60, however, the symbol for 1 was again used, just as we use a '1' symbol in writing 10; and similarly for 60×60, and $60 \times 60 \times 60$, and so forth. Although a zero appears late, and there was no symbol corresponding to our decimal point, these minor limitations did not detract seriously from the great merits of this 'place value' system of writing numbers, which enabled the Babylonians to perform elaborate arithmetical calculations with ease. Our division of the hour into 60 minutes each composed of 60 seconds (each composed of 60 thirds, and so on), and our similar division of the degree, reflect this Babylonian notation.

The early Babylonian skywatchers are often thought of as astrologers, but if astrology is to be understood in the Greek sense, as the study of the direct and unavoidable consequences for individuals that result from the configurations of heavenly bodies, this is a mistake. The Babylonians were on the alert for unusual occurrences in every aspect of nature – their examination of the entrails of sheep was one example of this – and when they found something unusual they interpreted this as an omen; not as the cause of a disaster about to happen, but as a warning sign to the king and his people of a disaster that might yet be avoided by the appropriate ritual.

These interpretations were based on past experience, and this experience was eventually codified in a series of some seventy tablets with 7,000 omens, known from its opening words as *Enuma Anu Enlil*. The collection, which took definitive form before 900 BC, was in effect a list of past signals that had been sent by the gods to the king, indicating pleasure or displeasure. Learned scribes would interpret new signals in the light of similar signals codified in the *Enuma*, and it was hoped that their advice would allow the king to take whatever avoiding action was necessary.

The celestial body most frequently cited

in the *Enuma* was the Moon; and as their calendar was lunar, the cycle of the Moon was doubly important. Calendars present every culture with an awkward challenge because the natural lengths of the day, the month, and the year are independent of each other: no simple multiple of days makes a month and no simple multiple of days or of months makes a year. We see the result of this today in the confusion of month-lengths that we employ, and in our complex system of leap years.

Since the length of the seasonal year was more than twelve lunar months and fewer than thirteen, from time to time the Babylonians had to supplement the normal year of twelve months with a thirteenth 'intercalary' month. For a long time these intercalations followed no rule, but around the fifth century BC the Babylonian astronomers realized that nineteen solar years are very close to 235 lunar months, and since $235 = (19 \times 12) + 7$, thereafter the intercalations followed a regular pattern of seven in every nineteen years.

The recognition of this so-called 'Metonic cycle' was made possible by the development in the eighth century BC of a practice whereby the scribes of the *Enuma*

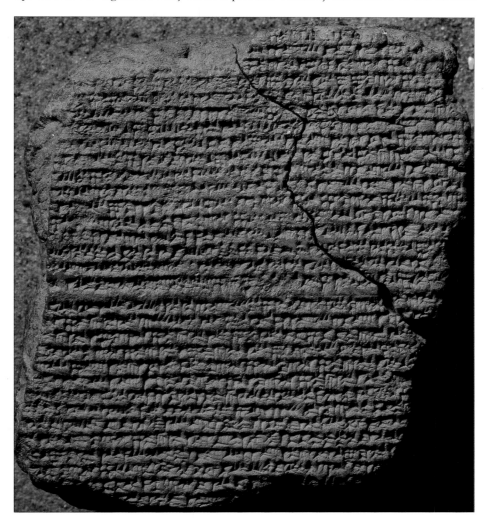

In 164 BC, Halley's Comet made a close approach to Earth, and would have been well placed for observation from September to November. This is one of two Babylonian tablets that record the comet's positions in the month from 20 October to 18 November, when the comet was in the region of the Pleiades and Taurus.

began systematically observing and recording astronomical (and meteorological) phenomena. Their purpose was to improve the efficiency of their prognostications, but the impact of their efforts on the history of astronomy was to be incalculable. For seven centuries the recording went on; and just as the *Enuma* was the expression of ominous experiences that had been repeated so often that they had come to be recognized as regularities in the cosmic order, so cycles – regularities – in the behaviour of the Sun, Moon, and planets were gradually identified and confirmed.

The need of astrologers to have tables ('ephemerides') of the future positions of the Sun, Moon, and planets was, until the seventeenth century, a driving force behind the study of the movements of the heavenly bodies. Equipped with ephemerides, the astrologer could ply his trade in fair weather and foul. The Babylonian scribes learned to do this, by using their sophisticated numerical system to take full advantage of the cycles revealed by their observational records.

For example, the speed of the Sun in its year-long journey among the background stars constantly varies, and the Babylonian techniques for coping with this illustrate the ingenuity of their methods. For half the year the Sun's speed gradually increases, until it reaches a maximum; thereafter, for the next half-year, it slows down until it reaches a minimum.

The Babylonians devised two ways of approximating to this. The first involved supposing that the Sun moved with a *constant* speed for half the year, and for a different constant speed for the other half of the year. The second, and more sophisticated, involved supposing that the Sun increased its speed *uniformly* for half the year, and then decreased it uniformly for the remaining half of the year. Clearly neither could have been intended as other than a highly artificial approximation to reality; but the resultant calculations were straightforward because of their efficient numerical notation, and the results were good enough.

Similar techniques were used to give control of the movements of the Moon and of the five planets, and this allowed the compilation of tables of extraordinary complexity. Some of these have come down to us, in the form of elaborate tablets with columns of numbers, whose underlying constructions historians patiently labour to unravel.

Whether the scribes had in mind any model of the universe, we do not know. What they handed on to astronomers who wrote in Greek were arithmetical

A representation in modern terms of the second Babylonian approximation to the speed of the Sun against the background stars, with the values found in a tablet for 133/132 BC. In each of six successive months the Sun was imagined to increase speed by the same amount, and thereafter its speed was imagined to decrease by the same amount for six months. In the tablet, the column of monthly velocities shows fluctuations between a calculated minimum of 28°10'39"40''' (28 + 10/60 + 39/60² + 40/60³ degrees) and a maximum of 30°1'59", even though these actual numbers are not listed for any given month.

Sun's speed slows to 28°10'39"40''' per month

Sun's speed increases to 30°1'59" per month

Month

12th   1st   2nd   3rd   4th   5th   6th   7th   8th   9th   10th   11th   12th relationships involving time and angular distances. This was just what was needed to turn the Greek speculative cosmologies into geometrical models from which ephemerides of high accuracy could be calculated.

GREEK ASTRONOMY: THE HEAVENLY SPHERES

The questions the Greek astronomers asked themselves, and the methods they used to answer them, were quite different from those we find represented on the Babylonian tablets. In addressing the calendric problem of when new moons would occur, the Babylonians were using arithmetic to investigate special configurations of heavenly bodies – the times when Earth, Moon, and Sun were aligned. This was in sharp contrast to the Greeks, who used geometric models of the movement of planets in an effort to represent their observed positions at all times. Again, Greek philosophers might debate the status of a geometrical model that represented the motions of a planet – for example, whether it was to be considered true/false/possible, or merely accurate/inaccurate. But, as far as we can tell, the Babylonian astronomers would have been judged simply by results – either their predictions were good enough for their purpose, or they were not.

The first cosmologists known to us are from the prosperous Greek colonies in Ionia. Thales of Miletus (c. 625–c. 547 BC) is said to have taught that there was a material unity (which he identified as water) underlying the transient phenomena that our senses register – and, this being so, nature was more intelligible than the superficial appearance of endless variety might have suggested.

Anaximander (c. 610–c. 545 BC), also of Miletus, attempted to explain the form of the heavenly bodies in the context of his vision of worlds constantly coming into being from the Infinite, only to perish and be reabsorbed into the Infinite. If we can rely on accounts written many centuries later, he thought of the stars as wheel-like condensations of air filled with fire, with openings through which flames were discharged. The Sun was the highest (that is, most remote) of the heavenly bodies, with the Moon next below it, then the 'fixed' stars (those unchanging in their positional relations), and finally the planets. The Earth he believed to be a cylinder, on one of whose end surfaces lived mankind; it rested in the middle of the universe, remaining where it was because it had the same distance from everything. The limitations of this cosmology are evident, but a fundamental shift had occurred: earlier mythologies had been replaced by a nature in which an impersonal law was at work.

Not long after Thales had suggested that there was a material unity in nature, the members of a religious sect in the Greek colonies of southern Italy were finding the underlying unity of nature in structure. The sectarians looked on Pythagoras as their founder. Later myths were to obscure his personal achievements; but a significant legend has it that when listening to the sounds made by a blacksmith striking his anvil, Pythagoras was led to a perception of the relationship between arithmetic ratios and harmonic intervals in music – between abstract numbers and

Meridies

b
Eclip citate lune

c
Vmbᵃ terre

a
Oriens

Ro ma

Parisius

Occidēs

¶ a. b. hore eclipsis lune aꝑ ciui
tatē orientaliorē ⁊ sic. 6. hore
noctis. c. b. ore eclipsis lune
ciuitatis occidentalioꝛ ⁊ sūt
4. hore g̅ ꝑus incipit nox
ꝑ una hoꝛa aꝑ oriētalioꝛ
et b' minoꝛ tᵉᵃ e.

oculis
i sūmi
tate mal'.

sol

Ocul' nõ uidēs.

Ocl's uidēs.

ocl's.

Rex uisa

a phenomenon of nature. What is certain is that the Pythagoreans, generalizing on this, perceived number as the basis of all things.

These same Pythagoreans attained a very remarkable insight into the natural world: they recognized that the Earth is a sphere. The arguments they used are not recorded, but the proof given later by Aristotle – that the Earth's shadow cast on the Moon during eclipses is always circular – was cogent. In confirmation he pointed out that one sees different stars as one travels north or south, which showed that the Earth was in fact a sphere of no great size. For the rest of Antiquity and throughout the civilizations of the Middle Ages and the Renaissance, anyone with any pretensions to education knew that the Earth was round.

More generally, the Pythagoreans believed the natural world to be a *cosmos*. This term implied a rational order, but had in addition overtones of symmetry and beauty, and of the harmony that existed in a healthy organism. This intuition that the universe must be harmonious was to be a powerful driving force in astronomy in the Renaissance, when a Pythagorean stance became popular once more following the recovery of the writings of Plato, who had visited southern Italy and been much influenced by the doctrines he found there.

Plato (427–348/7 BC) was the second of the three great philosophers who graced Athens in the late fifth and the fourth centuries BC. His teacher, Socrates, wrote nothing, but lives on as a character in the dialogues of Plato. His pupil, Aristotle, who was a naturalist as well as a philosopher, wrote much, and a vast quantity of his writing survives; by chance, nearly all of this was available to the universities of the later Middle Ages (unlike most of Plato's dialogues), and this encouraged there an attitude to nature that emphasized the naturalist's sense of function and purpose.

Plato, in contrast to Aristotle, focused on the certainty to be found in mathematical reasoning, and to later civilizations he was to represent the mathematical view of nature: the Pythagoreans' insight, though shorn of their extreme mysticism. But Plato and Aristotle agreed that the world was a cosmos, a macrocosm whose parts corresponded to the organs of the individual living body, or microcosm. It was this correspondence that would provide a theoretical basis for astrology and especially for astrological medicine (see page 73).

Earlier philosophers had seen two fundamental and contrasting pairs of qualities in the objects people encountered in the natural world: hot versus cold, and moist versus dry. According to Aristotle, bodies that were cold and dry were mostly made of earth; those that were cold and wet, of water; those hot and wet, of air; and those hot and dry, of fire. The elementary earth was to be found mainly in the (roughly spherical) Earth that was at the centre of the cosmos; around this was a shell of water (the seas), around this a shell of air (the atmosphere), and around this a shell of fire that terminated just this side of the Moon.

Within all this region – which constituted the terrestrial, or sublunary, world – there were life and death, coming to be and passing away. Any given body had a natural place – a natural height, or distance from the centre of the Earth –

Proofs that the Earth is a sphere. Opposite: In a thirteenth-century manuscript, an eclipse of the Moon is seen at different (local) times by observers at different longitudes; and a sailor at the top of a ship's mast may see a signal on shore that is invisible to a sailor on deck. Below: Sixteenth-century woodcuts illustrating Aristotle's argument that the circular shadow invariably seen on the Moon during a lunar eclipse proves that the Earth is a sphere (as in the left-hand drawing). If the Earth were triangular (right), its shadow would have a triangular shape.

depending on the proportion of the elements in its makeup; and unless prevented, the body would move towards this natural place. This was why stones fell down and flames moved up. Such 'natural' motions, we note, took place in a straight line, either towards the centre of the Earth or away from it; and they were transient, ending when the body reached its natural place or got as near to it as circumstances permitted.

Granted that the Earth was a sphere, the Greeks found there to be strong evidence that this spherical Earth was at the centre of a universe bounded by a much greater sphere, the sphere of the 'fixed' stars. For how else would people invariably see half the heavens?

There was, according to Aristotle, a fundamental contrast between the terrestrial and the celestial regions, between the imprecision and impermanence of what people saw about them here on Earth and the geometrical perfection of the eternal heavens with their points and circles of light. In the heavens, there was no life or death, no coming to be or passing away. Instead, the heavenly bodies cycled around eternally; they were all formed of a single fifth element, or 'quintessence', and they expressed the permanence of their nature by the permanence of their uniform circular motions. (Comets, which did indeed come to be and pass away, presented no difficulty: their behaviour showed them to be part of the terrestrial world – indeed they were supposedly caused by exhalations from the Earth, and Aristotle discusses them in his *Meteorology*.)

It is important to recognize that this Aristotelian cosmology drew strength from being an intellectual formulation that reinforced common sense, in contrast to modern science, which contradicts what seems self-evident. Aristotle tells us that the Earth is at rest beneath our feet, that we are on *terra firma*, whereas today's astronomers would have us believe that we are hurtling through space at an almost unimaginable speed. Aristotle confirms that there is indeed a fundamental contrast between Earth and sky, whereas post-Copernican astronomy insists that we must disregard the plain evidence of our senses, and accept the Earth as a heavenly body like Venus and Mars and the rest. The history of cosmology is not the easy story of the rejection of absurd ideas in favour of what (perhaps after a little thought) is seen to be patently true, but the heroic saga of the hard-won rejection of the patently true in favour of the absurd. It is this that gives the history of cosmology its fascination.

We have an example of this in the physical proofs that the Earth was at rest rather than spinning on its axis. These proofs were to prove convincing to nearly everyone who considered the question, from the philosophers of Athens in the fourth century BC to Tycho Brahe in his observatory on the Baltic island of Hven two millennia later, and even today they retain a certain plausibility. Let an archer, for example, fire an arrow vertically into the air; and suppose that while the arrow is in flight, the Earth moves, and therefore the archer with it. If so, then by the time the arrow has climbed vertically into the air, paused, and fallen vertically down again,

Aristotelian teaching illustrated in sixteenth-century woodcuts. Top: Since we see exactly half the heavens, the Earth must be at the centre of the universe. Bottom: The terrestrial or sublunary region of the elements earth, water, air, and fire, is surrounded by the unchanging heavens composed of the fifth element.

Atlas carrying the world, as portrayed in 1559. At the centre are the elements earth, water, air, and fire, after which come the celestial spheres in the order established in the second century AD by Ptolemy: Moon, Mercury, Venus, Sun, Mars, Jupiter, Saturn, and the fixed stars. Outside the fixed stars (the 'firmament') are two additional spheres invented in the Middle Ages for theological reasons (see page 78). Also shown is the band of the zodiac, together with the celestial equator, the tropics of Cancer and Capricorn, and other circles.

the archer will no longer be underneath it. Yet experience shows that in fact it is hazardous for an archer to fire arrows vertically. *Ergo*, the Earth does not move.

But if the stability of the Earth was not in doubt, the status of the heavens as a cosmos where law prevailed was open to question, and would remain so until the movements of the 'wandering' stars, or 'planets', had been shown to be lawlike and regular. With seven exceptions, the innumerable heavenly bodies behaved in rational fashion; maintaining their positions relative to one another, the 'fixed' stars wheeled around the central Earth with perfect mathematical regularity, night after night, year after year. But the seven 'planets' – Sun, Moon, Mercury, Venus, Mars,

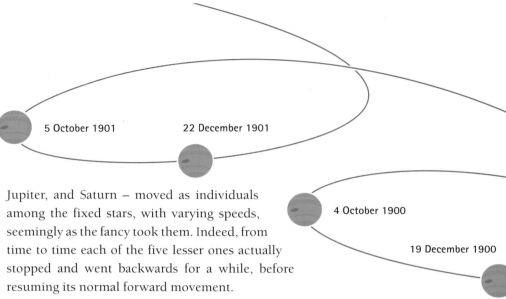

5 October 1901   22 December 1901

4 October 1900

19 December 1900

The path of Neptune as seen from Earth around the beginning of the present century (with vertical scale exaggerated). Because Neptune takes no less than 165 years to orbit the Sun, the Earth overtakes Neptune every one year and two days, and about this time Neptune appears from Earth to be moving backwards ('to retrogress'). Neptune is too faint to be seen with the naked eye and so was unknown in Antiquity, but similar (though much less pronounced) retrogressions in the planets then known earned them the name of 'wandering stars'.

Jupiter, and Saturn – moved as individuals among the fixed stars, with varying speeds, seemingly as the fancy took them. Indeed, from time to time each of the five lesser ones actually stopped and went backwards for a while, before resuming its normal forward movement.

Plato (according to a late commentator) put the challenge to his contemporaries, to show that the planetary motions were in fact as regular as the rest (though not of course as simple). And the terms of an acceptable answer were clear: the stellar motions were uniform and circular, and so the planetary motions must be shown to be similar in nature; that is, to result from combinations of uniform circular motions.

It was Eudoxus of Cnidus (c. 400–c. 347 B.C.), Plato's younger contemporary and one of the finest mathematicians of Antiquity, who first offered a solution to Plato's challenge. The occasional periods of backward ('retrograde') motion indulged in by the five lesser planets were the most extreme examples of non-uniformity, and Eudoxus's masterstroke was to achieve the representation of such an effect simply by means of a pair of concentric spheres. The astronomer was to imagine the planet as located at the equator of a sphere that was spinning uniformly. Projections protruded from its two poles, and these were embedded into a second sphere, outside the first and concentric with it. This outer sphere was also spinning but about a somewhat different axis, and as it spun it carried round with it the inner sphere. In consequence, the movement of the planet reflected the spinning of both spheres. Eudoxus realized that if the two spins were equal in speed but opposite in direction, and if the two axes were not very different, then the planet would

The hippopede produced by a pair of Eudoxan spheres. The planet is located on the equator of the sphere (shown in blue) that has the inclined axis, whose poles are embedded in the sphere (shown in green) with the vertical axis, and are carried round by it. The two spheres rotate with equal and opposite speeds. As a result of the combined motions, the planet moves on a figure-of-eight (which is in fact the intersection of the spheres with a cylinder – yellow – that touches them internally).

Planet

move back and forth in a figure-of-eight – a 'hippopede', or horse-fetter, named from the hobble placed round the front feet of a horse to prevent it straying.

In each of the Eudoxan models of these five lesser planets, the outer of the two spheres was itself carried around by a third sphere located outside it, whose spin was chosen so as to reproduce the average (mean) west-to-east motion of the planet along the ecliptic (the apparent path of the Sun); and that sphere in turn was carried round by a fourth and outermost sphere that generated the east-to-west daily journey of the planet around the Earth. These two outer spheres thereby reproduced the principal movements of the planet, while the two inner spheres served to replicate – qualitatively if not quantitatively – the occasional retrogressions.

For the planet Moon Eudoxus proposed a nest of three spheres. The outermost again generated the daily east-to-west journey of the planet around the Earth. The speeds of

Above: The motion of the planet Moon, according to Eudoxus. The Moon is on the equator of the innermost sphere, which spins once a month. The poles of this sphere are embedded in the middle sphere, which spins once every 18.6 years, a cycle familiar from eclipse records; and this in turn has its poles embedded in the outermost sphere, which spins daily.

spin of the other two were responsible respectively for the lunar month and for the 18.6-year cycle familiar from eclipse records. The various quantities, or 'parameters', could be chosen so as to represent the Moon's motion with some success. The nest of three spheres that Eudoxus proposed for the planet Sun, however, was less satisfactory.

With four spheres for each of the five lesser planets, three each for the planet Sun and the planet Moon, and one for the fixed stars, Eudoxus required a total of twenty-seven spheres in all; eight of the spheres – one from each set – simply supplied the identical daily movement, so that overall the complexity was far from excessive.

We are not told how Eudoxus viewed the status of his spheres, but it is probable that they were a mathematician's solution to the problem of the wandering stars.

They were the equivalent of equations that set out to describe how these bodies moved. There was need for further refinement – Eudoxus's younger contemporary, Callippus, gave the system greater flexibility by increasing the numbers of spheres – but Eudoxus's nests provided a reasonable basis for believing that the world was indeed a cosmos, that law prevailed even among the seven planets. This was after all what Plato, himself a mathematician, had called for.

But Aristotle was a naturalist who lived in the real world, and he insisted on knowing the physical causes of the celestial motions. Fortunately he saw an easy way of achieving this, by adapting Eudoxus's nests of spheres appropriately. He made the spheres physically real, and combined the various distinct nests of three or four spheres into an integrated whole, with the nest for the Moon (the planet believed to be nearest the Earth) at the inside and the nest (or more exactly, the single sphere) for the fixed stars at the outside.

But this was not satisfactory as it stood, for in any such physical system of spheres, the motions of each sphere would be transmitted down through the system and so would affect every one of the spheres within. There was indeed one motion – the daily spin – common to every planet, and so Aristotle was able to eliminate all of the twenty-four-hour spheres except the outermost; this one sphere would be sufficient to impose the appropriate motion on all the seven planets within. But most of the spherical motions applied only to the planets for which they had been designed. Aristotle therefore interpolated appropriate new spheres between any given nest and the next inner nest. Each such additional sphere moved about the same axis and with the same speed as one of the spheres of the given nest, but in the opposite direction; its function was to cancel the motion of the sphere in the given nest, and so prevent a motion that was peculiar to that planet alone from being transmitted down to all the planets within.

These interpolated spheres served a physical purpose only. Although they brought about a nominal increase in the total number of spheres in the system, they did not add to its mathematical complexity. Aristotle's coherent system of concentric spheres, each spinning with a uniform angular speed, was to be a cornerstone of natural philosophy as taught in the later Middle Ages; and the attractive simplicity of his conception of the cosmos would contrast with the daunting complexity of the geometrical models developed by later mathematical astronomers.

These same astronomers, on the other hand, saw grave defects in concentric spheres, whether they were viewed as mathematical conceptions or as physically real. They denied astronomers the flexibility they needed. In particular, nothing that could be done by way of adding spheres to a nest would permit the planet to vary its distance from the centre of the Earth. Yet some of the lesser planets altered in brightness in a manner that strongly suggested that they were varying their distances from Earth, while the Sun and Moon actually altered in apparent size and so demonstrably varied their distances.

HELLENISTIC ASTRONOMY: THE MERGING OF TWO TRADITIONS

Such quantitative considerations became more pressing after the conquests of Alexander the Great resulted in the spread of Greek culture into the Near East. This brought speculative Greek geometric astronomy into contact with the pragmatic and observation-based arithmetical astronomy of the Babylonians. Few astronomical writings survive from the early centuries of this 'Hellenistic era'; but we need no books to tell us why the Eudoxan models, for all their geometrical elegance, would have been unacceptable to astronomers who had learned from the Babylonians the fundamental importance of quantitative observation.

Aristarchus and the motion of the Earth

Aristarchus (c. 310–230 BC) was born on the island of Samos, just off the coast of Asia Minor, but little is known of his career. There had already been a number of speculations that the Earth is in motion. According to one tradition, Philolaus, a Pythagorean living in south Italy in the second half of the fifth century BC, held that the Earth, together with an anti-Earth, the Sun and Moon, and the five lesser planets, orbited around a central fire, 'the hearth of the universe'. Another tradition tells us that Hicetas of Syracuse, a contemporary of Philolaus, believed the Earth was spinning on its axis, and the same teaching is ascribed to Heraclides of Pontus (c. 390–post 339 BC), a pupil of Plato. Aristarchus, however, anticipated Copernicus by teaching that the Earth is in orbit about the Sun. According to Archimedes (c. 287–212 BC), 'His hypotheses are that the fixed stars and the Sun are stationary, that the Earth is borne in a circular orbit about the Sun, which lies in the middle of its orbit...'.

But if Earth-based astronomers observed the stars from a moving platform, why did not the stars appear to move, in reflection of the Earth's motion? Because, Aristarchus suggested (and as Copernicans were later to maintain), the radius of the Earth's orbit was negligible compared to the distances of the stars, whose apparent movements were in consequence too small for astronomers to detect.

We know of only one person who was convinced by Aristarchus's speculation, the Babylonian Seleucus of Seleucia, who lived in the mid-second century BC and who tried to prove the hypothesis true. This lack of support is not surprising: Aristarchus's proposal, however perceptive it can seem to modern eyes, belonged to a tradition of speculative cosmology that was about to give place to an astronomy that was concerned with quantitative observations, and preoccupied with devising geometric models that would replicate these observations.

Part of a page from the manuscript of Nicholas Copernicus's *De revolutionibus* (1543), in which he argued that the Earth is a planet that orbits the Sun. In the draft he wrote: 'Philolaus believed in the mobility of the Earth, and some even say that Aristarchus of Samos was of that opinion', but in the end he decided to omit the whole passage, probably because it formed an awkward digression. As the sentence indicates, Copernicus had no clear information as to Aristarchus's beliefs.

Aristarchus and the distances of the Sun and the Moon

In the only treatise of his that is extant, Aristarchus correctly shows how (in principle, at least) we may calculate the relative distances of the Moon and Sun. When the Moon is half full ('at quadrature'), the angle Earth–Moon–Sun is a right angle. If we then measure the angle Moon–Earth–Sun, we learn the shape of the triangle joining the three bodies, and hence the ratio of any two sides.

In practice, the time of quadrature of the Moon is very hard to determine accurately, and it is also difficult to measure the small difference that then exists between the angle Moon–Earth–Sun and a complete right angle.

MOON ('at quadrature')   SUN

Moon–Earth–Sun angle   Aristarchus takes the angle to be 3° short of a right angle, whereas the true difference is only 1/18th of this. As a result, his conclusion that the Sun is

EARTH

some nineteen times further than the Moon is less than 1/20th of the true value. But his treatise is modelled on works in pure mathematics, and the 3° may have been no more than a convenient value with which to illustrate his method.

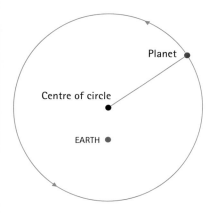

In an eccentric circle, the planet moves as usual with uniform speed around the circle; but since the Earth is not at the centre of the circle, the planet appears from Earth to move with varying speed (slower when at the top of the circle and faster when at the bottom).

On the other hand, belief in uniform circular motions as the key to the secrets of the cosmos remained strong in Greek minds. The way forward was to keep faith with the tradition of circular motions, but to employ them in a more flexible manner, and with parameters taken from Babylonian sources or calculated from Babylonian observations.

The man who developed more flexible forms of circular motions was Apollonius of Perga, famous for his mastery of the geometry of conics. Of his life little is known; he flourished around 200 BC, and lived for a time in Alexandria. In the first of the geometrical devices that he employed in astronomy, the planet moved around the Earth uniformly on a circle; however, the Earth was not at the centre of the circle, but to one side. Accordingly, a planet moving on one of these *eccentric* circles varied its distance from Earth, and therefore its apparent speed across the sky.

In the second device, the planet moved uniformly on a small circle, or *epicycle*, whose centre was carried uniformly on a large 'carrying' circle, or *deferent*, that had the Earth at its centre. If the motion of the planet on the epicycle was sufficiently fast in relation to the motion of the epicycle on the deferent, then the planet would appear from time to time to move backwards. It was not difficult to see that motion on an eccentric circle is mathematically (and observationally) equivalent to a special case of the motion produced by epicycle and deferent; which would one day lead to the philosophical question of how one could decide which of two such models was 'true', if the observed effects were necessarily identical.

With hindsight we can see that the introduction of epicycles and deferents set Greek astronomy onto a most promising path. For since in fact Mercury, Venus, Mars, Jupiter, and Saturn all circle round the Sun (in near-circular ellipses), and the Sun appears to circle round the Earth (also in a near-circular ellipse), a combination of two circles (epicycle and deferent, each doing duty for a near-circular ellipse) could provide an encouraging first approximation to the observed motion of one of these five planets.

Yet these planetary orbits are not circles but ellipses, and so no combinations of epicycles and deferents could ever reproduce the observed movements perfectly. Astronomers – whether in Antiquity, the Middle Ages, or the Renaissance – who limited themselves to circles would never finally close the gap between prediction and reality, try as they might. And so the struggle continued down the ages, until at length in the seventeenth century the unprecedented observational accuracy that had been achieved by Tycho Brahe would force his assistant and successor, Johannes Kepler, to abandon circles and try alternatives.

We know little of the use Apollonius attempted to make of these devices, for his astronomical works are lost. So too are all but one of the writings of Hipparchus, who died some time after 127 BC. Fortunately for the modern historian, Ptolemy made

Epicycle centre

Planet

Epicycle

EARTH

Deferent circle

Backward path of planet as it circles rotating epicycle centre

How an epicycle might explain the apparent backward motion of a wandering star. The Earth is assumed to be at the centre of the large circle or *deferent*, on which travels the centre of the small circle or *epicycle*, which carries the planet. Each circle rotates with uniform speed. The diagram shows how the apparent backward (retrograde) motion of a planet could be produced.

Eratosthenes and the size of the Earth

Eratosthenes (c. 276–c. 195 BC) was born in Cyrene (in what is now Libya), and after studying in Athens spent the rest of his life in Alexandria, where he was in charge of the famous library that was part of the Museum. Outstanding in several branches of mathematics and geography, he was responsible for a remarkable estimate of the circumference of the sphere of the Earth.

At noon on the day of the summer solstice, in the town of Syene in Upper Egypt, the Sun was directly overhead: a vertical pointer cast no shadow, and the Sun's rays reached to the bottom of a well that had been dug for the purpose. Eratosthenes, who was then in Alexandria, measured the

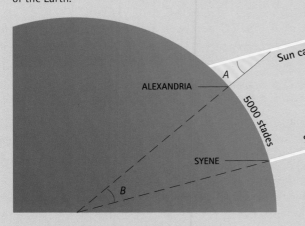

Sun casts shadow with angle A at Alexandria, 5000 stades north of Syene

Sun directly overhead at Syene casts no shadow

ALEXANDRIA

5000 stades

SYENE

A

B

shadow cast at noon by a vertical pointer, and found it to be 1/50th of a circle. Believing Alexandria to be some 5,000 stades due north of Syene, Eratosthenes concluded that the circumference of the Earth was fifty times 5,000 stades, or 250,000 stades.

Although the data used by Eratosthenes are not quite accurate, and although the modern equivalent of 250,000 stades is a matter of debate – some put it at 29,000 miles – there is no doubt that the fit between Eratosthenes's value for the circumference of the Earth and our modern value of nearly 25,000 miles is good, and possibly excellent.

Eratosthenes found by measurement that the angle *A* was 1/50th of a circle and, since *A* and *B* were equal, he concluded that the circumference of the Earth was fifty times 5,000 stades, or 250,000 stades.

much use of Hipparchus's work and was happy to acknowledge his debt to his great predecessor. From Ptolemy's great synthesis, the *Almagest*, we can reconstruct both Hipparchus's model for the motion of the Sun (which Ptolemy was happy to take over largely unchanged), and his much more complex model for the Moon (which Ptolemy improved and elaborated). The lesser planets Hipparchus left to posterity.

It is in Hipparchus's model for the motion of the Moon that we begin to experience the extraordinary transformation that was taking place in Greek geometrical astronomy, through its assimilation of the centuries-old Babylonian commitment to the facts of observation. In the model, the Moon moved on an epicycle whose centre moved around the Earth on a deferent circle. The parameters Hipparchus calculated from eclipse records, and from relationships that we find in Babylonian astronomical tablets.

Some numbers will illustrate the formidable technique at his command. He made use of the facts that in 126,007 days plus one hour, there occurred 4,267 synodic months (the time between successive conjunctions of the Sun and the Moon), 4,573 returns of the Moon to the same velocity, and 4,612 returns of the Moon to the same point on the ecliptic (less $7\frac{1}{2}°$). It is hard to believe that little more than 200 years had passed since the speculative cosmologies of Eudoxus and his contemporaries.

Hipparchus

Hipparchus was born in Nicaea in north-western Asia Minor, and Ptolemy cites observations he made at Rhodes between 141 and 127 BC. Aside from this, we know almost nothing of his life, and with one minor exception his writings are lost. Most of what we know of his astronomy we owe to the many references Ptolemy makes to him in the *Almagest*.

Hipparchus was a systematic observer, and he compiled a catalogue of the stars, apparently because he suspected that one of the stars may have moved, and he wished to bequeath to his successors data against which any future suspected movements might be tested. He also provides us with the first demonstration of how immensely fruitful was to be the union of the contrasting Babylonian and Greek traditions in astronomy. He used astronomical records and parameters derived from Babylonian sources to develop quantitative geometrical models of the motions of the Sun and the Moon.

Hipparchus's single most important discovery was that of the precession of the equinoxes, the slow movement from east to west among the stars of the equinoctial points (the places where the celestial equator is crossed by the path of the Sun or 'ecliptic'). Precession in fact results from a wobble of the Earth's axis. The spring equinoctial point is used by astronomers in defining the celestial frame of reference, and the discovery of precession showed that this point in fact is steadily moving, so that the measured position of a star varies with the date of the measurement. According to Ptolemy, Hipparchus thought precession amounted to 1° per century (though in fact it is nearer to 1° in seventy years).

Hipparchus's model of the motion of the Sun

Hipparchus's elegant and simple model of the motion of the Sun worked well enough for Ptolemy to retain it unaltered, and so it has been handed down to us. In the model, the Sun orbited the Earth on a circle, moving about the centre of the

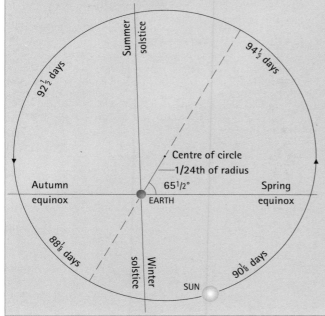

circle with a uniform angular speed that took it round in 365¼ days. But Hipparchus knew that the seasons are of different lengths: in particular, as seen from Earth, the Sun moved the 90° from spring equinox to summer solstice in 94½ days, and the 90° from summer solstice to autumn equinox in 92½ days. Since both these intervals were longer than one-quarter of 365¼ days, the Sun appeared to move across the sky with a speed that was not uniform but varied.

The Earth therefore could not be located at the centre of the circle, but must be 'eccentric'. By how far, and in which direction? To generate these longer intervals, the Earth had to be displaced from the centre of the circle in the opposite direction, so that the corresponding 90° arcs seen from Earth were each more than one-quarter of the circle, and it therefore took the Sun intervals of more than one-quarter of 365¼ days to traverse them. Hipparchus's calculations showed that the eccentricity needed to be 1/24th of the radius of the circle, and that the line from Earth to centre had to make an angle of 65½° with the spring equinox.

Granted these parameters, the model was completely defined. Luckily it reproduced the remaining two seasons well enough, and so this single eccentric circle was sufficient to reproduce the motion of the Sun.

The three centuries that separate Hipparchus and Ptolemy form a dark ages of astronomy, a period of which we know little. We would know even less were it not for the limited aims and ultra-conservative nature of such astronomy as was practised in India, for writings have been handed down in Sanskrit that embody elements from the Greek works of this period from which they derive. But with Ptolemy, the historian is on secure ground at last, for his works were cherished by later generations and much of his voluminous output has survived.

Ptolemy wrote a major work on astrology, the *Tetrabiblos*, and his studies in astronomy were motivated in part by his need, which he shared with fellow astrologers, for tables of planetary positions that were accurate and yet could be computed without excessive labour. The techniques of eccentre, epicycle, and deferent he had inherited from Apollonius and Hipparchus. But he found that to compute planetary positions both accurately and conveniently, he needed to resort to another device, the *punctum aequans,* or equant point.

Suppose the Earth is located at a distance from the centre of a given circle. Ptolemy defined the equant point as the mirror image of the Earth, on the opposite side of the centre and at an equal distance from it. This point he then used to define motion on the circumference of the circle. A point on the circle was required to

Ptolemy

About Ptolemy's career we know little. The observations he reports in the *Almagest* date from AD 127 to 141, so that he cannot have been born much later than the beginning of the century; and he wrote several major works after the *Almagest*, so that his death may have been around 165 or 170. The only location he gives for his observations is Alexandria, and he very probably spent his adult life in that great centre of Hellenistic civilization.

The original title of the *Almagest* is *Megale syntaxis*, or *Mathematical* (that is, astronomical) *Compilation*; it became known in late Antiquity as *The Greatest Compilation*, and Arabic translators rendered 'the greatest' as *al-majisti*, which became *almagestum* in medieval Latin. It is indeed a magisterial work, one that provided geometrical models and related tables by which the movements of the Sun, the Moon, and the five lesser planets could be calculated for the indefinite future. It also contained a catalogue of over a thousand stars arranged in forty-eight constellations, with the longitude,

latitude, and apparent brightness (magnitude) of each. Some historians have argued that Ptolemy simply took the positions from Hipparchus's catalogue and reduced each longitude by $2^2/3°$ to allow for precession in the interval between the two men. Whole forests have been sacrificed to the resulting dispute, and it is now evident that straightforward plagiarism by Ptolemy is altogether too simplistic an explanation.

Ptolemy later published a modestly revised version of the tables with an introduction explaining their use, under the title *Handy Tables*. He also published a digest of the *Almagest* which he called *Planetary Hypotheses* and in which he added a physical dimension to the geometrical models of the *Almagest*.

Ptolemy wrote major treatises on a wide range of other mathematical sciences, most notably astrology, optics (including discussion of refraction), and geography. Few if any mathematical works of comparable quality would be written in Greek after his time.

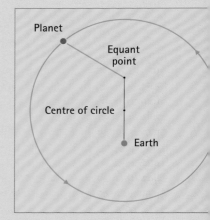

Above: In Ptolemy's circle, the Earth is located to one side of the centre, and the equant point is symmetrically positioned on the opposite side. The planet moves with a variable speed in such a way that it appears from the equant point to be moving uniformly. The planet therefore moves slower when at the top of the figure and faster when at the bottom.

Ptolemy, from a fifteenth-century manuscript in Greek of Ptolemy's *Geography*. He is shown using an astrolabe. Ptolemy the astronomer was confused in the Middle Ages with the Ptolemies who ruled Egypt, and so the astronomer is here wearing a crown.

Ptolemy's equant point

We must take advantage of our modern knowledge of planetary motion if we are to understand why the equant (seen opposite) was so convenient that to use it, Ptolemy was prepared to sacrifice the most fundamental principle of astronomy – that circular motions must be uniform. The first of the two laws published by Johannes Kepler in 1609 tells us that a planet moves around the Sun in an elliptical orbit with the Sun at one focus. Kepler's second law prescribes the speed of the planet in its orbit: a line (the 'radius vector') from the Sun to the planet traces out equal areas in equal times.

In obedience to the second law, the planet moves through space with a slower velocity than usual when furthest from the Sun (the radius vector then being longer than usual), and faster when nearest to the Sun (the radius vector then being shorter). How would the planet's movements appear to a hypothetical observer located at the focus of the ellipse *not* occupied by the Sun – the 'empty focus'. When the planet in its orbit is furthest from the Sun (and moving through space more slowly than usual), it is nearest to this observer at the empty focus; the slower velocity is therefore masked by the proximity of the planet to the observer. Similarly, when the planet is nearest to the Sun (and moving through space more quickly), this too is concealed from the observer because the

planet is now furthest from him. As a result, viewed from the empty focus, the planet appears to move around the sky with almost uniform angular velocity.

The consequence is that the motion of the planet in its orbit about the Sun – it matters little whether this orbit is a nearly circular ellipse or a true circle – is well represented by a model in which the planet moves with uniform angular velocity as seen from the empty focus (in the case of an elliptical orbit); or, if the orbit is taken as circular, from an equant point similarly located, that is, on the opposite side of the centre to the Sun and equidistant from the centre. Since this is true in particular of the Earth, it follows that the same is true of the motion of the Sun relative to the Earth.

It was of course the motion of the planet relative to the Earth (rather than to the Sun) that Ptolemy wished to represent; but this is compounded from the motion of the planet about the Sun, and the motion of the Sun relative to the Earth.

With the benefit of hindsight, then, we can see that Ptolemy's equant point was so useful because it was closely related to the empty focus of a Keplerian ellipse.

Kepler's first two laws tell us that a planet orbits the Sun in an elliptical path with the Sun at one focus, and that the line from the Sun to the planet traces out equal areas (shown in red) in equal times. As a result, the planet moves fastest when nearest the Sun and slowest when furthest from the Sun. The equant was a useful tool for Ptolemy because of the analogy between the equant point and the empty focus of the ellipse. (In the figure the eccentricity of the ellipse has been greatly exaggerated: with the exception of Mercury, which is difficult to observe, the planetary orbits are in fact nearly circular.)

Empty focus

Centre of ellipse

SUN

Planet

move, not with uniform speed, but with a speed that varied in such a way as to appear uniform to an observer at the equant point. The equant, therefore, was a device that violated the centuries-old principle that the heavenly motions were uniform. No doubt Ptolemy was more concerned with accuracy and mathematical convenience, than with questions of truth.

Yet, as time would show, Ptolemy had created a serious point of philosophical tension, one that disturbed many Islamic astronomers, and became a regular matter of debate in the universities of the medieval West. It would be difficult enough for university students of Arts to reconcile eccentres and epicycles with what they were taught in their natural philosophy about the uniform motions of the Aristotelian concentric spheres; to reconcile the equant was quite impossible. For this very reason, Copernicus in the sixteenth century would see it as a matter for self-congratulation, that his models needed no equants.

Ptolemy, in his monumental *Mathematical Compilation*, or *Almagest*, written about AD 145, provided geometric models that would, with fair accuracy, predict the motions of every one of the seven wanderers. He showed how the all-important parameters in his models could be derived from observations, though his general strategy was to work in the opposite direction: to present the models and then to 'test' them by comparing deductions from them against observations. With the help of the *Almagest*, mathematical astronomers (and astrologers) would be able to calculate tables of the positions of a planet in longitude and latitude for the indefinite future.

There were of course problems. The model for the planet Moon represented well enough the position of the body in the sky – that is, the Moon's direction as seen from Earth. But the epicycle carrying the Moon was unusually large in relation to the deferent that carried this epicycle about the Earth, so much so that in the model, the height of the Moon above the Earth varied between 33 and 64 Earth radii. This ought to have resulted in its apparent diameter varying by a factor of nearly two; yet such variations were not in fact observed in the real Moon. How much this bothered Ptolemy's readers depended on whether they were looking for something more than accurate tables of position.

Another problem stemmed from a curious fact of observation. Two of the lesser planets – Mercury and Venus – are never to be seen far from the Sun: they rise and set with the Sun, unlike Mars, Jupiter, and Saturn which may be seen in the sky at any time of night. Ptolemy replicated this in his models for Venus and Mercury by aligning the centres of their epicycles with the 'mean' Sun, so that all three had the same period of one year. Some of his readers, however, were unhappy with such an *ad hoc* device.

A model from the *Almagest*

The technical ingenuity that Ptolemy displays in the *Almagest* is illustrated by his model of the motion in longitude (that is, motion around the Sun's path or 'ecliptic') of any of the planets Mars, Jupiter, or Saturn. The Earth is at *O*, and *M* is the centre of the large (deferent) circle. *C* is a point on the deferent circle, and is itself the centre of the epicycle on which travels the planet *P*. *E* is a point (the equant point) on the line *OM* (extended), such that *OM* = *ME*. *C* moves around on the deferent circle with a variable angular velocity, such that, viewed from the equant point *E*, it *appears* to be travelling uniformly. *P* moves on the epicycle in such a way that *CP* is always parallel to the line from *O* to the 'mean' Sun (reflecting the fact that the Earth-based observer is in reality in motion around the Sun). The ratio of the radii of the circles, the 'eccentricity' *OM*, and the two angular velocities, are then chosen to fit the motion of the particular planet.

Epicycle centre

P Planet

C

E Equant point

M Centre of circle

O EARTH

MEAN SUN

But these were details. Ptolemy's *Almagest* achieved what it set out to do: to provide a set of geometrical models for calculating accurate tables of the future positions in the sky of each of the seven planets. It marked the culmination of centuries of effort.

Ptolemy made no attempt in the *Almagest* to present an integrated 'Ptolemaic System' of the heavens, but he did permit himself a few sentences on the order of the planets, in terms of their respective heights above the central Earth. It seemed plausible to make the fixed stars the outermost (as indeed Eudoxus and Aristotle had assumed half a millennium before), and to place nearest to these stars those planets whose motions most closely imitated the motion of the stars. On this basis Saturn – whose motion differed from that of the stars by only one circuit every thirty years – was the highest of the planets, with Jupiter (one in twelve years) and Mars (one in two years) next. At the other extreme, the Moon (one each month) was to be placed closest to the Earth.

That left him with three planets, Sun, Mercury, and Venus, whose order was as yet unassigned. Unfortunately, they kept company with each other as they wheeled about the central Earth, and so all their motions differed from that of the stars by the same amount: one circuit in one year. Which then was nearest the Earth and which furthest?

As the Sun was of unique status, Ptolemy decided to follow earlier astronomers in placing the remaining planets above it and below in equal numbers. Indeed, in the Latin Middle Ages the Sun would be seen as a king on a royal progress through his dominions attended by three planetary courtiers on either side. Since the previous argument had placed Saturn, Jupiter, and Mars above the Sun and only the Moon below, Ptolemy evened things up by locating Venus and Mercury below the Sun. The order he selected for these two, putting Mercury below Venus, was based on little more than the toss of a coin. And so, by reasoning that varied from the plausible to sheer guesswork, Ptolemy finally arrived at the order: Moon, Mercury, Venus, Sun, Mars, Jupiter, Saturn, fixed stars.

A fifteenth-century figure showing the Earth at the centre surrounded by the planets in the order chosen by Ptolemy: Moon, Mercury, Venus, Sun (which *non habet epicentrum* – 'does not have an epicycle'), Mars, Jupiter, and Saturn. Beyond Saturn is the firmament of the fixed stars.

The model for the motion of Mercury proposed by Georg Peurbach in the fifteenth century. Peurbach popularized an idea transmitted by Islamic astronomers, that the details of the planetary motions could be realized within a system of solid spheres. Here the epicycle carrying Mercury is indicated by the small, triple circle at the top of the figure, and this epicycle moves in the unshaded groove, about the central Earth. Being the innermost planet of the solar system, Mercury was to ancient observers invisible most of the time, lost in the glare of the Sun. Ptolemy made assumptions to supplement the observations available to him, and these implied that the planet came close to Earth, not once but twice in each orbit. As a result, the Ptolemaic model for Mercury was more complex than those of the other planets.

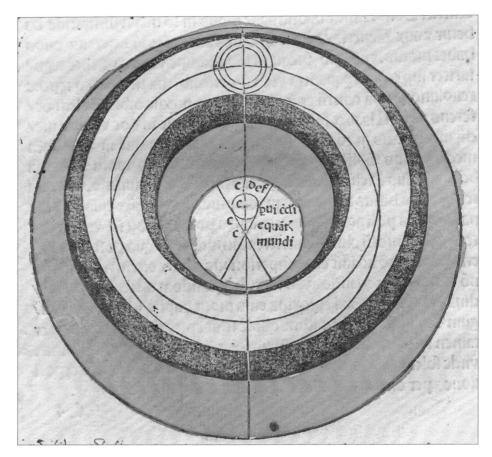

This enabled Ptolemy, in a later work called the *Planetary Hypotheses*, to combine his planetary models into a unified and physically real system, one that later craftsmen attempted – imperfectly, of course – to imitate in metal. Knowing the order of the planets, he made the assumption that all the possible heights in the heavens were shared out among them; that each planet had its own range of heights that it occupied from time to time, and that these ranges neither overlapped nor left any gaps.

This being so, he could exploit the fact that (according to the *Almagest*) the maximum height of the Moon was 64 Earth radii. Because the range of heights occupied from time to time by the Moon adjoined the range occupied by the next planet, Mercury, he inferred that the minimum height of Mercury was also equal to 64 Earth radii. Knowing the ratio of Mercury's epicycle to its deferent, he could now calculate the maximum height of Mercury. This he put equal to the minimum height of Venus; and so on, until finally he put the fixed stars at the maximum height of the outermost planet, Saturn.

By this means Ptolemy arrived at a universe whose radius was 19,865 times the radius of the Earth, or about 75 million miles. Some modern writers dismiss this figure as hopelessly wrong, pointing out that it is less than the true distance from the Earth to the Sun. But to think in this way is unhistorical. Rather, it was in the work of Ptolemy that the universe first became too large for the human mind truly to comprehend.

Ptolemy's astronomical writings brought to a triumphant conclusion the opening, Greek phase of the campaign to devise geometrical models that would replicate the motions of the seven planets. Future planetary positions could now be predicted encouragingly well, and there seemed no reason to doubt that adjustments made by later astronomers would bring prediction and observation ever closer. This programme was to preoccupy astronomers for the next fourteen centuries – indeed, the radical revision of Ptolemaic astronomy carried out in the Renaissance by Copernicus was inspired by a commitment to the uniform circular motions of the Greek natural philosophers that was even greater than Ptolemy's.

Circular motions would finally succumb to the precision of Tycho Brahe's observations. But meanwhile, after the collapse of the *pax romana*, astronomers of civilizations unborn in Ptolemy's day would struggle to master the intricacies of the *Almagest*, against a background of philosophical disquiet at the devices Ptolemy had found it necessary to employ.

The Ptolemaic system in outline. Note that every distance from the Earth is occupied from time to time by just one planet. Note also that the centres of the epicycles carrying Mercury and Venus are aligned with the Sun, while the radii of the epicycles of Mars, Jupiter, and Saturn are parallel to the line from the Earth to the Sun. Following Copernicus, we realize that this reflects the fact that we Earth-dwellers observe the planets from a platform in orbit about the Sun.

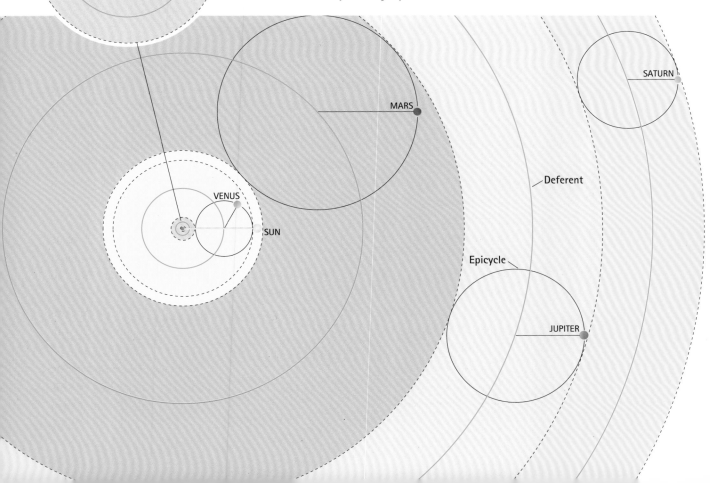

The traditional Chinese term for the inhabited world was *tianxia*, 'That which is below Heaven'. In astronomical terms the Chinese were under the same sky as everybody else, but the ways in which they interpreted what they saw in the sky were often unique.

There was in China no single subject of 'astronomy'. On the one hand there was *lifa*, 'calendrical methods', a discipline that aimed to master the regularities of celestial phenomena through careful measurement, record-keeping, and mathematical calculation. On the other there was *tianwen*, 'celestial patterns', whose practitioners watched for unpredictable and transient celestial phenomena and tried to interpret their significance for the world of human beings.

There was, however, one important factor linking *lifa* and *tianwen*: both were regarded as of vital importance to the state, and the practitioners of both disciplines were nearly all members of the imperial civil service. From around 1000 BC both the moral and the natural order of the cosmos were seen as embodied in a somewhat impersonal entity known as *Tian*, 'Heaven', whose will it was that the world below should be well governed. If the ruler misconducted his government or his personal life, this would be expected to cause disturbance in the natural world as an expression of Heaven's displeasure. Such disturbances might include flood, famine, and plague, but disorder in the sky was the most ominous of all and might be a warning of other troubles to come. It was therefore prudent for the imperial government to maintain a staff of sky-watchers, whose job it was to record, report, and try to interpret all unusual celestial events. This was the province of specialists in *tianwen*. Meanwhile, specialists in *lifa* worked to detect and codify all discernible regularities in celestial phenomena, with the aim of providing the emperor with all predictable phenomena, so that he could demonstrate his own effectiveness in maintaining the orderliness of the cosmos by promulgating an accurate astro-calendrical almanac.

Detailed records of the activities of *lifa* practitioners go back to 104 BC, when a new system of mathematical astronomy was inaugurated by Emperor Wu of the Han dynasty. This was to be followed by forty-seven other new systems up to the seventeenth century. All such systems had as their core requirement the need to run a luni-solar calendar, in which a civil year of twelve lunar months had to be kept in step with the cycle of the seasons by the addition of extra 'intercalary' months, *run yue*, at appropriate intervals. Lunar months began with conjunction (the 'astronomical' new moon), and the basic mark point for the seasons was the winter solstice. But in addition to predicting conjunctions and solstices, it became customary to include in officially promulgated astronomical tables methods for predicting the apparent motions of the planets, and for predicting lunar eclipses. Solar eclipses could never be accurately predicted by the available methods, and hence remained to a large extent within the domain of *tianwen*. The increasingly sophisticated mathematical methods used by *lifa* practitioners were numerical rather than geometric and exploited the recurrence of long-term cycles without laying stress on the motions of celestial bodies in three-dimensional space. Unlike the ancient Greeks and medieval Europeans, Chinese *lifa* specialists were not much concerned with cosmographical disputes. For most of the imperial period it was found satisfactory to assume the inhabited world was a small central region near the centre of a flat Earth, whose surface was a diametric plane of a celestial sphere rotating about an inclined axis. The celestial bodies moved on the inner surface of this sphere; the means by which they did so were not discussed.

The activities of *tianwen* specialists can be traced back to the late second millennium BC, when inscribed 'oracle bones' are

Left: A Chinese star chart from 940 AD, showing the Great Bear on the left.

Opposite: An armillary sphere, made in 1437. Armillary spheres contained metal circles that represented the zodiac, the celestial equator, and other circles in the sky. They were used for instruction or, as here, for the measurement of celestial co-ordinates.

thought to mention an observation of a nova (a new star). From around 200 BC a continuous stream of data is available from official sources. Apart from novae, officials recorded meteor showers, comets, solar eclipses, sunspots, and phenomena such as unusual clouds or the aurora borealis. In many cases such records are detailed and accurate enough to be useful to modern astronomers interested in such matters as identifying supernova remnants or tracing early appearances of Halley's Comet. Like the specialists in *lifa*, the observers of such phenomena had access to large and well-made armillary spheres and other graduated instruments that enabled them to record the apparent positions of celestial bodies to a degree of naked-eye accuracy not surpassed in Europe until the work of Tycho Brahe.

CHAPTER 3   *Islamic Astronomy*

It had taken Greek geometrical astronomers nearly half a millennium to achieve their goal of predicting with tolerable accuracy the positions of the seven 'planets' – Sun, Moon, Mercury, Venus, Mars, Jupiter, and Saturn – which moved as individuals against the background of the 'fixed' stars. By the time Ptolemy's *Almagest* was written, in the second century after Christ, the golden age of Athenian culture was a distant memory, and the power of Rome itself was showing signs of decline. Some elements of Babylonian and Greek astronomy had already percolated into India, whence they were eventually to be restored to the mainstream tradition by Arabic authors. But although many Greek works (including the *Almagest*) would survive into the later Middle Ages in the original language, as manuscripts gathering dust in libraries in Constantinople and elsewhere, study of the natural world was to continue within the cultural context of a religion not yet born, that of Islam.

In the mid-fourth century AD, a school was founded in Edessa in Mesopotamia by St Ephrem. The language spoken there was Syriac, but Greek was studied and some writings of Aristotle and of Ptolemy's distinguished medical contemporary, Galen, were translated into Syriac. In 489, the school was closed by the Emperor Zeno, and some of the teachers moved further east, to Jundishapur. There they translated further medical and other writings from Greek into Syriac: modest in quantity, but providing a sample of the intellectual riches on offer to those who could gain access to works originally written in Greek.

The year 570 saw the birth of the prophet Muhammad at the trading centre of Mecca. Soon after his death in 632, the religion he had founded began to spread with astonishing speed throughout the Middle East and across North Africa and into Spain. In 762 his successors in the Middle East founded a new capital, Baghdad, by the river Tigris at the point of nearest approach of the Euphrates, and within reach of the Christian physicians of Jundishapur. Members of the Baghdad court called on them for advice, and these encounters opened the eyes of prominent Muslims to the existence of a legacy of intellectual treasures from Antiquity – most of which were preserved in manuscripts lying in distant libraries and written in a foreign tongue. Harun al-Rashid (caliph from 786) and his successors sent agents to the Byzantine empire to buy Greek manuscripts, and early in the ninth century a translation centre, the House of Wisdom, was established in Baghdad by the Caliph al-Ma'mun. There Syriac- and Arabic-speaking scholars, under the leadership of the Christian Hunayn ibn Ishaq al-Ibadi (808–73), collaborated to translate the Greek works, either from the original or from a Syriac version, into Arabic.

That Arabic should emerge as an international scientific language is itself surprising. Hitherto it had been the language of poetry, and of the Koran and of studies related to the Islamic religion. Wherever Islam ruled, Arabic would be understood, and the translation of these Greek works into Arabic ensured their

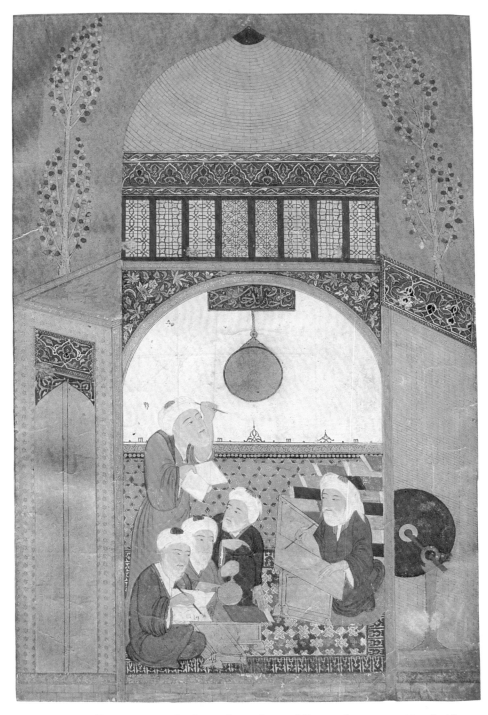

Islamic astronomers at work, from a manuscript of the early fifteenth century. The astronomer standing appears to be using a very large (and therefore, it was no doubt thought, accurate) astrolabe-type instrument to measure the altitude of a heavenly body. Four colleagues consider the implications of the measurement with the aid of a smaller astrolabe and written documents.

widespread diffusion, not only throughout the Middle East but across North Africa and into Islamic Spain. Eventually, mainly as a result of the Christian advance in Spain against the Moors in the twelfth century, these works would come into Christian hands, and many would be translated into Latin. Although Greek writings that made their way into Latin through these circuitous routes (possibly via Syriac, Arabic, and Castilian, for example) would arrive somewhat the worse for having been subjected to successive translations, they nevertheless provided an important continuity in traditions of astronomical thought.

ASTRONOMY AND ISLAMIC PRACTICES

Most of the 10,000 or so astronomical manuscripts written in Arabic, Persian, or Turkish that have come down to us continue to rest undisturbed on the shelves of libraries. However, it is clear that a large number of these writings on astronomy are Islamic in character, in the sense that they were the work either of religious scholars dealing with folk astronomy, or of astronomers dealing with mathematical techniques, each group addressing the same problems of Islamic practice. Other extant astronomical writings are relatively few in number, though of considerable historical significance; some attempted to perfect the planetary models of the *Almagest* (which was among the works translated, and later retranslated, in Baghdad), while others continued the debate between philosophers and astronomers over the nature of the cosmos.

Long before translations began, a rich tradition of folk astronomy already existed in the Arabian peninsula. This merged with the view of the heavens presented in Islamic commentaries and treatises, to create a simple cosmology based on the actual appearances of the sky and unsupported by any underlying theory. Meanwhile religious practices generated three specific challenges to which mathematical astronomers attempted solutions, the complexity of which often went beyond the actual needs of the community.

The first challenge arose out of the lunar calendar inherited in a modified form from pre-Islamic times. (The usual twelve lunar months in the year had previously been supplemented at intervals with an extra, or 'intercalary', month to keep the year roughly in step with the seasons; but Muhammad apparently taught against such intercalations, so the Muslim 'year' was, and still is, consistently some eleven days short of the seasonal year. One result of this is that Ramadan, the sacred month of fasting, may – unlike Lent – occur in any season.) Each month began with the new moon – understood not astronomically (when the Sun, Moon, and Earth were shown by calculation to be in a straight line), but practically, when the lunar crescent was first sighted in the evening sky. On the appropriate evenings witnesses would be sent to suitable locations with instructions to watch the western sky; if the lunar crescent was not seen one night, they would try again the following night.

This simple procedure led to difficulties. The skies were not always clear; and even if they were, the watchers of one town might see the new moon on an evening when those of the next had been unsuccessful, so that the two towns would begin their months on different days. Early Muslim astronomers followed a criterion they found in Indian sources, whereby the new moon would be seen if the difference in the setting times of the Sun and Moon on the local horizon was at least forty-eight minutes. Ptolemy's theory of the lunar motion was tolerably accurate around the time of new moon, but it described the Moon's position with respect to the Sun's path (the ecliptic), and this is inclined to the horizon. To relate this to motion with respect to the horizon was therefore a problem in spherical geometry. Later Muslim astronomers devised more elaborate conditions and compiled sophisticated tables to assist in the resulting calculations, leading to the production of almanacs with information on the possibility of sightings around the beginning of each month. But even today the problem remains a challenge to the Muslim world.

The second religious requirement that involved astronomy concerned the times of prayer, the number of which was established as five: sunset, nightfall, daybreak, midday, and afternoon. The timing of the latter two, plus that of a voluntary midmorning prayer, corresponded to the ends of the third, sixth, and ninth of the (variable) hours of daylight, and an approximate formula of Indian origin was used to relate these 'hours' to increases in shadow lengths. The problem of the translation of these rules into uniform hours and minutes aroused the interest of astronomers, though the muezzins who actually made the calls to prayer were more likely to follow a simple folk astronomy. Early in the ninth century a member of the House of Wisdom, al-Khwarizmi (a corruption of whose name gives us the word *algorithm*), compiled prayer-tables for the latitude of Baghdad, while the first tables for finding the time of day from the altitude of the Sun or the time of night from the altitudes of bright stars appeared in Baghdad soon after. Solving any of these problems involved finding the unknown sides or angles of a triangle on the celestial sphere from the known sides and angles. Thus the time of day could be derived from a triangle whose vertices were the zenith and the celestial North Pole (so that the line joining them was in the meridian), and the Sun's position. The observer had to know his geographical latitude (and hence the angle between his zenith and the Pole), and the Sun's position on the ecliptic for the time of year. He measured the altitude of the Sun, and the time was then given by the angle between the meridian and the Sun's hour circle (the arc from the Pole to the Sun). The necessary formulae were derived from two-dimensional projections of the celestial sphere, using techniques adapted from Indian sources. Later, spherical trigonometry was also used.

The method Ptolemy used to solve spherical triangles had been devised around the end of the first century AD by Menelaus of Alexandria. It involved a clumsy procedure whereby five segments had to be known in order to determine the unknown sixth quantity, but it sufficed to solve all the basic problems of spherical trigonometry. Unfortunately, however, this meant that telling the time from the Sun's altitude involved several applications of the theorem.

By the ninth century, the medieval equivalents of the six modern trigonometric functions had been recognized, whereas Ptolemy had operated

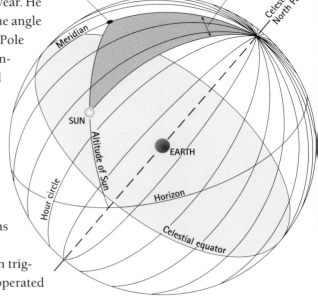

Calculating the time of day from the altitude of the Sun. The procedure involved finding the angle at the Pole in the triangle (shaded) whose vertices are the Sun, the observer's zenith, and the celestial North Pole. The zenith–Pole distance is the observer's latitude, the zenith–Sun distance is the complement of the Sun's altitude, and the Sun–Pole distance is (by definition) the complement of the Sun's 'declination' on that particular date. The angle at the Pole is proportional to the time.

with only a single function, one involving chords of circles. The concept of the sine of an angle (which, in a right-angled triangle, equals the side opposite the angle divided by the hypotenuse) was introduced into Islam from India, as were the tangent and cotangent functions important in calculations involving shadow lengths. Islamic astronomers discovered basic trigonometric identities that greatly simplified calculations involving triangles on the celestial sphere.

Eventually, the office of *muwaqqit,* or timekeeper, was instituted for mosques. This at last gave competent astronomers an institutional haven within one of the central structures of society, and the development resulted in a rapid increase in the quantity and quality of astronomical writings. Because of the hostility of Islam to astrology, astronomers who became muwaqqits were denied the financial rewards that could be earned by astrologers, but they were compensated by their secure and respected position in the community.

The third challenge posed by Islam to astronomy stemmed from the requirement that many acts of a religious nature, and more especially the orientation of mosques, should be directed towards the sacred shrine at Mecca known as the Kaaba. In the first two centuries of Islam many mosques, from Spain in the west to central Asia in the east, paid token respect to this requirement by facing south, on the grounds that when at Medina (which is north of Mecca) the Prophet had faced south when praying. In other places the *qibla,* or sacred orientation, was assumed to be the direction taken by pilgrims leaving for Mecca. Others again adopted an orientation of the rectangular shrine itself, whose major axis faced the rising of the star Canopus and whose minor axis was aligned with the summer sunrise and winter sunset. But in later centuries, as muwaqqits and other professional astronomers applied their minds to the problem of how to determine the qibla mathematically using the available geographical data, formulae in spherical trigonometry were developed and tables calculated from them. One outstanding achievement, which seems to date from the eleventh century, was the development of cartographic grids for Mecca-centred world-maps, from which one could read off the qibla and distance to Mecca directly. The culmination of this activity was a table prepared at Damascus in the fourteenth century by the muwaqqit al-Khalili, which gives the qibla for each degree of latitude from 10° to 56° and for each degree of longitude from 1° to 60° east and west of Mecca, with the vast majority of the entries correctly computed from a complicated (and accurate) formula.

THE EMERGENCE OF OBSERVATORIES IN ISLAM

The Koran declares that 'Nobody but God can know the future', and Islamic religious leaders were as resolute as their Christian counterparts in condemning astrology – and as ineffective. Ruler and people alike saw astrology as of great practical use, and they were prepared to pay for the information they wanted. Astrologers who practised in the market-place were little more than fortune-tellers,

Opposite: A late sixteenth-century Ottoman manuscript showing a huge armillary sphere (a representation of the principal circles in the heavens), supported on a wooden frame. The man in the middle is using a plumb-bob to adjust the position of the meridian (north–south) circle, which we see almost edge-on. Above, three astronomers are, somewhat precariously, taking observations, which are recorded by the assistant at the desk.

but in princely courts and elsewhere were to be found astrologers who were
also astronomers and who based their astrological predictions on tables of
planetary positions.

Early Arabic works on planetary astronomy had been eclectic, drawing on
pre-Ptolemaic writings in Greek as well as importations from Persia and India.
The translation of the *Almagest* transformed the situation by demonstrating the
incomparable superiority of the Alexandrian astronomer. However, the effectiveness of Ptolemy's planetary models depended not only on the geometric configurations themselves but also on the accuracy of the parameters employed; and by
now several centuries had passed, and improvement in many of these parameters
was clearly necessary. Ptolemy had shown his readers how to derive the parameters
from observations, and this was a lesson his Islamic successors learned well: we
find that the observations they record are mostly designed for the improvement
of the values used for such quantities as the eccentricity of the solar orbit and the
obliquity of the ecliptic plane (the angle at which it is inclined to the celestial
equator). These observations called for instruments of increased precision.

At first the instruments in question were small and portable, but in time the
quest for greater accuracy led to a call for large and fixed instruments. Here and
there, a ruler or other powerful patron would pay for the construction of such
instruments; as these were no longer portable they required permanent homes, and
these institutions marked the beginnings of astronomical observatories. But
hostility to astrology on the part of the religious authorities made it all the more
likely that the death of a patron, or even his loss of nerve in the face of criticism,
would bring observing to an end.

Two of the larger observatories were actually demolished. In Cairo in 1120 construction of an observatory was begun on the order of the vizier of the Fatamid
caliph, and when the vizier was murdered the following year work continued under
his successor. But in 1125, when the instruments were constructed but the building
was not yet complete, the new vizier was killed by order of the caliph, his alleged
crimes including communication with Saturn. The observatory was demolished,
and the personnel forced to flee for their lives.

A similar fate befell the observatory built in Istanbul for the astronomer Taqi al-
Din by the Sultan Murad III. Its construction began in 1575 and was completed in
1577, which made it an exact contemporary of the first major observatory of
northern Europe, that of Tycho Brahe (see pages 104–12). In addition to the main
building, there was a smaller observatory, where, as the artist records at the top of
the picture opposite:

> fifteen distinguished men of science were in readiness in the service of Taqi al-
> Din. In the observations made with each instrument five wise and learned men
> co-operated: there were two or three observers, and the fourth was a clerk, and
> there was also a fifth person who performed miscellaneous work.

Building work finished just in time for observations of the bright comet of 1577. Taqi al-Din interpreted the apparition as boding well for the Sultan in his fight against the Persians, but the success of the Turks proved to be far from complete, and other misfortunes befell the Sultan, including an outbreak of plague and the deaths of several notable figures. In 1580 religious leaders persuaded him that attempts to pry into the secrets of nature would only bring misfortune, and he ordered the observatory to be destroyed 'from its apogee to its perigee'. Tycho's observatory did not long outlive it: in 1588 Tycho's royal patron died, and soon his observatory was in decline.

Only two Islamic observatories enjoyed more than a brief existence. The first was at Maragha, the present-day Maragheh in northern Iran. It was built for the great Persian astronomer Nasir al-Din al-Tusi (1201–74) by the Mongol ruler of Persia, Hulagu, who was addicted to astrology.

Construction began in 1259. The observatory (whose foundations still survive) was built on the flattened top of a hill, and included an extensive library; no doubt it was there that the students were given part of their systematic instruction. The instruments were set up in the open air. They comprised a mural quadrant (an instrument fixed in the north–south direction, for measuring altitudes) no less than 14 feet in radius, an armillary sphere (a representation of the principal astronomical circles, used for other measurements of position) with circles of about 5 feet radius, and numerous lesser instruments. With their help, the team of astronomers (known to history as 'the Maragha School') completed in 1271 a *zij*, or collection of astronomical tables with instructions for their use, composed in the tradition of Ptolemy's *Handy Tables*.

Three years later al-Tusi left Maragha for Baghdad, where he died. His departure from the observatory marks the end of its creative period, though observations continued into the next century.

The other major Islamic observatory was built at Samarkand in central Asia, by Ulugh Beg (1394–1449), who was a provincial governor long before he succeeded to the throne in 1447. For once there had been no need to petition for royal patronage: the most enthusiastic, and perhaps most knowledgeable member of staff was Ulugh Beg himself. Construction of a three-story building took place in 1420. The major instrument was mounted out of doors, in a trench, between two marble walls aligned north–south and separated by some 20 inches. The instrument itself was a form of sextant (for measuring altitudes of celestial bodies over a range of one-sixth of a circle), its range being chosen so that the Sun, Moon, and other planets could all be observed. The radius of the sextant was over 130 feet, an illustration of Islamic astronomers' mistaken conviction that increased size would inevitably lead to increased accuracy.

The great achievement of Ulugh Beg's observatory was a set of astronomical tables that included a catalogue of over 1,000 stars. Many of its star positions had been established by the Samarkand astronomers themselves, making this the one important star catalogue of the Middle Ages. Ulugh Beg was murdered in 1449, by which time the observatory had enjoyed three decades of existence. As so often with the death of the patron, the end of observing soon followed.

ARABIC PLANETARY ASTRONOMY

An early example of a zij had been composed in the House of Wisdom at Baghdad by al-Khwarizmi. It derived from a Sanskrit astronomical work that had been brought to the Baghdad court around 770 by a member of an Indian political mission, and most of the parameters used in the zij derived from Hindu astronomy, as did many of the computational procedures. In a version by a later astronomer, it was to be translated into Latin by Adelard of Bath in the twelfth century, and so became one of the ways in which Indian methods reached the medieval West.

The Andromeda constellation, from a medieval Islamic copy of drawings by the tenth-century Persian astronomer Abd al-Rahman al-Sufi, who revised the star catalogue in Ptolemy's *Almagest*. The drawing shows (directly above this caption), very remarkably, the Andromeda Nebula, which is visible to the naked eye only under very favourable conditions and which had to be rediscovered telescopically in the West in the seventeenth century. The fish represents an alternative depiction from pre-Islamic Arabia.

One of the many astronomers who respected the Ptolemaic models but amended certain of his parameters was Muhammad al-Battani (c. 850–929), who spent most of his working life at al-Raqqa on the Euphrates. His zij, with an improved treatment of the orbit of the Sun in relation to the Earth, reached Christendom via Muslim Spain. The invention of printing eventually gave it wider circulation, and it was much used by Copernicus, who mentions its author in *De revolutionibus* no fewer than twenty-three times.

An astronomer whose work, by contrast, was unknown in the West in the Middle Ages was Abd al-Rahman ibn Yunus, who lived in Cairo in the late tenth century. He composed a major astronomical handbook called the *Hakimi Zij*, which he prefaced with a series of more than one hundred observations, mostly of eclipses and planetary conjunctions. His timekeeping tables were still in use in Cairo in the nineteenth century. The *Toledan Tables* prepared by the eleventh-century Moorish astronomer al-Zarqali (Azarquiel to the Latins), on the other hand, were translated early and had wide circulation; they became the model for the *Alfonsine Tables* that,

The constellation Perseus, after Abd al-Rahman al-Sufi. It is depicted as if on a globe; that is, the mirror image of the way the stars appear in the sky. The brightest star in the severed head of Medusa is labelled 'al-ghul' ('Algol'), an abbreviation of the Arabic 'ra's al-ghul', 'the Demon's Head' (for Ptolemy's 'Gorgon-head'). Algol is a variable star and this was recognized in the West in the seventeenth century (see page 203), but there is no convincing evidence that the earlier use of the word 'demon' implies that Ptolemy or his Islamic successors knew of this variability.

from early in the fourteenth century, were to dominate this aspect of Latin astronomy into the Renaissance.

The first revision of the star catalogue in the *Almagest* was due to an astronomer who worked in Persia and in Baghdad, Abd al-Rahman al-Sufi (903–86). In his *Book on the Constellations of Fixed Stars* he gave improved magnitudes, and Arabic versions of the identifications, but left both the stars themselves and their (often inaccurate) relative positions unchanged. This is a symptom of the lack of commitment to observation that was common among Islamic astronomers: despite their industry in applying mathematics to astronomy, their workplace was often the study rather than the open air. Thus they were almost entirely oblivious of the supernova explosion of 1054 that gave rise to the Crab Nebula; only a single reference to this is known from Arabic sources. They had a versatile and convenient observing and calculating instrument in the astrolabe (see pages 64–7), but a single measurement with it was sufficient for most tasks. After all, astrology depended on the positions of planets, and for this the *Almagest* provided an excellent basis.

Imperfections in Ptolemaic astronomy of a more conceptual nature became clear from a comparison of the abstract, geometrical models of the *Almagest* with the same author's physical view of the cosmos as embodied in his *Planetary Hypotheses*. Already in the ninth century we find Thabit ibn Qurra (836–901), a Syriac-speaking scholar who worked in Baghdad, drawing attention to such inconsistencies, and by the tenth century texts appeared regularly whose subject matter was *shukuk*, or 'doubts', concerning Ptolemy. While we look in vain in Islam for a rebel who challenged the basis of the Earth-centred Aristotelian/Ptolemaic cosmology, there was philosophical unease over some of the aspects of the received planetary models, as there had been in Antiquity and as there would be in the Christian Middle Ages and Renaissance. The most obvious target for criticism was Ptolemy's device of the equant (see pages 42–3), which led the planet to slow down and speed up and so violated a fundamental rule of Greek astronomy, that motions must be uniform. Eccentres (see page 38) were also objectionable in that they involved circular motions about centres other than the centre of the Earth, and even epicycles (see page 39) fell foul of Aristotelian fundamentalists.

The debate over eccentres and epicycles (which took place mainly in Islamic Spain) was essentially between philosophers and mathematical astronomers. An influential spokesman for the philosophers was the greatest Islamic exponent of Aristotle, the Andalusian Muhammad ibn Rushd (1126–98), who became known to the Latins as Averroës, or simply The Commentator. He accepted that the predictions from the Ptolemaic models might indeed match the observed motions, but in his view only concentric spheres could compose the true universe. Most mathematical astronomers realized that to attempt planetary models on this principle was hopeless, but Averroës's contemporary and fellow Andalusian, Abu Ishaq al-Bitruji (Alpetragius), was bold enough to try. Not surprisingly, the results were grossly unsatisfactory; Saturn, for example, found itself deviating from the ecliptic by a monstrous 26° instead of the correct maximum of only 3°. Yet his attempt struck a sympathetic chord among Christian natural philosophers after his work was translated into Latin early in the thirteenth century.

Most attempts at reform of the layout of the planetary models were the work of Eastern mathematical astronomers, and were driven by theoretical considerations rather than the need to improve predictions. One of those who rejected the Ptolemaic equant was Ibn al-Haytham (965–c.1040), later known to the Latins as Alhazen, who worked in Cairo. In his *On the Configuration of the World* he tried to adapt the planetary models of the *Almagest* so that they could take on physical reality. He saw the heavens as formed of concentric spherical shells, within whose thicknesses other shells and spheres were located. In this way he tried to assign a single spherical body to each of the simple motions of the Ptolemaic models. In the thirteenth century his work was translated into Castilian in the court of Alfonso the Wise, and not long after from Castilian into Latin. The concept of separate celestial spheres for each component of the planetary motions was to gain currency in the later fifteenth century through the work of Georg Peurbach.

The equant was enough to arouse misgivings even among observational astronomers. In the thirteenth-century the Maragha astronomer al-Tusi, in his *Tadhkira*, or *Memorandum*, succeeded in devising a satisfactory substitute that used only uniform circular motion: two small epicycles were to be added to each of the planetary models, a complication that seemed a price worth paying.

Eccentres, though less shocking than equants, still represented a violation of Aristotelian doctrine. The most competent attempt to purge the Ptolemaic models of all their objectionable features and to replace them with models that were acceptable both philosophically and observationally was made around 1350 by Ibn al-Shatir, muwaqqit of the Umayyad mosque at Damascus. Ibn al-Shatir's lunar model avoided the gross variation in the apparent diameter of the Moon that Ptolemy's model had implied (see page 44); his solar model was based upon new observations of the solar diameter; and all his planetary models were free not only from the equant but also from eccentres. True, he could not do without epicycles; but the existence of individual stars demonstrated that Aristotle had gone too far in his insistence that the heavens were composed throughout of uniform matter, and so even an Aristotelian ought to keep an open mind over the possible existence of epicycles.

Though not the last mathematical astronomer to write in Arabic, Ibn al-Shatir represented the culmination of a movement that had been maturing for half a millennium. But the moment had passed when Arabic astronomers could influence the future course of planetary theory. The enthusiasm in the Latin world for translations from the Arabic had long since abated; the Latins were developing their own astronomical tradition, and the writings of al-Tusi and Ibn al-Shatir seem to have been virtually unknown in the West.

But if so, historians have to accept a strange coincidence; for when Ibn al-Shatir's work was rediscovered in modern times, it was realized that it employed geometrical devices similar to those later used by Copernicus, who was likewise scandalized by the Ptolemaic equant. Copernicus took the radical step of turning the Earth into a planet, but when it came to developing detailed models, many of the problems he encountered were not so very different from those confronted by his predecessors. In the *Commentariolus*, a preliminary sketch of his Sun-centred theory that circulated in manuscript in the early years of the sixteenth century, Copernicus used an arrangement equivalent to Ibn al-Shatir's in order to eliminate the equant and generate the intricate changes in the Earth's orbit. In the fully developed *De revolutionibus* (1543), Copernicus reverted to the use of eccentric

A diagram from a manuscript now in Oxford that sets out to explain the model of the motion of the Moon proposed in the fourteenth century by Ibn al-Shatir. It is an example of his use of a pair of epicycles in order to avoid the equant of Ptolemy, which he considered unacceptable in principle. With the values chosen by Ibn al-Shatir for the radii and speeds of rotation of the various circles, not only is the motion of the Moon across the sky well reproduced, but the changes in the distance of the Moon from the Earth are moderate. This avoided a striking defect in Ptolemy's lunar theory, in which the distance of the Moon (and hence its diameter as it should appear from Earth) varied by a factor of two.

orbits, but he used a model that was the Sun-centred equivalent of one developed at the observatory founded by al-Tusi at Maragha in Persia. No Latin translation of these Arabic works has been found, nor is any Latin account of them known. A Greek translation of some of al-Tusi's writings found its way to Italy in the aftermath of the fall of Constantinople in 1453, and Copernicus studied in Italy from 1496 to 1503 and acquired a knowledge of Greek. But whether he owed to others his technique for replacing the equant, or whether he developed the method independently, is for the present moot.

By far the most sophisticated (and historically important) astronomical instrument of the Middle Ages was the planispheric astrolabe. It was one of four forms of astrolabe, but two of the other forms, the linear and the spherical, were very rare, and the mariner's astrolabe (see page 84) was a crude tool for use at sea and seemingly developed only towards the end of the Middle Ages.

The origins of the astrolabe go back to Greek Antiquity; it reached maturity in Islam (where it was to retain its popularity into modern times), and then received further refinements in the West. The basis of the instrument is a disc of brass which can be suspended from a ring. The back of the disc is essentially an observing instrument, being fitted with an observing bar, or *alidade*, that rotates about a pin in the centre of the disc and is used for measuring the altitude of celestial bodies. The observer simply suspends the instrument from the ring so that it hangs vertically, looks along the alidade towards the body in question, and reads off its altitude from a scale engraved around the circumference of the disc.

Also on the back of the instrument in its Western form are two circular scales which together give the position of the Sun on the ecliptic (its path against the background of the stars) for any date in the year. One scale is engraved with the days of the year, and the other shows the corresponding position of the Sun; the alidade can be used to line up corresponding points.

The front of the astrolabe is a calculating device that embodies representations both of the heavens (individual stars, ecliptic, equator, and tropics), and of the local system of co-ordinates (angular altitude above the horizon, angle of 'azimuth' around the horizon, usually from south) in which the user will be making measurements of the positions of heavenly bodies.

To make an astrolabe, we need to transfer the sky onto a sheet of brass. Fortunately, we know from geometry how to set up a 'projection' between the points of the celestial sphere and the points of an (infinite) plane, so that each point on the sphere 'projects' into exactly one point on the plane. We simply imagine the line joining the celestial South Pole to any other celestial point, and take the projection of the point to be the place where this line intersects the plane that contains the celestial equator. Such a projection has the valuable and surprising property that angles between curves on the heavenly sphere are unchanged after projection, so that problems in spherical triangles can be converted into more tractable problems in plane trigonometry.

The brass disc of our astrolabe is a physical replica of the plane through the celestial equator, onto which we have projected (that is, represented) the celestial sphere. Unfortunately the disc is limited in size, and this means that in practice not all of the celestial sphere can be represented on it. But as the Muslim and Christian users of astrolabes never saw the skies

Below: The back of an astrolabe. The observer holds the instrument suspended from the ring at the top, and measures the altitude of a celestial body by sighting along the observing bar and reading the angle on the outermost scale. The bar can also be used to determine the current position of the Sun on the ecliptic: the user aligns the bar with the current date on the circle engraved with the days of the year, and the position of the Sun is shown by the intersection of the bar with the circle engraved with the signs of the zodiac.

Right: Observing with the astrolabe in thirteenth-century France.

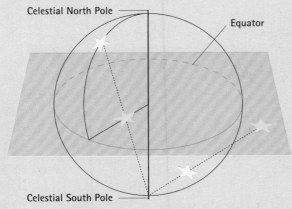

Celestial North Pole

Equator

Celestial South Pole

So far, every feature of the representation we have mentioned is fixed and static. But we now need to project the individual stars of the heavenly sphere (other than those too far south, of course); and since the heavens rotate, the projection must rotate. Had transparent plastic been available in the Middle Ages, no doubt the projected stars would have been engraved on a plastic disc that rotated about the central point that represents the celestial North Pole; through the plastic the user would have been able to see the co-ordinate circles

Above: The front of the astrolabe is a replica in brass of the plane through the celestial equator, onto which we have projected the heavenly sphere from the celestial South Pole. From this Pole we draw a line to any point on the heavenly sphere, and the 'projection' or representation of the point is the point where the line meets the plane. The celestial equator, since it already lies in the plane, projects into itself. Points (or stars) in the northern hemisphere (as seen to the left in the figure) project downwards into points inside the equator, while points (or stars) in the southern hemisphere (as seen to the right) project upwards into points outside the equator.

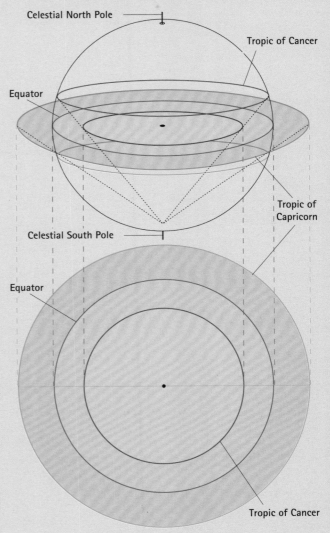

Celestial North Pole

Tropic of Cancer

Equator

Tropic of Capricorn

Celestial South Pole

Equator

Tropic of Cancer

near the celestial South Pole, no harm was done if the southern skies were left unrepresented. This was the reason behind the choice of the South Pole as the centre of projection.

At the very centre of our disc of brass is the representation of the celestial North Pole. Around it, three concentric circles represent respectively the Tropic of Cancer, the equator, and the Tropic of Capricorn. We choose to make this latter circle the outer limit of our disc, so that it is the skies that lie south of the Tropic of Capricorn that are not represented.

The observations that the user makes with the alidade of the astrolabe are of the angular altitudes of heavenly bodies, and these are angles within a co-ordinate system in which the horizon is 0° and the zenith 90°. To enable us to make convenient use of these observations, circles of equal altitudes (0°, 10°, up to 90°) must be represented in the projection, along with circles of equal azimuth. The problem is that these circles depend upon the latitude of the place from which the observations are being made, and so therefore do their projections. The solution adopted was to supply the astrolabe with discs for a range of latitudes, each engraved with the projection of the appropriate co-ordinate circles. These *climates* were stored in the instrument one above the another, and the user would simply select the most appropriate one available and place it at the top of the pile.

Step 1 (of 5) in making the front of an astrolabe: the projection onto a brass disc of the celestial North Pole, the tropics, and the equator. Because the astrolabe was of finite size and the skies to the far south were of no practical interest, the skies south of the Tropic of Capricorn were not represented.

engraved on the climate below. Instead, the medieval maker took another disc of brass (the *rete*), and marked on it the projections of the important moving features of the sky (the ecliptic path of the Sun and the positions of the principal stars). Retaining these, he cut away as much of the rest of the brass as possible, thereby exposing the climate below with its co-ordinate grid.

The heavens spin: in modern parlance, the heavenly sphere has just one degree of freedom. Therefore, only a single observational fact is sufficient to locate all the stars in the positions they occupy at the time of the observation. If the astronomer, for example, has measured the altitude of Sirius with the back of his astrolabe (and has verified whether the star is rising or setting), he may then retire to his study and rotate the rete

Step 2: the projection of the circles of equal altitude that are part of the observer's local co-ordinate system. These are the angles, ranging from 0° at the horizon to 90° at the zenith, that are directly measured with the back of the astrolabe. In this figure the observer's latitude has been taken to be 50°, but the astrolabe would have brass discs engraved with similar projections for a range of latitudes.

Step 3: the projection of circles of equal azimuth (angles around the horizon). On astrolabes, azimuth was usually measured clockwise from the south, although some makers followed other conventions. Also part of the observer's local co-ordinate system, the lines of equal azimuth converge at the zenith.

until the representation of Sirius is appropriately located over the corresponding altitude circle in the co-ordinate grid. This done, he now has not only Sirius correctly located, but all the other stars as well; he can determine their current positions, he can tell which of them are about to rise and which have just set, and so on.

A basic use of the astrolabe was to tell the time. The pair of scales on the back of the astrolabe tells the user the position of the Sun for the day in question, and so he knows where to mark its position on the projection of the ecliptic. In a Western astrolabe an outer circle on the front was engraved with twenty-four equal divisions, to represent the hours taken by the heavens to complete one rotation (Islamic astrolabists had to manage without this device). An *index* bar allowed the user to align the projected position of the Sun with this outer circle and so read off the time. In other words, the astrolabe is a twenty-four-hour clock that permits the time to be determined from a single observation — of the Sun in the day-time, or of a star at night.

Alternatively, the astrolabe can be used to predict the time when an astronomical event will happen. So, for example, if the user wishes to know the hour when the Sun will rise, he need only rotate his rete until the Sun is on the eastern horizon, and he can then use the index to read off the time when this will happen.

These are only a few of the applications of this very remarkable instrument, which in the Islamic and the Christian Middle Ages was fundamental to astronomy, astrology, and astrological medicine alike.

Rete

Index bar (incomplete)

Hour angle lines

Step 4: 'Hour angle lines' are added to the left of the combined projections of the principal fixed circles and of the circles that form the observer's local co-ordinate system. The hour angle lines were used to tell the time with the period of daylight divided into twelve equal 'hours' and the period of dark divided into twelve equal 'hours' (normally different from the daylight 'hours').

Step 5: the rotating 'rete' is now superimposed on the disc which, as a result of Steps 1–4, is engraved with hour angle lines and with projections of the principal fixed circles and of the circles that form the observer's local co-ordinate system. The complete circle that is present in the rete represents the ecliptic path of the Sun, and the various 'pointers' of the rete indicate the positions of bright stars. The actual position of the Sun on the ecliptic for the date of the observation could be found by consulting scales on the back of the instrument. If the 'index' bar (which is broken at one end in the astrolabe above) was aligned with the Sun's position on the ecliptic, the bar acted as the hand of a twenty-four-hour clock. Accordingly, some Western astrolabes have a twenty-four-hour scale around the outer edge.

CHAPTER 4   *Medieval Latin Astronomy*

In the second century BC Roman soldiers had conquered the Greek city states, but it was Greek culture that had triumphed over the Roman mind. Works written in Latin scarcely featured in our discussion of astronomy in Antiquity, because within the *pax romana* the language of cultivated men was Greek: among writers on astronomy Latin was mainly reserved for low-level didactic works.

The final disintegration of the *pax* came about in the second half of the fifth century AD, and thereafter everyday life in the regions of western Europe became increasingly precarious. For the few to whom scholarly writings still mattered, access to such works became ever more difficult, and an ability to read the Greek in which they were written became a rarity.

One scholar made a desperate attempt to render the outstanding works of Greek philosophy into Latin before it was too late: Anicius Manilius Severinus Boethius (c. 480–524/5), a high official in the Roman Gothic kingdom. His most ambitious plan was to 'translate and comment upon as many works by Aristotle and Plato as I can get hold of'. Sadly, he angered his master Theodoric by his refusal to stand silently by while injustices were perpetrated, and he was tortured and executed, his grand task scarcely begun. But what he did achieve was to prove significant. In particular, he translated or paraphrased a number of Greek writings on logic and assembled them along with logical writings by Latin authors such as Cicero, so providing students living centuries later and in happier times with a corpus of secular writings that they might compare and contrast, discovering in the process how to display an independence of mind. Logic would become an obsession in the medieval universities, most of whose Arts teachers would be more concerned with the status of Aristotelian spheres or Ptolemaic equants than with the effectiveness of the planetary models constructed out of them.

By his own writings and by his translations and commentaries, Boethius codified the existing tradition of mathematical studies into what would later become the standard university format of the *quadrivium*: arithmetic (the abstract theory of discrete quantities); harmony (the same discipline applied to nature); geometry (the abstract theory of continuous quantities); and astronomy (the same discipline applied to nature). However, it so happened that his own quadrivial writings that survived were mainly on arithmetic and harmony, rather than astronomy.

THE 'DARK' AGES

One definition of the Middle Ages is, the period between the fall of Rome and the fall of Constantinople (in 1453). Another is, the period when few Western scholars knew Greek. For the time being this ignorance of Greek scarcely mattered, since Greek works were rarely accessible in the original. Sadly, neither were they available in translation. Aside from the legacy of Boethius, the most significant translation from the Greek was due to Calcidius (it is a sign of his disturbed times that we are not sure if he lived in the fourth or the fifth century). He rendered into Latin two-thirds of Plato's cosmological myth, the *Timaeus*, and wrote a lengthy and influential commentary. Remarkably, even in the later Middle Ages, when the whole of the vast Aristotelian corpus had been translated into Latin and was dominating the learned world, Calcidius's version of the *Timaeus* was still the most influential of the tiny handful of Plato's dialogues available.

Writings composed in Latin in the middle centuries of the first millennium AD bear testimony to the dire state of secular studies. The best Latin treatise on astronomy from this period appeared in *The Nuptials of Philology and Mercury* by Martianus Capella of Carthage (c. 365–440). This work was an allegory of a heavenly marriage in which seven bridesmaids presented a compendium of astronomy and the other six Liberal Arts. In particular, it contains a statement of the theory, nowadays often ascribed to Heracleides of Pontus (c. 390–c. 310 BC), that Venus and Mercury always appeared close to the Sun because they circled the Sun while all three circled the Earth. In the Middle Ages, the Capellan tradition became entangled with another from Pliny, sometimes with the result that diagrams showed the orbits of Mercury and Venus curiously interlinked as they circled the Sun. In any event, diagrams and passages such as these would ensure that in the medieval

A thirteenth-century manuscript showing a curious variant of the system of Martianus Capella (c. 365–440). Venus and Mercury are carried around the Earth by the Sun, but the paths along which they circle the Sun are interlinked. Notice that the three superior planets (Mars, Jupiter, and Saturn in the upper part of the diagram) have the customary Ptolemaic epicycles.

Publius
Cornelius

Sapio.
Affricanus.
Publij Jumons.

Quaue
Corneli?
Sapio Jumor
cus numatis.

paulus sapio
pater publij

Primum mobile

Spem

Speu

Iue

Spem scu
te

omhg

ROMA

CARTAGO

TERRA

Spera lune

Spem mercurij

Spem Ienerio

Spem Solis

Spem Martis

Spem Jouis

P.C. SCIPIO

universities, dissenting voices would always be heard, and Martianus Capella was to be cited with approval (albeit some puzzlement) by Copernicus.

Ambrosius Theodosius Macrobius, who was apparently also from North Africa and who lived in the early fifth century, was the author of a commentary on the *Dream of Scipio* of the Latin author, Cicero. Macrobius took the opportunity to expound a cosmology deriving ultimately from Plato and the Pythagoreans, in which numbers underlay all things. The commentary also incorporated a popular handbook on astronomy. A spherical Earth lay at the centre of the spherical universe, encircled by seven planetary spheres. At the outside was a starry sphere; this rotated daily from east to west, dragging along the planetary spheres, though each of these also had its own motion in the opposite direction. Significantly, Macrobius left the order of the planets vague, because his sources – Plato and Cicero – differed on the place of the Sun in the sequence. Like Martianus Capella, Macrobius related the theory of Crates of Mallus (second century BC) that the Earth was divided by oceans into four quarters, each of which was inhabited, and he reported Eratosthenes's value (see page 39) for the circumference of the Earth.

As in Islam, so in Christendom religious requirements posed questions whose answers demanded some study of the heavens. Monasteries needed to know the time at night so that the community could rise and sing the office psalms; this was often done by observation of the stars, the monks having to resort to more mundane methods – water clocks or rhythmic chanting – when the stars could not be seen. Of wider significance was the determination of the appropriate date for the great festival of Easter; in conformity with its origins in Jewish observances, Easter was linked with the first full moon after the spring equinox, so that its dating depended on both the lunar month and the solar year. Church authorities needed to publish Easter dates for some years ahead, so that the feast could be celebrated on the same day throughout Christendom. One possibility would have been for the bishop of Alexandria to invite local astronomers, as the inheritors of accurate values for the month and year that had been handed down from the Babylonians, to carry out calculations of the date of the spring equinox, year by year. Instead, the authorities took the more practical course of looking for a cycle in which a certain number of solar years would equal (with sufficient accuracy) a number of lunar months: the dates of Easter would then be calculated within the cycle, and simply repeated in succeeding cycles. From the third century onwards various cycles were tried ranging from eight years to eighty-four, until the Metonic cycle (see page 27) of nineteen years was recognized as the most suitable: nineteen years differ from 235 lunar months by only 1/12th of a day, and so such a cycle departs from nature by only one day in 12×19 years, or more than two centuries.

The definitive treatise on the Easter problem was written in 725 by the Venerable Bede (672/673–735), monk of Yarrow in the north of England. In his *On the Divisions of Time*, he set out for the date of Easter a cycle of $19 \times 28 = 532$ years. Under the Julian calendar, a leap year occurred every fourth year (without exception).

Opposite: The frontispiece to a fourteenth-century manuscript of Macrobius's commentary on Cicero's *Dream of Scipio*. The Earth is shown at the centre, with buildings of Rome and Carthage. Surrounding the Earth are the spheres of the seven planets, and at the outside is the sphere of the fixed stars. Interestingly, the Milky Way is shown as composed of stars; this explanation for the milky appearance in the sky had been suggested in Antiquity, but was not confirmed until after the invention of the telescope. Scipio himself is asleep at the foot of the picture, and the other figures are ancestral souls encountered by Scipio in his dream.

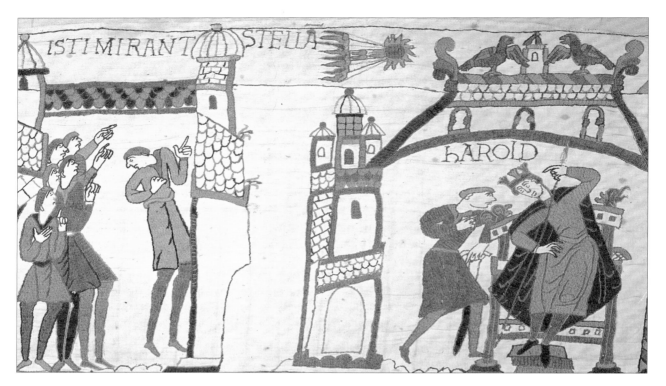

The Bayeux tapestry was woven after the victory in October 1066 of William of Normandy over the English King Harold, which opened the way to the Norman conquest of England. At the top is Halley's Comet, which was visible in England in the spring of that year. On the left are men marvelling at the star (*isti mirant stellā*), while on the right Harold is informed of the evil omen.

Then as now, the weekday on which a given day of a given month fell normally advanced by one day each year, but by two when the extra day of a leap year intervened. Accordingly, in four years, the weekday invariably advanced a total of five days, and in $4 \times 7 = 28$ years it advanced by 5×7 days or exactly five weeks, thereby repeating itself. Since the Metonic cycle of nineteen years dealt adequately with the solar/lunar component of the Easter dating, and any cycle of twenty-eight years repeated the day of the week, a compound cycle of 19×28 years also took into account the requirement that Easter Day be a Sunday.

THE REVIVAL OF ASTRONOMY IN THE WEST

Writers such as Macrobius had helped preserve fundamental concepts of Greek astronomy such as the sphericity of Earth and its location at the centre of a spherical universe, and the distinction between the seven planets and the 'fixed' stars. But compared to the sophistication of the *Almagest*, knowledge of astronomy among the Latins in the second half of the first millennium was primitive in the extreme. The tenth century, however, witnessed the first contacts with Islamic centres of learning, and monasteries on the southern slopes of the Pyrenees became centres for the transmission of Arabic culture. One influential scholar to visit Spain and study there was Gerbert of Aurillac (later Pope Sylvester II; c. 945–1003). After his return, Gerbert was given charge of the cathedral school of Rheims, which soon began to attract students from far and wide. Gerbert may well have brought from Spain knowledge of the astrolabe, that sophisticated and versatile instrument for both measurement and calculation. Certainly the instrument itself had appeared in the West by about 1025, and soon thereafter two Latin treatises on it were composed (or adapted from the Arabic) by an Austrian monk, Hermann the Cripple (Hermannus Contractus, 1013–54).

The astrolabe allowed astronomy to become once more a mathematical science, for with it an astronomer could measure the angle between the horizon and the position of a heavenly body. As in Islam, astrology had always retained a certain popularity despite forthright expressions of disapproval from religious leaders, and this popularity would increase with the passage of time, most notably in the demoralization that followed the terrible Black Death of 1348–50. Nor was astrology wholly without a basis in reason. So, for example, the Aristotelian analogy between the macrocosm (the world in the large) and the microcosm (the individual living body) enabled medical treatment of a patient's organs to be planned in the light of the favourable or unfavourable disposition of the corresponding heavenly bodies. The teachers in university medical faculties would include instructors in astronomy/astrology, and some knowledge of the influence exerted by the planets, and therefore of the celestial configurations that generated this influence, would be expected of every practising physician. The English poet Geoffrey Chaucer says of the physician among the pilgrims on their way to Canterbury, that 'he was grounded in astronomy,' and so 'well could he fortunen the ascendent'.

Astrology therefore generated an incentive to master the movements of the planets, one that went far beyond the disinterested pursuit of astronomy for its own sake. Accordingly, the tables that had been compiled by Islamic astronomers and the rules for their use – and the theories that underlay them – aroused keen interest among the translators attracted to Spain in the twelfth century, as the tide of Islam receded and left in Christian hands manuscripts that embodied the philosophical and scientific achievements of both Greece and Islam. The fall of the great intellectual centre of Toledo in 1085 opened the floodgates. Gerard of Cremona (c. 1114–87) was one of the translators who flocked there, and his companions were later to list no fewer than seventy-one astronomical and other works translated by him. Surprisingly, modern scholars agree that even this list is not exhaustive, and credit Gerard with translating additional works, including the rules (*Canons*) for the use of the *Toledan Tables* of al-Zarqali. These tables were adapted for various longitudes in western Europe and enjoyed an immense success. Their availability stimulated the study of astronomy, for it was now possible to compute planetary positions for any moment in time, and to

The Aristotelian analogy between the world in the large and the individual living body enabled physicians to use astrology as a guide to the treatment of patients. This fifteenth-century figure instructs them in the correspondence between the various organs of the human body and the planets and 'houses' of the zodiac.

compare these predictions with observations. However, the underlying theories remained a mystery until the *Almagest* itself became available – though to minds accustomed to astronomical treatises of the level of Boethius and Martianus Capella, it was at first wholly indigestible. In Sicily, which had been ruled in turn by Greeks, Muslims, and Latins, translations likewise got under way. Communications with Constantinople were restored, and from the libraries there came another copy of the *Almagest*, this time in the original Greek.

It is hard for us today to comprehend the impact of the translations of works from Antiquity and the Islamic Middle Ages, as they descended upon the newly fledged universities of the Latin West in the twelfth century. When a teacher in a Faculty of Arts first laid hands on the *Almagest*, it must have represented a level of technical virtuosity exceeding by several orders of magnitude any astronomical writings he had previously known.

In this fifteenth-century miniature, the Moon is depicted as a young woman standing on two 'wheels of fortune'. Between her legs is the zodiacal sign for Cancer, the 'house' in which the Moon exerted her greatest influence. Ptolemy in his book on astrology assigned to the Moon the power of humidifying, while other writers thought the Moon the cause of tides; these associations are symbolized in the seascape.

Paris University soon established itself as the great intellectual centre of Christendom. In a period when all educated men could write and speak Latin, there were few obstacles (other than financial) to prevent the best teachers, and the best students, gravitating towards a recognized centre of excellence, and Paris in the thirteenth century attracted an astonishing galaxy of talent. At Paris, as elsewhere, the role of the Faculty of Arts was simply to provide students with the basic education that was all that most of them would need. The Faculty's teaching was structured initially around the seven Liberal Arts: the literary *trivium* of grammar, rhetoric, and logic, and the mathematical *quadrivium* of arithmetic, harmony, geometry, and astronomy. The students of Arts were mostly young – much younger than is normal today – and the Arts teachers in Paris were acutely conscious of the prestige of the higher faculties (medicine, law, and especially theology), to which the more ambitious students might progress after becoming Masters of Arts. They were therefore exhilarated to find that most of the non-medical works that were now arriving in translation clearly belonged to the Faculty of Arts. In particular, Aristotle's writings owed nothing to the Hebraeo-Christian revelation that was the subject matter of the Faculty of Theology. The new works therefore provided the Masters of Arts with a weapon to use in their struggle to improve their status *vis-à-vis* the theologians.

This political dimension added to the inevitable difficulties of assimilating a pagan Aristotelian world-picture into a culture that had been dominated for a millennium by Christian theology, and as a result the University of Paris was in turmoil for much of the thirteenth century. The synthesis eventually achieved there by the Italian Dominican friar, Thomas Aquinas (1225–74), propagated an immensely powerful cosmology, in which the teachings of Aristotle concerning the natural world – so many of them plain common sense anyway – seemed to acquire an aura of revelation.

The Christianized cosmos as represented in the *Nuremberg Chronicle*, a world history published in 1493. At the centre is the region of the four elements, surrounded by the seven planetary spheres, after which come the firmament, the crystalline heaven, and the First Mover postulated long ago by Aristotle. At the outside we see God enthroned, with the nine orders of angels.

A woodcut of 1504 depicting the organization of studies in Arts after the assimilation of works translated from the Greek. At the lower levels, grammar and rhetoric are represented by the Roman authors Donatus, Priscian, and Cicero ('Tullius'), while Boethius represents arithmetic, Aristotle logic, Pythagoras music, Euclid geometry, and Ptolemy astronomy. At the top are the three philosophies: moral philosophy (Seneca), natural philosophy (unnamed, perhaps because Aristotle is present elsewhere), and, topmost of all, metaphysics or theology, represented by Peter Lombard, whose *Sentences* was the basic text in theology.

Most of the Aristotelian works were eventually incorporated into the last years of the Arts course, where they were divided between the three 'philosophies' of metaphysics, moral philosophy, and natural philosophy. This meant there was now less time for the quadrivium, where a very rudimentary astronomy had long been taught; but this loss was more than offset by the teaching, in the courses on natural philosophy, of Aristotelian material relating to astronomy. As to the *Almagest* and the major Arabic works in mathematical astronomy, no young student in Arts could think for a moment of attempting to master them; manuscripts, even if available, were rare, prohibitively expensive, and subject to copyist's errors, and in any case the conditions under which students worked precluded the study of anything other than the most elementary treatises of astronomy.

Furthermore, research – the extension of the frontiers of knowledge – was not the function of a university, whose duty was rather to educate. This was done through the systematic study of approved texts of recognized quality, and through intensive training in the skills of logical reasoning. The typical medieval Arts student was no mathematical astronomer, and certainly there was no place for observation in his astronomy course; but on the other hand, once introduced to epicycles or equants, he was likely to form the most decided opinions about their reality or validity.

The quality of the translated works in astronomy soon made it abundantly clear that the introductory texts in use were no longer acceptable. The first to fill the gap was John of Holywood, better known by the Latinized form of his name, Sacrobosco, who taught in Paris in the mid-thirteenth century. He provided students with three short works that together supplied their needs in astronomy: a *Compotus,* or introduction to time-reckoning, an *Algorismus* that taught the arithmetic needed in astronomical computation, and a *Tractatus de sphaera* (or *Sphere*).

Sacrobosco's *Sphere* contained four 'books'. Book I dealt with the heavenly sphere, its revolution, the sphericity of the Earth, and its location at the centre; Book II defined the celestial equator, the ecliptic, the zodiac and zodiacal constellations, the meridian, the altitude of the Pole, and the division of the Earth by means of the tropics and polar circles; Book III dealt with the rising and setting of celestial bodies, and the length of day and night at different latitudes and 'climates'; and Book IV gave a brief description of the motion of the Sun, Moon, and planets, and an outline theory of eclipses.

Astrologia (left) directing the attention of Sacrobosco to Ptolemy (right) who is observing with an astrolabe.

While the first three books were of a tolerable standard, if only just, the fourth was hopelessly inadequate, and teachers needed something better. The most popular of the texts written to fill the gap was the anonymous *Theory of the Planets*, written in the second half of the thirteenth century. It began with Hipparchus's theory of the Sun, and went on to the Ptolemaic theories of the Moon and the outer planets. The author then had chapters on Venus and Mercury, and on the direct and

The heavenly spheres and their moving spirits

The incorporation of the Aristotelian spheres into the Christian world picture resulted in a cosmology based on both sacred and profane learning, and consequently of great strength. However, there were problems over the region that lay outside the spheres of the seven planets. How did the multiple motions of the sphere of the fixed stars come about? What was the relationship between this sphere and 'the heaven' created on the first day according to the story in Genesis — and between that and 'the firmament' brought forth on the second day and made visible on the fourth? And what was to be made of 'the waters above the firmament'?

Everyone agreed that in addition to their daily motion, the fixed stars performed the slow movement that in fact results from the wobble of the Earth's axis ('precession'). Many astronomers wrongly believed that the rate of precession itself varied, and that this involved a third movement known as 'trepidation'. To generate these three movements, writers often thought it necessary to postulate three spheres. Albert of Saxony (c. 1316–90), for example, a leading figure first at Paris and then at the University of Vienna, is typical in assigning precession to the eighth sphere, trepidation to the ninth, and the daily motion to the tenth.

But how did these astronomical spheres relate to the heavenly realms described in Genesis? The 'firmament' was frequently identified with the eighth sphere, in which lay the fixed stars themselves. 'The waters above the firmament' were often thought of as being as hard as crystal, and as forming the ninth sphere (or the ninth and tenth). 'The heaven' created on the first day was the outermost (usually the eleventh) of all the spheres, the motionless *Empyreum*, which served a purely theological purpose. The Empyreum was the ultimate container of the universe, and the dwelling place of God and the elect.

Aristotelian natural philosophy required all motions to have a cause. What was it that caused the motions of the planetary and starry spheres? The heavens were all made of a single substance, the fifth element, or 'quintessence', and so the motions were, in effect, frictionless and therefore effort-

less; but nevertheless there had to be some cause of their motion. The ultimate source was God, the Prime Mover, who acted directly on the outermost of the moving spheres. But God had also assigned to each of the other moving spheres an immaterial, spiritual intelligence, or angel, which was able, voluntarily and inexhaustibly, to move its sphere with the appropriate uniform circular motion.

For a more satisfactory explanation of the continued motion of the heavenly spheres, Jean Buridan in fourteenth-century Paris adapted the concept of *impetus*, which had been devised as the cause of the continued motion of projectiles. There was, he pointed out, no mention of the angelic intelligences in Scripture; and if God had impressed an impetus on each sphere at the Creation, then in the absence of friction this impetus would itself have been sufficient to keep the sphere in motion indefinitely.

Opposite: A theologian and an astronomer discuss the heavenly spheres. Besides the basic complement of eight spheres (one for each of the seven planets and an eighth for the fixed stars), astronomers needed an additional sphere to account for precession – two if they believed the rate of precession itself varied – while theologians needed further spheres to accommodate the descriptions of the heavens in Genesis.

Above: God imparting motion to the outermost of the heavenly spheres, in Giovanni di Paolo's *Expulsion from the Garden of Eden*. This panel was painted in about 1445 for the Guelfi Chapel in the church of San Domingo in Siena.

Right: Angels at work, generating the rotations of the heavenly spheres.

The 'old' quadrant (quarter-circle in brass) was known in Europe in the first half of the thirteenth century. It could be used to measure altitudes: the observer looked along the edge of the instrument towards the target object, whose altitude was shown by the position of the plumb-bob on the scale at the circumference. But the sophistication of the instrument lay in the engraved, curved hour lines and adjustable scale, which enabled it to be used as a sundial for any latitude.

retrograde motion of planets. He described the geometrical models briefly but well, with definitions clearly stated. He was, however, less satisfactory on the other topics he tackled, such as eclipses.

In time other authors were to write treatises to make good the shortcomings in the *Theory*, just as the *Theory* made good shortcomings in the *Sphere*. The historian of medieval astronomy is able to monitor the steady improvement in the technical quality of teaching in astronomy, by studying the works that were grouped together in surviving manuscripts to form the set of treatises that a student needed: the quality of the sets steadily improved as the less satisfactory works dropped out and were replaced by better ones.

Other works that we find in such collections are treatises on the calendar, and the *Alfonsine Tables* of the planetary motions, named after the thirteenth-century patron of astronomy, King Alfonso X of Castile (though some modern historians believe the tables are in fact of French origin). These tables soon replaced the old *Toledan Tables* inherited from the Arabs. Modern computer analysis has shown that they were calculated from Ptolemaic models of the planetary motions, with only occasionally modified parameters.

Another striking feature of the manuscript collections of the time is the new interest in instruments. The 'old quadrant' (shown left) is described in an encyclopedia compiled by Alfonso's astronomers, but it was supplemented at the end of the thirteenth century by the much more complex 'new quadrant' (a combination of quadrant and astrolabe) of the Montpellier astronomer Profatius the Jew (Jacob ben Mahir, c. 1236–c. 1304). Another Jewish astronomer, Levi ben Gerson (1288–1344) of Provence, invented the cross-staff, which allowed the angle between two objects (such as stars) to be measured.

In the Latin West, however, purpose-built observatories were unknown and would remain so until the time of Tycho Brahe in the late sixteenth century. Any observations were taken from private homes and with portable instruments. So, for example, an astronomer of Roskilde in Denmark in 1274 used an astrolabe to measure the noon altitude of the Sun every day of the year; from this could be calculated the length of daylight.

Criticism of Ptolemy came from natural philosophers concerned to learn the truth about the cosmos, rather than from the mathematical astronomers, for whose

A navigator learns how to use a cross-staff to measure the angle between the Moon and a particular star. Elsewhere in this woodcut of 1533, surveyors are using the instrument to determine the height of a tower. To use a cross-staff to measure the angle separating two objects, the observer slides the cross bar towards or away from himself until each object is in line with one or other end of the bar; the angle can then be read off from the scale on the central member. A version of this instrument was later in widespread use among seamen, although it suffered from the defect that in moving the cross bar, the observer affected the alignment of both objects: it was not feasible to align one object and then adjust for the other.

calculations the Ptolemaic models (only rarely with updated parameters) were indispensable. Most Latin philosophers agreed with their Arabic counterparts, that the non-uniform motions implied by equants could not occur in the real world. In fact, in the eyes of many, epicycles and eccentres were likewise unacceptable: the motions involved were indeed uniform and circular, but they did not take place around the Earth, that spherical aggregate of earthy bodies gathered at the centre of the cosmos.

But what of the possibility that the Earth itself was spinning on its axis? Such a motion is of course inherently implausible to us humans who believe ourselves to be on *terra firma*, but the arguments against came primarily from consideration of the motion of projectiles such as arrows – not from astronomy, nor from the

Above: Richard of Wallingford as Abbot of St Albans, with the clock he constructed for the abbey. Richard suffered from 'leprosy' and is depicted with a spotted face.

The introduction of clockwork

The mechanical clock is recorded with certainty only from the fourteenth century, and historians have been intrigued and puzzled by the fact that it appeared, not in a simple and primitive form, but as a highly complex and sophisticated mechanism. Not surprisingly — since timetelling was derived from astronomy — the more sophisticated of these clocks were astronomical. Some were mechanical astrolabes, while others were mechanizations of astronomical demonstration instruments. Historians have been hampered in their investigations of earlier and simpler mechanisms by the fact that the Latin word *horologium* can refer either to a clock or to a sundial.

Richard of Wallingford (c. 1292–1336), Abbot of St Albans in England, built one of the most celebrated of these astronomical clocks in the early fourteenth century, and left a detailed technical account of its construction. The clock was a replica of the medieval universe. Its features included the phases and eclipses of the Moon (using differential gears), and an oval wheel gave an accurately varying velocity to the Moon as it moved around an astrolabe dial.

Even more famous in its time was the 'astrarium' of the professor of astronomy at Padua, Giovanni de' Dondi (1318–89). The machine, which he completed in Pavia in 1364, is also fully documented, and modern reconstructions have been made. It had a seven-sided frame with one dial for each of the planets (Sun, Moon, Mercury, etc.), each mechanism being in effect a geared Ptolemaic diagram. Below there were dials or displays showing the twenty-four hours, the times of sunrise and sunset, and much else. The astrarium was one of the marvels of the age; Regiomontanus (see pages 88–9) visited it in 1463 and he records that prelates and princes flocked to see it as if they were to witness a miracle.

The large astronomical public clock became popular in the fifteenth and sixteenth centuries, with famous examples installed at Strasbourg (see page 138), Padua, Prague, and many other centres of civic and ecclesiastical pride.

In this fifteenth-century manuscript of the *Horloge de Sapience*, a work on the practical aspects of mysticism by the German Heinrich Suso (1300–66), we see Sapientia (Wisdom) instructing the author, in a room full of time-telling devices. Behind Suso is a mechanical clock with a single hand and a dial marked with twenty-fours (I–XII, repeated). From the front of the clock hangs an astrolabe. To the left of Sapientia is a bell-ringing mechanism whose precise nature is uncertain. On the table are a horizontal dial, a spring-driven clock movement, and a universal equinoctial dial (a sundial termed 'universal' because it could be used for any latitude, 'equinoctial' because it measured equal hours rather than dividing daylight and nighttime each into twelve equal parts). Hanging from the table is an 'old' quadrant (see page 80).

Latitude and the time at night on board ship

As the Middle Ages drew to a close, European navigators began to venture ever more frequently out of sight of land. Mistakes in position could easily lead to shipwreck, and so navigators usually kept to basics: they followed a course that brought them to the required latitude well to the west (or east) of their destination, confirmed their latitude by observation, and then travelled east (or west) along the parallel of latitude.

Measurement of latitude was therefore of great importance. In the daytime, latitude could be determined from the altitude of the Sun at noon (that is, when it was highest in the sky). To measure this, the navigator would use an instrument such as a cross-staff or a mariner's astrolabe. He would then consult a table that gave the position of the Sun above (or below) the celestial equator for the time of year. Combining his noon observation with the angle given in the table, he obtained the altitude of the celestial equator as seen from his ship, and hence the ship's latitude.

Had the Pole Star been located *exactly* at the celestial Pole, the navigator at night could simply have used the altitude of the star as the ship's latitude. Unfortunately this was not the case – in 1500 the star was some 3° from the Pole – and some adjustment was necessary.

The present position of the Pole Star in relation to the Pole could be inferred from the configuration of the other stars rotating around the Pole. In practice navigators normally used the 'Guards', two stars in the constellation of Ursa Major (the Plough or Big Dipper) that happened to be in line with the Pole Star. This line turned in the sky like the hand of a twenty-four-hour clock. There was, however, a complication: the clock ran on star time ('sidereal' time) rather than Sun time. We see the Sun and stars go round every day, but because the Sun also circles once a year among the background stars, the 'day' of a star is some four minutes shorter than that of the Sun, and this difference accumulates steadily throughout the year.

An ingenious instrument called a 'nocturnal' allowed the navigator either to calculate the required adjustment in altitude, or simply to tell the (conventional) time at

A mariner's astrolabe made in 1616. To reduce interference by the wind, the framework has been reduced to a skeleton, and to encourage stability, the instrument is weighted at the bottom.

relevant Scripture passages (to which the medievals were able to take a more measured approach than would later be possible in the post-Reformation period). As Aristotle had taught, the fact that an arrow fired vertically upwards on a calm day would fall to ground where the archer was standing, proved that the ground below had not moved during the period of flight.

night. It was fitted with two concentric circles. The outer circle, which was attached to the handle, was graduated with an annual calendar. The inner circle, which rotated freely, was graduated into twenty-four hours.

The navigator rotated the inner circle until the noon mark was aligned with the date on the outer circle; this took care of the difference between sidereal and Sun time. He then held the instrument above him so that it was roughly in the plane of the celestial equator. He looked at the Pole Star through a hole in the centre, and he then took hold of an index bar, which rotated about the centre and projected beyond the edge of the circles, and turned it until the bar touched the two stars. The time was then shown by where the index bar crossed the hour circle.

To find out what adjustment to make to the altitude of the Pole Star in order to get the altitude of the true Pole, and hence the

latitude of the ship, the navigator simply consulted a scale on the reverse of the instrument. Otherwise tables known as 'The Regiment of the North Star' yielded the required adjustment from an estimate of the orientation of the Guards.

Above: The nocturnal being used to find the time at night (in this case, 1.30 a.m.). This nocturnal has 'teeth' around the edge, so that if the navigator merely wishes to know the approximate time, he can 'read' the instrument in the dark simply by feel.

Left: A nocturnal made at the beginning of the eighteenth century. This instrument is designed for use with stars from either Ursa Minor (Little Bear or 'LB') or Ursa Major (Great Bear or 'GB').

Opposite: A miniature showing Nicole Oresme at work in his study, with an armillary sphere. It comes from a fifteenth-century manuscript (which once belonged to the Duke of Berry) of Oresme's *Traité de l'Espere*, which is related to his commentary on Aristotle's astronomical treatise, *On the Heavens and the Earth*. This was one of a number of works that, at the request of King Charles V, he either wrote in (Old) French or translated, thereby helping to formulate a new scientific terminology in that language. In his commentary to *On the Heavens and the Earth*, Oresme argued that there was no proof that the Earth was at rest; in other writings he proved himself a determined critic of astrology.

However, the Aristotelian theory of projectiles was itself in disarray, for his explanation of why it was that the arrow continued to rise in the air after it had lost contact with the bowstring was clearly absurd. In continuing its upwards movement, the arrow, although composed mainly of the element earth, was getting still further from its natural place at the centre of the Earth; and this unnatural (or 'violent') movement must be the result of the action of an external force. Anxious to identify the source of this external force, Aristotle saw only one possible candidate, the surrounding air, for this was the only material body in contact with the arrow. Somehow, the air contrived for a time to push the arrow upwards, so preventing its beginning its fall to the ground the instant it left the bowstring.

Parisian masters of the mid-fourteenth century were not the first to point out the absurdity of this – for if Aristotle was right, how was it possible to fire an arrow into the teeth of a gale? – but they were the ones who elaborated an alternative theory, and then applied it to the possible spin of the Earth. Jean Buridan (c. 1295–c. 1358) and Nicole Oresme (c. 1320–1382) both agreed with Aristotle that *some* force must be at work, but both saw the absurdity of making the air the origin of this force, and both proposed instead that the projector – in this case, the archer with his bow – might impart to the projectile an 'incorporeal motive force', which they termed *impetus*.

Buridan then used this concept to explain other puzzling phenomena, such as the acceleration of a freely falling body: impetus was responsible for the continuance of the downward motion of the body, while the natural downwards tendency produced the acceleration. Oresme meanwhile pointed out that if the Earth were to rotate, then the archer would be moving sideways with the Earth. Consequently, merely by holding the arrow prior to firing it, the archer would share with the arrow his sideways motion and hence impart to it a sideways impetus. This impetus would cause the arrow while in flight to keep pace with the sideways movement of the archer below, and hence to fall beside the archer. In other words, an arrow fired vertically would fall onto the archer, whether or not the Earth was in motion: one therefore could not decide whether or not the Earth was at rest, simply by firing arrows into the air. Nor, declared Oresme, could the question be decided by other arguments from natural philosophy, or by astronomical observation, or by the study of quotations from Scripture. He himself believed that the Earth was at rest; but there was no proof either way.

Oresme's discussion was but one example of the growing self-confidence of Masters of Arts of the fourteenth century, and of their readiness to contradict Aristotle. The twelfth century had been that of the translations, the thirteenth, that of assimilation; now the learned world was on the move forward. In the coming century a new technology was to become available that would revolutionize the study of the mathematical sciences, and especially geometry and astronomy: the invention of printing.

La figur et la disposicion du monde le nombre et ordre des elemens et les mouuemens

des corps du ciel apptienment a sa uoir a tout home qui est de france condicion et de noble engin Et est bele chose et delectable profetable et honeste et aueqes ce est neccessaire

pour sauoir plue et p especial po astrologie Et pce afin que engin humain peust plus leggeremet tele chose comprendre les sages anciens composerent et les uns un instrument qui est appelle espere matiel ou artificiel le ql on peut ptgarder tout entour ci muoir et tourner et y considri en ptie la disposicion et le mouue ment du monde et du ciel aussi

PRINTING AND THE RISE OF HUMANISM

Books are the most effective means whereby the knowledge of one generation is passed to the next. Before printing, books existed in the form of manuscripts, copied – one from another, at great expense of labour – by scribes. If the subject matter was expressed in words – if the text was by Aristotle, for example – then the scribe would have some grasp of what he was copying; he might even be able to correct slips of his predecessors that he found in the manuscript before him. But if the subject matter was expressed in symbols, as in a treatise on geometry or astronomy, then the scribe would seldom have any hope of noticing when one of his predecessors had introduced an error. These earlier errors would therefore be repeated, and no doubt in making this fresh copy our scribe would himself introduce new ones of his own. As a result, a would-be astronomer wishing to master the *Almagest* – even when he had managed to lay hands on a copy – would find himself struggling to make sense of an unreliable text.

The mid-fifteenth-century invention of printing with movable type transformed this situation. It now made sense for the scholarly editor of an ancient treatise to devote months, or even years, to establishing a text faithful to the intentions of the original author; for he could then correct printer's proofs and supervise the making of woodblock illustrations, and, when everything was to his satisfaction, hundreds of identical copies would be run off and distributed throughout Europe for sale at reasonable cost.

Nowhere was this development more significant than in astronomy. Furthermore, the invention coincided with an increasing appreciation of the achievements of the Greeks and Romans. Antiquity was now coming to be seen as a Golden Age of artistic, literary, and philosophical achievement that had been followed by the decline of the 'Middle Ages'. This 'Humanist' movement was primarily concerned with the works of man rather than with the understanding of nature, but all ancient writings, even the scientific, were legacies to be treasured.

A leading figure of the Humanist movement was the Greek scholar Johannes Bessarion (c. 1395–1472), who had been educated in Constantinople. He had come to Italy to take part in moves to reunite the Greek and Latin churches; and when these proved unsuccessful, he had remained in the West, becoming a cardinal in 1439. In 1460 Bessarion arrived in Vienna as papal legate, and there he met the Austrian court astrologer, Georg Peurbach (1423–61), and his young collaborator in astronomy, Johannes Müller of Königsberg (1436–76), known as Regiomontanus from the Latinized name of his birthplace. Peurbach had already written a *Theoricae novae planetarum* (*New Theories of the Planets*) designed to supersede the thirteenth-century textbook of similar name. Regiomontanus was to see this work into print in about 1474,

The invention of printing permitted the duplication of books with cardboard models ('volvelles') that could be used to compute positions of planets. By far the most sumptuous of these books was *Astronomicum Caesareum*, published in 1540 by Peter Apian (1495–1552), professor at Ingolstadt. The detail shows the model for the longitude of Mercury, set for 23 February 1500, the birthday of Apian's patron, the Emperor Charles V. The threads were used, when required, to align corresponding points on the component dials.

1502	1502	1502
finſter der Sunnē	finſter des mōdes	finſter des mōdes
30 19 24	19 12 20	29 13 30
Des Derbſtmōdes	Des Weinmōdes	Des Dornungs
Halbe werung	Halbe werung	Halbe werung
1 8	1 1	1 26
Czehen punct	Drei punct	

Part of a page from the *Kalendar* published by Regiomontanus in 1474 and taken by Columbus on his fourth voyage. At right is the forecast of the eclipse of the Moon of 29 February 1504, which Columbus used to frighten the natives of Jamaica. Believing that the prayers of Columbus had restored the Moon, the natives brought much-needed food for Columbus and his crew.

some years after Peurbach's early death, and it became a best-selling textbook, appearing in dozens of editions well into the seventeenth century. One feature of the work was its detailed descriptions of physically real representations of Ptolemaic planetary models in terms of solid spheres (see page 46), representations whose shortcomings may have been the motivation that led Copernicus to take up planetary astronomy. Eventually, Tycho Brahe's demonstration that the comet of 1577 had passed without hindrance through the planetary regions (see page 101) was to show that no such solid spheres existed.

Bessarion was himself an outstanding scholar and collector of manuscripts, and he was anxious to see the *Almagest* made more accessible to students, through the preparation of a systematic abridgement. This project appealed to the two astronomers, who set to work; within a year, however, Peurbach died, with the abridgement completed of only the first six 'books'. On his deathbed he made his friend promise to complete the project, and this Regiomontanus did while with Bessarion in Italy. The *Epitome* of the *Almagest*, half the length of the original and a model of clarity, was printed in 1496; and as copies spread throughout Europe, Ptolemy's mathematical astronomy at last became truly accessible. There were even prospects that the *Almagest* itself might one day be superseded.

In 1471, after some years spent partly with Bessarion, Regiomontanus settled in the prosperous commercial centre of Nuremberg. The best instrument-makers were on his doorstep, and his home became not only his observatory but also his printing house, where he could produce astronomical books that commercial publishers might find unattractive: Peurbach's *Theoricae novae planetarum* was followed by an *Ephemerides* giving the positions of the heavenly bodies for each day from 1475 to 1506, the first such to get into print. Columbus took a copy of one of Regiomontanus's books with him on his fourth voyage (1502–04) and used its prediction of the lunar eclipse of 29 February 1504 to overawe the hostile natives of Jamaica.

In 1475, however, Regiomontanus left Nuremberg for Rome, apparently to help with the calendar reform, and there he too met an untimely death. He had, however, found a collaborator in Nuremberg, the merchant Bernhard Walther (1430–1504), whose own observations began five days after Regiomontanus left the city and continued for nearly thirty years. Walther achieved a new level of accuracy in astronomical measurements, and his observations were to be used extensively by Copernicus, Tycho Brahe, and Kepler.

COPERNICUS AND THE CULMINATION OF THE GREEK PROGRAMME

Nicolaus Copernicus (1473–1543) was born in Toruń in Poland and studied at the University of Cracow. Although remote from the main cultural centres of Europe, Cracow University had a distinguished tradition in astronomy, including teachers who made no secret of their dissatisfaction with the way that equants violated the principle of uniform circular motion.

In 1496 Copernicus went to Italy to study. Constantinople was no longer in Christian hands, for in 1453 the city where Greek culture had survived for so long had at last fallen to the Muslims. In the decades preceding this traumatic event, Greek scholars had begun making their way west, and especially to Italy, where in Florence a school had been established in 1462 by Marsilio Ficino in explicit imitation of Plato's own Academy. Greek was now widely studied in universities, and the elegant dialogues of Plato – most of which were unknown in the Middle Ages – had become available and were proving profoundly influential. Although the Platonic

Copernicus

Nicholas Copernicus was born in Toruń, Poland, in 1473, and entered the University of Cracow in 1491. In 1496 he enrolled at the University of Bologna as a student of canon law, eventually taking his doctorate at the University of Ferrara in 1503. While in Italy he enhanced his love of astronomy, making his first recorded observation in March 1497, and acquiring some reputation in the subject: in 1500 he is said to have lectured on astronomy in Rome to a large audience. Meanwhile, through the influence of his uncle, a bishop, he was elected a canon of the cathedral chapter of Frombork (Frauenburg), an administrative position that did not require him to be ordained, and it was with the approval of the chapter that he returned to Italy in 1501 for two years of medical studies. In 1503 he went back to Poland, where he was in the service of the Frombork chapter until his death in 1543.

Opposite: A horoscope for Copernicus's birth. In the central square are his name and the date, hour and minute of his birth, and around the square are twelve triangles, one for each of twelve portions ('houses') of the zodiac. The fate of the person depended on which planets were in the various houses, which signs of the zodiac the various planets occupied, whether a planet was in one of the signs with which it was especially associated or was in the place on the ecliptic where it exerted its greatest (or least) influence, and how the positions of the planets were interrelated.

tradition was towards abstraction and away from routine observation of the natural world, it was imbued with a mathematical view of nature, and demanded of any theory of the cosmos that it reflect the cosmic harmony and symmetry.

In Italy Copernicus studied canon law and medicine, but he also learned Greek and developed his interest in astronomy, his future disciple Georg Joachim Rheticus (1514–74) telling us that around 1500 in Rome he lectured on astronomy before experts. Astronomy was in a far from satisfactory state. The calendar – society's basic requirement from astronomers – was clearly getting out of step with the seasons (and would remain so until the Gregorian reform of 1582 – see page 108). To all students in Arts, the presence of equants in the Ptolemaic planetary models was a cause for scandal, so that Rheticus would describe the equant as 'a relation

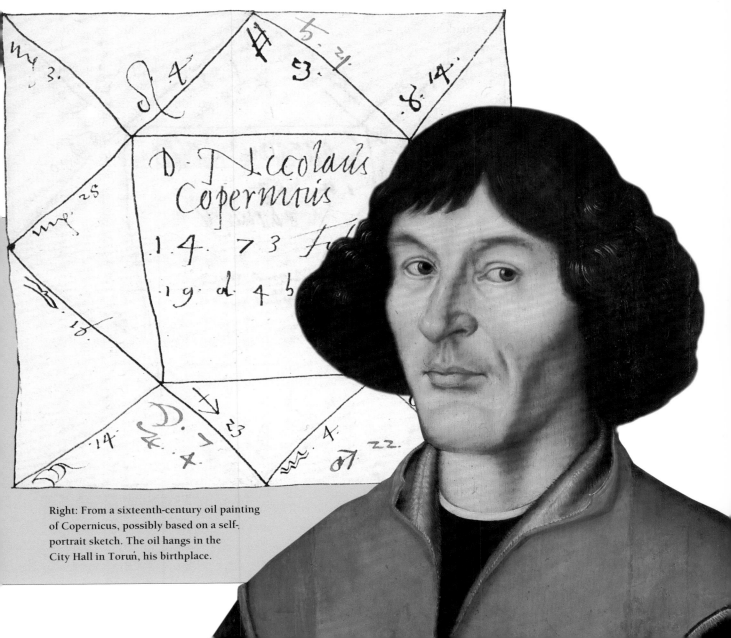

Right: From a sixteenth-century oil painting of Copernicus, possibly based on a self-portrait sketch. The oil hangs in the City Hall in Toruń, his birthplace.

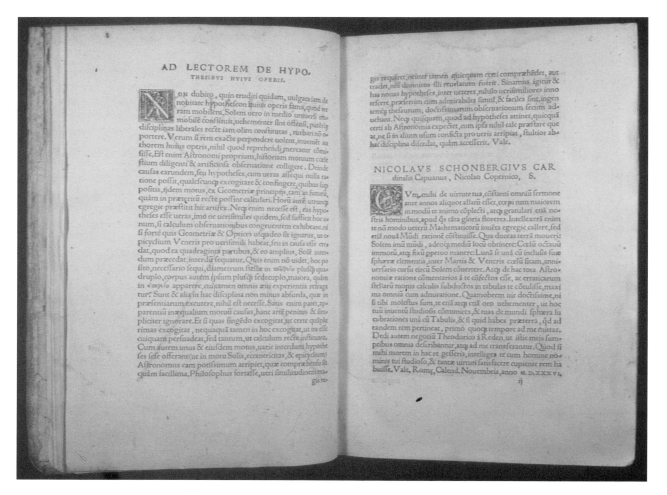

A copy of Copernicus's *De revolutionibus*, in which the unauthorized and unsigned introduction inserted by Osiander has been cancelled by an early owner. The true identity of the author of the passage became generally known only after 1609, when Kepler publicized the deceit in his *Astronomia nova*.

that nature abhors'. Those who delved further into the technicalities of mathematical astronomy were disturbed because the Ptolemaic model of the Moon's motion implied that the Moon's apparent size varied greatly (see page 44), which it clearly did not. Ptolemy's *Planetary Hypotheses* had been lost, and with it his exposition of his integrated system of the planets; in the current Platonist climate, the fact that in the *Almagest* Ptolemy treated each planet independently, without any attempt to develop a coherent cosmic system, made the work deeply unsatisfactory. Past astronomers, Copernicus would later write,

> have not been able to discover or to infer the chief point of all – the structure of
> the universe and the true symmetry of its parts. But they are just like someone
> taking from different places hands, feet, head, and the other limbs, no doubt
> depicted very well but not modelled from the same body and not matching one
> another – so that such parts would produce a monster rather than a man.

On the other hand, the geometrical models from which planetary tables were calculated were seen as reasonably satisfactory for this purpose: both the level of complexity of the models and the accuracy of the resultant tables were acceptable. Modern legend has it that by the sixteenth century, the Ptolemaic models had been elaborated in a desperate search for greater accuracy, to the point where the number of circles required was beyond all reason. The legend goes on to claim that on the Sun-centred hypothesis the number of circles was greatly reduced, and that the motion of the Earth was considered an acceptable price to pay for this simplification.

In this legend there is no truth: Copernicus's detailed models are every bit as complicated as Ptolemy's. Indeed, it is obvious that the legend cannot be true; for the supposed elaboration of the Ptolemaic models could only have been necessitated by new observations, observations of such precision that mere adjustments in the parameters were not enough. But observations of this accuracy not only were not, but could not be made at this period – the necessary instruments simply did not exist. Sadly, the legend will doubtless persist; which is a shame, because it offers a pedestrian motivation for what is in fact one of the greatest of the intellectual leaps known to the history of science.

For those (like Copernicus) who saw other reasons for a reform of astronomy, there were several clues to the direction such a reform might take. Some were to be found in Aristotle's own writings. In discussing the stability of the Earth, for example, the Philosopher followed his usual custom and reported the views of his predecessors before demolishing them; in consequence, every student in a medieval or Renaissance university heard of ancient Greeks who believed the Earth to be in motion. To take a single example: Albert of Saxony, who was Rector of Paris University in the 1350s, expounded these ancient opinions in his *Questions* on Aristotle's *On the Heavens and the Earth*, concluding, '*Ita dixerunt solem non moveri circa terram sed magis terram circa solem*' – 'Hence they said that the Sun does not move around the Earth but rather the Earth about the Sun'. It is no surprise to find that Albert concluded in favour of Aristotle; but his *Questions*, with their account of this alternative view, was reprinted no fewer than six times during the lifetime of Copernicus.

Another clue came from the mystery as to why the annual period of the Sun had found its way into the model of each planet. So Peurbach had written: 'It is clear that each of the six planets in its motion shares something with the Sun, and the Sun's motion is, so to speak, the common mirror and measure for their motions.' With hindsight we know the reason: relative to us on Earth, the Sun orbits once a year taking with it all the other planets, so that satisfactory representations of their motions relative to us – which is what Ptolemy attempted – must in every case involve the Sun's period of one year.

In 1503 Copernicus returned to his native Poland, where he had meanwhile become a canon of the cathedral of Frombork (Frauenburg), where his uncle was bishop. As a canon (but not a priest) he was an administrator, and a private astronomer. We know little of the development of his thinking in the years that followed, but before long a manuscript by him later entitled *Commentariolus*, or *Little Commentary*, began to circulate. In it he briefly stated his dissatisfaction with existing planetary astronomy, singling out equants for special mention. He then set out the postulates of a Sun-centred alternative (in which the Earth became a planet with the Moon as its satellite), showed how this approach allowed an unambiguous order to be assigned to the planets (now six in number), and went on to develop equant-free outline models for the planets and the Moon.

Little more was heard of the Polish canon until Rheticus, then a teacher of mathematics in the University of Wittenberg, decided to visit him in 1539. Copernicus agreed to allow Rheticus to publish a *First Report* of his work, and the slim volume appeared the following year. No violent controversy ensued, and perhaps it was this that persuaded Copernicus to allow Rheticus to take his complete work, *On the Revolutions of the Heavenly Spheres*, for printing in Nuremberg. *De revolutionibus*, as it is always known, was published in 1543, the year of its author's death.

Rheticus's teaching duties had not permitted him personally to oversee the passage of the manuscript through the press. This task fell to a Lutheran clergyman, Andreas Osiander (1498–1552). No doubt with the good intention of shielding the author from controversy, Osiander inserted in the front of the work an introduction that was unsigned (and so apparently by Copernicus himself), declaring that the author was not maintaining that the Earth truly moved around the Sun, only that this was a convenient hypothesis on which to base efficient mathematical models of the planetary motions.

Not surprisingly, this served to obscure Copernicus's message. The author's true position should have been clear to anyone carefully studying the work as a whole; but the readers of *De revolutionibus* were mainly technical astronomers preoccupied with computing tables of planetary positions and eager to plunder the work for any help it could give them in this. In 1551, for example, Erasmus Reinhold (1511–1553), professor of astronomy in the University of Wittenberg, based his *Prutenic Tables* on *De revolutionibus*; he made minor adjustments in the parameters in Copernicus's models, but the models themselves he followed slavishly, seemingly uninterested in the truth or otherwise of the Sun-centred hypothesis.

Copernicus's treatise attempted simultaneously to serve two quite different purposes. In Book I he demonstrated the attractive consequences of the proposition that the Earth was an ordinary planet orbiting the Sun: the planets would form a coherent and integrated system, and many otherwise puzzling facts of observation would become natural and to be expected. Book I was a treatise on the cosmos, and although Copernicus could offer no compelling proof of what we might term his cosmovision, Book I was concerned with the fundamental structure of the world in which we live. The remaining 'books', however, served a quite different purpose. They were to provide 'justification by works' for the cosmovision, by showing that adequate planetary tables could indeed be calculated from geometrical models with the Sun at the centre. These books were concerned with quantity and with the close match between prediction and observation, not with truth and falsehood.

In Book I Copernicus showed that one natural consequence of his making the Earth into a planet in orbit about the Sun, was that the other planets would divide themselves into two groups, those whose orbits lay inside the path of the Earth and those whose orbits lay outside. It had been known since long before Ptolemy that Venus and Mercury are to be seen only around dusk or dawn, while Mars, Jupiter, and Saturn may be visible at any time of night. But what had been so mysterious a division on the Ptolemaic view, now became a natural and expected consequence of the Sun-centred hypothesis.

In Ptolemaic astronomy the Sun orbited the Earth with a period of one year; and since Venus and Mercury kept company with the Sun, they too inevitably had a period of one year. But if Copernicus was right and it was the observer on Earth who had the period of one year, it was a simple matter to make the appropriate allowance for this and to disentangle the true periods of Venus and Mercury. These turned out to be around seven and a half months and eighty-eight days, respectively. As a result, every planet now had its own individual period, and they could at last be ordered by increasing period in a well-defined sequence: Mercury, Venus, Earth, Mars, Jupiter, Saturn.

Erasmus Reinhold, one of the first great students of Copernicus's *De revolutionibus*, was especially interested in the various mechanisms that Copernicus proposed for eliminating Ptolemy's equant. Here his marginal notes tabulate and distinguish the circles used for the Earth's motion. In the text Copernicus states that he does not know which is correct 'but it must be one of them since they all give the same result'.

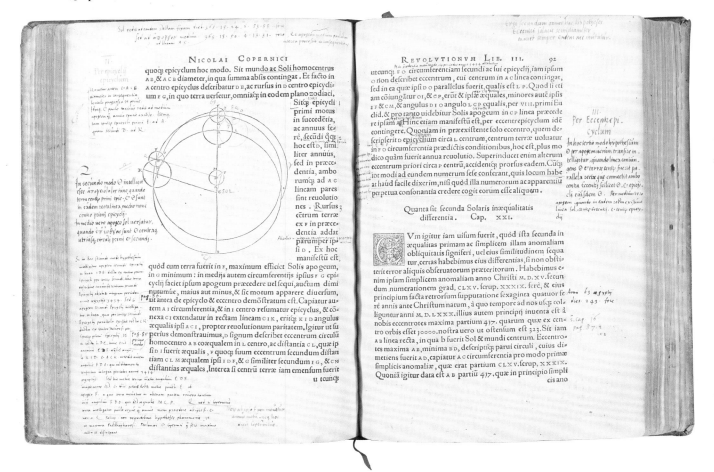

But it is not only the period of the Earth that enters into any observations we make of the other planets – so does the Earth's distance from the Sun (the 'astronomical unit'). Once this was recognized, this common distance scale could be applied to each of the geometric models of the planets, and it was then a simple matter to calculate the (relative) distance of each of them from the Sun. This allowed Copernicus to order the planets by a second criterion: their increasing remoteness from the Sun at the centre of the system.

To the delight of his Platonist soul, Copernicus found the two lists identical: 'Therefore in this arrangement we find that the world has a wonderful commensurability, and that there is a sure linking together in harmony of the movement and magnitude of the orbital circles, such as cannot be found in any other way.'

Copernicus's diagram of the solar system, from the manuscript of Book I of *De revolutionibus*, with (very approximate) periods. Note that the Earth ('*Tellus*' here in the genitive '*Telluris*') is anomalous in carrying a satellite around the Sun, as indicated by the caption: '5. The annual revolution of the Earth with the Moon.'

What of the occasional and very puzzling backward ('retrograde') motions that mathematical astronomers since the time of Plato had struggled so hard to reproduce in their geometrical models? Copernicus showed that these too were to be expected on the Sun-centred view, as no more than the natural consequence of our observing the other planets from a moving platform. For example, we on Earth orbit the Sun on the inside of Mars. Most of the time we see Mars moving around the Sun in the same direction as ourselves; but for the relatively short time that we are actually overtaking Mars, Mars appears to us to be moving backwards. The age-old mystery of the retrogressions that earned the planets the name of 'wanderers' had at long last been given a simple explanation.

If one looked a little more closely at the sizes of the orbits and the relative periods, one could easily calculate many of the finer details of the planetary retrogressions – such as where a planet would be when it appeared to come to a halt, for how long it would seem to move backwards, and where it would be when it appeared to resume its onwards movement. Copernicus triumphantly listed many such observed facts, pointing out that 'all these proceed from the same cause, which rests in the movement of the Earth'.

He reserved his greatest display of emotion for the passage where he reflected on how the Sun, symbol of the good in the Platonist tradition, was now in the middle of the universe:

> In the centre of all resides the Sun. For in this most beautiful temple, who would place this lamp in another or better place than that from which it can illuminate the whole at one and the same time? As a matter of fact, not inappropriately do some call it the lantern of the universe; others, its mind; and others still, its ruler. The Thrice-Great Hermes calls it a 'visible god'; Sophocles's Electra, 'that which gazes upon all things'. And thus the Sun, as if seated on a kingly throne, governs the family of planets that wheel around it.

The proclamation of Copernicus's cosmovision occupied no more than the first half of Book I of *De revolutionibus*. The remaining 95 per cent of the volume was dauntingly mathematical, offering proof that the Sun-centred hypothesis could deliver the goods – that it could be made the basis of an equant-free mathematical astronomy at least as effective as Ptolemy's *Almagest* in the computation of planetary tables. After thirteen centuries, Ptolemy had at last been beaten at his own game.

De revolutionibus was a book whose reception was affected by the complexity of its construction. It attempted within the covers of a single volume not only to provide mathematical astronomers with the geometrical machinery for the calculation of planetary tables to a new level of accuracy, but also to canvas support for a revolutionary cosmovision; and even the few pages devoted to the cosmovision were undermined by the anonymous and wholly misleading preface. The cosmovision badly needed a publicist; but for half a century, Rheticus apart, no publicist of stature would be forthcoming.

CHAPTER 5

From Geometry to Physics: Astronomy Transformed

The publication in 1543 of Copernicus's *De revolutionibus* marked the culmination of the campaign that had begun in the time of Plato in the fourth century BC, to develop geometrical models that reproduced the observed motions of the planets. The aim had been to reveal the regularities that underlay the planetary odysseys against the background of the 'fixed' stars, and so to permit predictions of their future positions. Interest was focused on how the planets moved, not what caused these movements.

The essential components of these geometrical models had been uniform circular motions. That is, the planetary motions were analysed into their component cycles; and since cycles endlessly reproduce themselves, the parameters could be improved without the need for observations of great accuracy.

EFFIGIES TYCHONIS BRAHE OTTONIDIS ÆTATIS SVÆ
ANNO 50 Completo.

355

Tycho Brahe

Tycho Brahe, seen left at the age of fifty, was born on 14 December 1546 in Skaane, which is now in Sweden but was then in Denmark. The son of a noble family, he did not need a profession, and so until he was 30 he studied at a number of universities, combining this with extensive travels and an increasing involvement in astronomy. In 1576 the king of Denmark offered him the lordship of the island of Hven, where he was to create the first major observatory in Christian Europe. There, for some two decades, Tycho developed instrumentation of ever-increasing precision, which he and his assistants used to compile observational records of a completeness and accuracy never before attempted. But after the death of the Danish king Tycho's position deteriorated, and in 1597 he left Hven, arriving two years later in Prague where he found a new patron in the emperor Rudolf II. He died near Prague on 24 October 1601, being succeeded by Johannes Kepler who had become one of his assistants the year before.

All this was to change in the century following Copernicus's death in 1543. He had been a traditionalist in purpose, methodology, and techniques, yet his *De revolutionibus* sowed the seeds of revolution – for how could the stable Earth not only spin, but hurtle through space, its passengers going about their lives in blissful ignorance of what is happening to them?

Indeed, how had the Earth come to be spherical in the first place? To Aristotle, the answer had been easy: any earthy body that had found itself away from its natural place in the centre of the cosmos would have moved naturally towards the centre, and so it was no surprise that the resulting aggregate approximated to a sphere. Copernicus could only suggest that earthy bodies belonged together and had assembled to form the planet Earth, just as venusian bodies belonged together and had assembled to form the planet Venus.

To explain the daily motion of the Earth, Copernicus argued that it is a natural sphere, and that natural spheres naturally spin. He may have thought that the Earth was itself embedded in a vast but invisible sphere, whose spinning carried the Earth in its annual orbit around the Sun; but his views on this are unclear. Evidently, he had advanced the solution of problems in kinematics – how the planets move – only to create new problems in dynamics – what are the causes of their movements.

In the transformation brought about in response to these unanswered questions, crucial roles were played by four men, of different countries and contrasting talents. Tycho Brahe of Denmark was an observer, who made accuracy and completeness his first priority; Johannes Kepler was a German mathematician, who transformed astronomy from applied geometry into a branch of dynamical physics; Galileo Galilei was an Italian physicist, who used a telescope to reveal celestial truths hidden since the Creation, and developed a new concept of motion to underpin Copernicus's claims; and René Descartes was a French philosopher, who conceived of an infinite universe, in which no position and no direction was special, and the Sun became merely our local star.

TYCHO BRAHE AND THE QUEST FOR ACCURACY

Copernicus had uncritically used the observations of previous astronomers when these were conveniently to hand; he himself did only what was necessary, content with the modest accuracy that the available instrumentation allowed. In the later decades of the sixteenth century, Tycho Brahe (1546–1601) was to bring about a revolution in the attitude of astronomers to observation.

Tycho was born into the Danish nobility, but he was fostered by the academically inclined family of his father's brother, and so was able to escape the restrictions of feudal life. Instead, he wandered from one university to another, exempt by the privilege of birth from the normal pressures of career. By 1563, when for the first time for twenty years Jupiter overtook Saturn as the two planets slowly moved against the background stars, the sixteen-year-old Tycho was already sufficiently interested in astronomy to make simple observations of this 'conjunction'. He

Right: Tycho's drawing of the new star of 1572 in relation to known stars of Cassiopeia. He measured the angles separating pairs of stars with his half-sextant (illustrated on page 106). Note how the caption identifies the stars descriptively, with reference to the features of Cassiopeia and her chair. Modern astronomers use the unambiguous Greek letters soon to be introduced by Johann Bayer (1572–1625), a German lawyer and amateur astronomer, who assigned to each star of a constellation a letter of the Greek alphabet (continuing with the Latin alphabet where necessary).

Below: Star-identification by Greek letters in a chart from Bayer's *Uranometria* (1603) showing the constellation Cassiopeia. The very bright star is the new star of 1572, which was long since gone by the time *Uranometria* was published. The star *A* (*caput Cassiopeae* – 'the head of Cassiopeia') on the right in Tycho's drawing has become Zeta Cas.

found that the prediction of the date in the thirteenth-century *Alfonsine Tables* (calculated from Ptolemaic planetary models) was a month out, while even the modern *Prutenic Tables* based on Copernicus's models were nearly two days wrong. This convinced him that there must be a reform of astronomy built on a solid foundation of accurate observations, and that such accuracy could come only from a combination of improved instrumentation and improved observing techniques.

The New Star of 1572 and the Comet of 1577

In the years that followed, Tycho increasingly sought the company of astronomers, both amateur and professional, and meanwhile he took his first serious steps as an observer. Then, in November 1572, Nature treated mankind to an astonishing spectacle: a star-like object, bright enough to be seen in the daytime, appeared in the constellation of Cassiopeia. Was it what it appeared to be, a new star? New heavenly bodies were unheard of, and there were compelling reasons for thinking such a celestial novelty to be impossible: the received Aristotelian cosmology was founded on the dichotomy between the unchanging heavens of the planets and stars, and the changeable central region of the four elements.

Had the 'nova' appeared even a decade earlier, it is possible that no consensus would have emerged from the clamour of contradictory arguments. In 1572 the astronomical instrumentation at Tycho's disposal was modest in both scale and quality. But he was able to satisfy himself that the object was either 'fixed' or nearly so. This showed that it was far above the Moon, and therefore celestial.

More importantly, he recognized the question-mark that the apparition placed against the accepted theory of comets. Because comets 'come to be and pass away', thus changing in the manner characteristic of the region of the four elements, it had been assumed since the time of Aristotle that they were terrestrial rather than celestial; their study belonged not to astronomy, but to 'meteorology'. Astronomers had seen little point in measuring the heights of comets, because the answer was already known. But Tycho now had proof that changes occurred in the heavens. He therefore promised himself that if a comet appeared one day, he would carefully measure its height and see whether or not it was indeed terrestrial.

Nature obliged: in 1577 a brilliant comet appeared, and Tycho's observations showed it was celestial. More precisely, it was located among the planets. Yet, if so, it was passing effortlessly through the invisible spheres that were thought to carry the planets around the central Earth. The implication eventually dawned on Tycho: these spheres did not exist after all.

The effects of this discovery on Johannes Kepler a generation later would be profound. Few had seen any great problem in explaining the spinning of the spheres that carried the planets: if angelic intelligences were no longer acceptable, perhaps Copernicus had been right and natural spheres naturally rotated. But if the spheres did not exist and the planets were isolated bodies moving in orbit, it would be altogether more difficult to explain the causes of their motions. As it turned out, to provide an answer, astronomy had to migrate from kinematics to dynamics, from geometry to physics.

Tycho had published his observations of the nova of 1572, but his tract had had little impact. On the question of the comet of 1577, he was determined to be heard, and heard clearly. While a hundred other writers rushed to cash in on the alarm

Aristotle's teaching on comets

In his *Meteorology*, Aristotle discussed the effects of the contact between the uppermost of the four terrestrial elements, namely fire, and the adjacent part of the rotating heavens. He thought of the fire (an unsatisfactory name for which he apologised) as a kind of fuel, 'so whenever the circular motion stirs this stuff up in any way, it bursts into flame at the point where it is most inflammable'. In his view, the whole of the element of fire, and much of the element of air below

it, was carried round by the rotation of the heavens, and 'in the course of this movement it often ignites wherever it happens to be of the right consistency', sometimes because of the motion of a particular star or planet. This resulted in shooting stars, and in comets of various shapes.

Aristotle's doctrine that comets were atmospheric rather than celestial survived without serious challenge until the time of Tycho.

Above: A 1531 woodcut showing an Aristotelian interpretation of the appearance that year of a bright comet. The comet (which was in fact Halley's) rounded the Sun on 26 August, having earlier in the month passed under the legs of the Great Bear (the constellation that includes the Plough or Big Dipper).

Right: The comet of 1577. In this picture, printed at Prague, the comet's tail is depicted extending above Saturn and the Moon. An artist is sketching it with the aid of a man who is holding the sketch block and illuminating it with a lantern.

aroused by the awesome apparition, Tycho prepared an elaborate analysis of his observations running to over 200 pages, to which he appended an even longer critique of the results arrived at by other observers. It was a block-buster of a book, and when in 1588 it finally appeared, the transfer of comets from meteorology to astronomy was assured; with the transfer came recognition that changes in the heavens are by no means unusual. In a second and this time exhaustive study on the nova of 1572, Tycho's contemporaries were again subjected to severe examination, though this work was not to appear until 1602, after Tycho's death.

These two treatises – which form the bulk of Tycho's literary output – served notice on other observers that new standards now applied. Tycho's own observations of the nova and the comet had been made with instruments commercially available; but by the time of the comet's appearance, Tycho was committed to a fundamental reform of both instrumentation and observational technique. For this he would need funding on a princely scale.

The Observatory on Hven

Astonishingly, the money became available. In 1575, Tycho had visited the landgrave (count) of Hesse, Wilhelm IV, who was himself an enthusiastic observer, and it may have been on the landgrave's recommendation that in the following year the king of Denmark granted Tycho the lordship of the island of Hven in the Danish Sound. There Tycho had the space, the time, and the financial resources to establish the first major observatory in Christian Europe.

On his island Tycho built for himself everything an observer could wish for: machine rooms for the design and construction of new instrumentation, purpose-built observing rooms, accommodation for himself and his assistants, even paper mills and a printing press so that Tycho could publish his results himself. When the first observatory, Uraniborg ('Heavenly Castle'), proved too small, Tycho built a second, Stjerneborg ('Castle of the Stars'), nearby. Year by year he constructed, tested, modified, and tested

The grounds of Uraniborg viewed from the east. The observatory itself is at the centre, and it is surrounded by ornamental gardens (with some 300 trees) and protected by high walls. At the east and west corners are gates, over which are kennels for watch-dogs. At the north corner are lodgings for servants, and at the south corner a printing office. Elsewhere on the island was a paper mill.

again his instruments of many different types, until they were capable of measuring angles to an accuracy of better than one minute (a sixtieth of a degree).

Meanwhile Tycho and – more especially – his assistants maintained an intensive observing programme, creating a treasure-store of data that were of an accuracy, reliability and completeness never before attempted. The bulk of their observations were made in the 1580s and early 1590s, during which time Tycho and his team were at work on about one night in four, mostly during the winter months. Whereas Tycho's predecessors had been content to make a single observation of a desired quantity, on Hven it was common to make multiple observations with a whole range of instruments. On the other hand, Tycho was often as much interested in the performance of the instrument as in the observations themselves, and so most of the measures were left in their raw state ('un-reduced'). Nevertheless, the Hven observers accumulated a rich store of data, to form a legacy with which Kepler would one day be able to test his various planetary hypotheses.

Tycho also compiled a catalogue of 777 stars, to supersede existing catalogues that derived from Ptolemy's *Almagest*. First he and his teams determined the positions of selected 'reference' stars as carefully as possible, and then

The observatory of Uraniborg from the east. At the very top of the building are eight bedrooms for assistants. On the principal upper floor are four observing rooms (with conical roofs), two at the southern end of the building and two at the north, and between them are various chambers including a summer dining room with a view out to sea. On the ground floor are the library (with the large globe), kitchen and winter dining room, and three spare bedrooms. In the basement is the alchemical laboratory. Construction was begun in 1576 and completed in 1580.

they measured the positions of other stars relative to appropriate reference stars. The accuracy of the catalogue therefore depended in the first place on the accuracy of the positions of the network of reference stars, and by the late 1580s Tycho had determined these angles to within half a minute or so of the true values. Yet although the places of the brightest of the non-reference stars are mostly correct to

Tycho's sextant

To measure the angle between two stars, late medieval astronomers used a wooden cross-staff (see page 81), which had the disadvantage that it was necessary to observe both stars simultaneously. Tycho overcame this problem by inventing the sextant, so-called because the arc is one-sixth of a circle. On the right we have a half-sextant, a hand-held instrument that Tycho used to position the new star of 1572 in relation to stars of Cassiopeia. The observer sights on one star along the left-hand (fixed) radius, and when he is satisfied that the radius is aligned correctly, he sights on the other star by moving the adjustable radius shown on the right; the angle is then read off from the scale.

After a further decade of experiment, Tycho had found how to make numerous improvements in such instruments. The mature example shown on the left is supported on a sphere (a kind of universal joint) which has a firm mounting. The direction of sighting has been reversed, so that the users now look from the arc to the centre. This allows two observers to collaborate (as we see in the picture of Stjerneborg opposite), one observing along the fixed radius and holding the instrument in position with the rods once the radius is correctly aligned, while the other then sights along the movable radius. It also allows the second observer to make adjustments with delicacy. The sights and the divisions have also been greatly improved, although these changes are not apparent in the diagram.

Stjerneborg: 'castle of the stars'

Stjerneborg, Tycho Brahe's satellite observatory (below), was a stone's throw from Uraniborg. Built about 1584 entirely of masonry, its rooms were below ground level, so that the instruments were shielded from the wind. Tycho attempted to link Stjerneborg to Uraniborg by an underground passage, but this was never completed. Instead, an observer walked across from Uraniborg and descended to the underground complex from the north, down steps (centre left) that led directly to the heating chamber. From this he could pass to any one of five chambers, each of which contained a single major instrument; observations were made through ground-level windows or through openings in a rotatable roof. Above ground the enclosure contained stable supports onto which other apparatus might be placed when necessary; so (foreground right) we see two observers using a massive sextant mounted on a sphere. Having two separate buildings in this way also enabled Tycho to ensure that the measurements made by different teams of assistants really were independent.

STELLÆBURGUM five OBSERVATORIUM SUBTERRANEVM, A TYCHONE BRAHE NOBILI DANO IN INSULA HVÆNA, EXTRA ARCEM URANIAM, EXTRVCTVM CIRCA ANNVM M D LXXXIIII.

Amfteladami. Joannes Blaeu excudebat.

The Gregorian reform of the calendar

Because the Christian Church in Alexandria was heir to a great tradition in astronomy, its date of 21 March for the spring equinox (when the Sun crosses the celestial equator) was eventually adopted throughout Christendom, and used in the various attempts at agreement on when to hold the great spring festival of Easter.

However, the calendar then in use was the Julian, introduced by Julius Caesar in 46 BC, and this prescribed a leap year every fourth year, without exception. As the solar ('tropical') year is in fact some 11 minutes short of 365¼ days, by the later Middle Ages this error had accumulated to the point where the spring equinox was occurring several days before 21 March, and astronomers and churchmen alike were calling for calendar reform.

Eventually, in the 1570s Pope Gregory XIII set up a commission to resolve the issue. The resulting papal decree, issued in 1582, laid down revised rules for the calculation of Easter, but to laymen the most noticeable changes came from the steps taken to restore the spring equinox to 21 March (by the omission of ten days from October 1582), and to keep it thereabouts

(by turning the years that are multiples of 100 into non-leap years unless they are also multiples of 400). This is the Gregorian calendar that we use today.

Catholic countries adopted the reform at once, and Queen Elizabeth of England wished to do the same, but her bishops would not have it. Confusion ensued. Isaac Newton, for example, was born on Christmas Day 1642 Julian ('Old Style'), but by that day in Catholic Europe, the year 1643 had already begun (in the 'New Style'). Newton was himself to be misled in calculations of the path of the comet of 1680–81, by a confusion of 'styles' in the dating of observations.

Britain finally introduced the Gregorian calendar in 1752. As the year 1700 was no longer a leap year under the new rules, the number of days to be omitted had grown to eleven. In 1753 Sweden followed suit, the last country of western Europe to adopt the new style. Russia resisted the Gregorian reform until the Bolshevik Revolution (by which time thirteen days had to be omitted), and even today the Orthodox Churches calculate Easter on the Julian calendar.

around the minute of arc that was his standard, the fainter ones are less accurately located, and there are many errors. It seems that Tycho may have lost enthusiasm for the work once the zodiacal stars – especially important as reference positions for future observations of planets – were completed. The catalogue eventually appeared posthumously, in his second publication on the nova of 1572.

The Tychonic System

Inside every astronomer there lurks a cosmologist, and Tycho was no exception. To us, his reform of observation is his outstanding achievement, for it gave to astronomy the respect for facts that we think of as characterizing modern science. But he himself was probably proudest of the cosmology expressed in the 'Tychonic system', which soon replaced the Ptolemaic as the most popular of the Earth-centred world pictures. Tycho was well able to appreciate the merits of Copernicus's cosmovision, but he was enough of a traditionalist to regard the price as too high.

October — 1582 — September

Positus Planetarum Diurnus.

Anni veteris	Anni Grega	Dies	☉		Moon		♄		♃		♂		♀		☿		☊	
1	1		17	29 45	11	27	4	52	12	26	14	30	12	22	8	30	4	28
2	2		18	29 12	26	28	4	49	12	27	14	57	13	35	10	9	4	25
3	3		19	28 41	11	29	4	46	12	27	15	23	14	48	11	49	4	22
4	4		20	28 12	26	23	4	43	12	28	15	49	16	1	13	29	4	18
5	15		21	27 45	11	2	4	40	12	29	16	15	17	14	15	10	4	15
6	16		22	27 19	25	21	4	37	12	31	16	40	18	27	16	52	4	12
C 7	17		23	26 55	9	18	4	35	12	33	17	5	19	40	18	34	4	9
8	18		24	26 33	22	52	4	33	12	35	17	30	20	53	20	17	4	6
9	19		25	26 13	6	2	4	30	12	37	17	54	22	6	22	0	4	3
10	20		26	25 55	18	52	4	28	12	40	18	18	23	19	23	43	3	59
11	21		27	25 39	1	21	4	26	12	43	18	42	24	32	25	27	3	56
12	22		28	25 25	13	33	4	24	12	46	19	5	25	45	27	11	3	53
13	23		29	25 13	25	32	4	23	12	49	19	28	26	58	28	55	3	50
C 14	24		0	25 3	7	20	4	22	12	53	19	50	28	11	0	40	3	47
15	25		1	24 54	19	1	4	21	12	56	20	12	29	24	2	24	3	43
16	26		2	24 47	0	39	4	20	13	0	20	34	0	37	4	8	3	40
17	27		3	24 42	12	17	4	19	13	4	20	55	1	51	5	52	3	37
18	28		4	24 39	23	58	4	18	13	8	21	16	3	4	7	36	3	34
19	29		5	24 38	5	46	4	17	13	12	21	36	4	18	9	20	3	31
20	30		6	24 39	17	45	4	16	13	16	21	56	5	31	11	4	3	28
C 21	31		7	24 42	29	58	4	16	13	21	22	16	6	45	12	47	3	24

Latitudo Planetarū ad diē

| | | | | 1 | 1 27 | 0 46 | 0 14 | 0 56 | 1 24 | |
| | | | | 2.1 | 1 26 | 0 46 | 0 27 | 1 25 | 0 53 | Mensis |

Incidit in hoc mense Octobris correctio anni, & restitutio Kalendarij Romani per decem dierum sublationem, facta per Sanctissimum D.N. GREGORIVM diuina prouidentia PP. XIII. anno eius Pontificatus X. vt accommodetur Aequinoctium vernum, & ad eandem sedem restituatur, in qua fuit tempore Niceni Concilij, ad diem nempe 21. Martij (quoniam iam ferè dies decem versus initium Martij retrocessit) ad hoc vt Pascha, & reliqua festa mobilia suis debitis temporibus congruant, & sic omittendo decem dies transeundum est à die 4. Octob. ad diem 15. eiusdem. Itaque ob hanc decem dierum detractionem litera Dominicalis mutatur hoc anno 1582. post diem 4. Octob. in C.

A page from an astronomical calendar ('ephemeris') for October 1582. The first column gives the day of the month in the Julian calendar, while the second gives the corresponding day in the reformed calendar, the 4th being immediately followed by the 15th.

Galileo would later admire Copernicus for 'committing such a rape upon his senses' as to move the Earth; Tycho's feet were solidly on *terra firma*. As a Protestant he saw difficulties for Copernicus in certain passages from the Old Testament. The ancient proof of the immobility of the Earth, based on the behaviour of projectiles – an arrow fired vertically in the air hit the ground at the place from which it had been fired, showing that the ground below had not moved while the arrow was in flight – seemed to him as valid as ever. And, even with the aid of his superb instrumentation, he could not detect the apparent annual movement of stars ('annual parallax')

that was to be expected if they were being observed from a moving platform on the orbiting Earth.

There were two possible explanations of his failure to detect annual parallax. Either Copernicus was wrong and the observer on Earth was in fact at rest, or the stars were so far away that their apparent movements were too small to be detected even with Tycho's instruments. Tycho estimated that, in that case, the stars would have to lie at 700 or more times the distance of the outermost planet. This would create an incredible gap between the planets and the stars – and the stars would have to be of colossal size in order to appear so large at so vast a distance. To Tycho, such a universe made no sense at all.

How then to retain the advantages of the Copernican cosmology without becoming entrapped in the absurdity of a moving Earth? With hindsight the compromise seems obvious. Any motions that we see a planet perform are motions *relative to us*, and any geometrical explanation of what we see is irrelevant to the question of whether it is the Earth or the Sun that is *absolutely* at rest. Therefore, if Tycho retained the geometry of relative motions as proposed by Copernicus, but declared the Earth to be the body that is absolutely at rest, he would have the best of both worlds.

This step may seem obvious now, but to Tycho it was far from obvious. By 1578 he had reached a half-way stage, the system described by Martianus Capella in the

The great mural quadrant

The 'great mural quadrant' at Uraniborg (left), was mounted onto a north–south wall. The instrument itself consists of a quarter-circle of brass, measuring over 6 feet in radius and divided at intervals of 10 seconds of arc. A fixed foresight at the centre of the circle is built into the wall at the left. There are two (alternative) movable backsights that slide on the arc. The observer (only partially visible at the extreme right), who is perhaps Tycho himself, adjusts one of the back-sights to the correct position as the star approaches the meridian, calls out the moment when the star is on the meridian, and reads the angle on the scale; the time is announced by an assistant at the clocks (which have only an hour hand), and the angle and the time are written down by a second assistant seated at the desk. Although the two clocks are prominent in the figure, we know that Tycho had little

confidence in mechanical timekeeping, and often determined the time from measurements of a star.

As decoration, the wall onto which the quadrant is mounted has a painting of Tycho with his faithful dog. In the background we see, on the top floor of the observatory, other assistants making observations with various instruments; below them, in the library, more students are working near the great celestial globe (see page 112); and the apparatus in the basement is for Tycho's alchemical experiments.

In a niche in the wall above the painting of Tycho is a brass globe that was fitted with wheels and could represent the daily spin of the Earth and the paths of the Sun and Moon, and much else besides. To either side of the globe are portraits of Tycho's royal patron and his consort, and above the globe is part of Tycho's library.

Tycho's celestial globe, the wooden structure of which was made not on Hven but in Augsburg, beginning in 1570. It measured nearly 5 feet in diameter. On visiting Augsberg in 1575, Tycho found it somewhat distorted. His solution was to fill in the cracks, using pieces of parchment to perfect the spherical shape. He then waited two years to see whether it would retain its sphericity and resist seasonal temperature changes. Satisfied over this, he had it covered with sheets of brass, on which were engraved the zodiac and the equator and their poles. On the globe he eventually inserted the positions (for the year 1600) of the thousand or so stars that are visible to the naked eye.

fifth century, in which Venus and Mercury were satellites of the Sun, while the Sun, along with the Moon, Mars, Jupiter, and Saturn, orbited the central Earth (see page 69). Six years later he was contemplating making all five lesser planets – Mercury, Venus, Mars, Jupiter, and Saturn – into satellites of the Sun. But there was a physical problem: the sphere that carried Mars would have intersected the sphere that carried the Sun.

At last it dawned on him that the unimpeded passage of the comet of 1577 through the planetary regions implied that these spheres did not exist after all, and now there was nothing to prevent him from making the five lesser planets into satellites of the Sun. In 1588, in the book on the comet, he published the resulting system in outline, along with detailed geometrical models for the motions of the Sun and Moon.

In the Tychonic system, the Earth was at rest at the centre. Around it orbited the Moon and the Sun. The other five planets were satellites of the Sun, and were carried with the Sun around the Earth. Just beyond the outermost position attained by any planet was a thin shell of space, centred on the Earth, and within this space the stars were located. The Tychonic universe was reassuringly compact, with a radius equal to some 14,000 Earth-radii. Even Ptolemy's universe had had a radius half as large again.

The End of a Dream

Tycho's 'Heavenly Castle' was always vulnerable, depending for its existence on the continuing favour of a princely patron. The death in 1588 of Frederick II brought no immediate change, for the regency that followed was dominated by Tycho's family and friends. But as the years passed and the young Christian IV began to take an increasing role in government, it became clear that the halcyon days of Hven were over. In 1597 a resentful Tycho quit Uraniborg, and after two difficult years he crossed Europe to Prague, and took service with a more appreciative patron, the emperor Rudolf II.

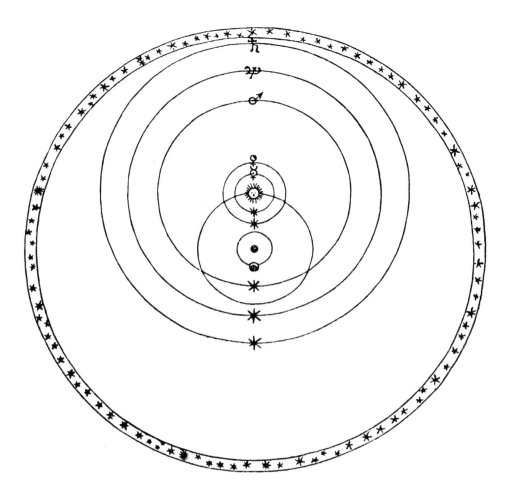

The Tychonic system. The Earth is at rest at the centre, and the Moon and the Sun orbit around it. The other five planets are satellites of the Sun and are carried with it about the Earth, while just beyond the planetary regions lie the stars. The orbit of Mars intersects that of the Sun; in Tycho's opinion, this was physically possible as the free passage of the comet of 1577 had shown that the celestial spheres did not in fact exist. The relative motions are the same as in the Copernican system, so that observations that concerned only relative motions – such as the phases of Venus observed by Galileo – were neutral in the debate between the Copernicans and the advocates of the Tychonic system.

Rudolf did his generous best, but by now Tycho had lost his appetite for observations. Four of his instruments were still in place on Hven, and the biggest of the remainder were in store at Magdeburg. His chief concern was for the publication of past research. Tycho now had only months to live; but in those months he renewed an earlier invitation to Johannes Kepler (1571–1630) to come and join him as his assistant; and this time Kepler accepted.

KEPLER AND THE INTRODUCTION OF DYNAMICS

Astronomers of the early and middle decades of the seventeenth century were often at a loss to know what to make of the writings of Johannes Kepler. He was that rarity, a astronomer who is frank and open about his mistakes: unlike his modern counterparts, who 'launder' their accounts before publication and give them an artificial simplicity, Kepler required his readers to share with him both triumphs and disappointments on the road to discovery. He made errors in the endless pages of calculations, and these sometimes led him to draw wrong conclusions. Because the mathematical techniques he needed did not yet exist, he resorted to procedures that we can recognize as flawed. In dynamics he was a transitional figure, whose

Like the medieval artist who made this image, Kepler believed that God was a geometer-architect who (in Kepler's words) had 'approached the foundation of the world according to order and rule'. Indeed, it was almost as though God had copied in advance the building methods of the human being he had not yet created. Throughout his astronomical career, Kepler attempted to identify the geometrical relationships that had found favour with the Creator when designing the universe.

idiosyncratic conception of motion (according to which a planet would instantly come to a halt unless driven on at every moment by an outside force) was to be repudiated by the next generation. And he was motivated by a religious mission, to penetrate the mind of God the geometer.

Kepler had none of the social advantages of Tycho Brahe. The son of a quarrelsome father and a mother whom he would later have to defend from charges of witchcraft, he was born in Weil der Stadt near Stuttgart and studied at the University of Tübingen. Although he intended to enter the Lutheran ministry, his studies at Tübingen began with courses that included astronomy. The professor was Michael Mästlin (1550–1631), an exceptionally competent mathematical astronomer, who – wherever his true preference lay – made sure that his students were aware of the merits of the Copernican hypothesis. Kepler began his studies of theology in 1591; but in his third year a mathematics teacher in Graz died, and the Tübingen authorities, asked to propose a replacement, nominated Kepler. Under protest, Kepler complied.

The Cosmographic Mystery

Established at Graz, Kepler began to reflect on the universe created by God whom, like so many in the tradition of Plato, he saw as a geometer. Copernicus had discovered the layout of God's universe – but not what had motivated God to choose this layout rather than another.

The immobile parts – the central Sun, the exterior sphere, and the space between – were easy to understand: they mirrored God the Father, God the Son, and God the Holy Ghost. The moving parts – the planets – were more difficult. Why were there six (as Copernicus had found), rather than, say, five or seven; and what had motivated God to locate a given planet at one distance from the Sun rather than another, and to give it one speed rather than another?

In retrospect, the answer to the question of numbers must have seemed obvious. As every student of Euclid knew, there were exactly five regular solids

Johannes Kepler

Johannes Kepler (seen, right, as a young man) was born of modest parentage in Weil der Stadt near Stuttgart on 27 December 1571. He studied at the University of Tübingen and intended to enter the ministry, but in 1594 he was nominated to a post as mathematics teacher in Graz. Early in 1600 he paid an extended visit to Tycho Brahe in Prague; later the same year, following religious persecution in Graz, he returned to become Tycho's assistant and, in 1601, his successor. From 1612 he lived in Linz where he held the additional post of District Mathematician, until 1626 when he abandoned the city in the aftermath of a military siege. There followed four unsettled years, before he fell victim to fever and died in Regensburg on 15 November 1630.

A diagram from Kepler's *Cosmographic Mystery* (1596) in which he reveals the geometrical relationships underlying God's scheme for the planetary orbits. Each orbit is represented by a sphere, and the separation between one sphere and another is defined by one of the five regular solids. The outermost sphere is that of Saturn. Inscribed within it is a cube, and within the cube is inscribed the sphere of Jupiter, so that the cube defines the ratio of the radius of Jupiter's orbit to that of Saturn. Similarly, the tetrahedron defines the ratio of the radius of Mars's orbit to that of Jupiter; and so on.

(tetrahedron, cube, octahedron, dodecahedron, and icosahedron), neither more nor less. There were therefore five appropriate figures for God the geometer to use to space out the pairs of adjacent planetary spheres: five figures, five spaces, and therefore six planetary spheres to be separated by these spaces.

The idea of a nest of spheres alternating with regular solids opened his way to the explanation of the distances of the planets from the Sun. After a little creative manipulation, Kepler found a particular nest in which the radii of the spheres were in fair agreement with the distances from the Sun derived by Copernicus from observation of planets in the real world.

There still remained the question of the speeds of the planets in their orbits. Copernicus had found with great satisfaction that the further a planet was from the Sun, the longer it took to complete a circuit. This was for him a demonstration of the geometrical harmony in the universe. But Kepler began to take astronomy along a decisively new path when he convinced himself that this harmony was brought about through the physical – that is, dynamical – influence of the Sun at the centre, as it drove the planets onwards in their orbits. Obviously this solar force would be less effective at greater distances, which was why the outer planets travelled more slowly.

Kepler's physical intuition was reinforced by his appreciation that the Sun was larger than the Earth or indeed any of the planets: it made sense to have the huge and therefore powerful Sun at the centre of the system. When the next generation of astronomers accepted *forces* as the key to understanding the heavens, and as the enormous size of the Sun was increasingly recognized, the days of Earth-centred cosmologies (such as Tycho's) were numbered.

Kepler had penetrated the mind of God and uncovered the geometrical motivation underlying the structure that God had chosen to give the universe. His proudly titled *Cosmographic Mystery* appeared in 1596, when Kepler was just twenty-five. Even without its hints of new physical insights, it was significant as being the first irredeemably Sun-centred book: there was no way in which the nest of spheres and solids could be adapted so as to have the Earth at its centre.

Kepler sent a copy of his book to Tycho. Tycho had by then developed his detailed models for the motions of the Sun and Moon, but had made no headway with the planets. Undeterred by the Copernican stance of Kepler's book, Tycho recognized it as the work of a very special mathematical talent, and he pressed Kepler to visit him on Hven. The remoteness of the Danish island, however, made such a visit impracticable, and Kepler had other commitments. But soon conditions for Protestants in Graz took a turn for the worse, and when Kepler learned that Tycho had left Hven and was in Prague, he decided to pay him an exploratory visit.

Kepler's explanation of how the Sun influences a planet. At A the 'magnetic threads' of the planet (indicated by the arrow) are such that their extremities are equidistant from the Sun, and the Sun's action is the simple one of pushing the planet around in its orbit. As the planet continues in its path, the 'threads' preserve their orientation, and so 'the region friendly to the Sun' begins to turn towards the Sun and the planet is attracted to the Sun. This continues until the planet reaches the neutral position E, after which the planet is increasingly repelled by the Sun, until it reaches A once more.

Warfare with Mars, and the New Astronomy

Kepler arrived at Tycho's observatory in February 1600, and for three months worked on the orbit of Mars. Except for Mercury, whose proximity to the Sun made it hard to observe, Mars was the planet whose orbit differed most from the circular, and it was therefore the most difficult to handle in terms of the traditional circular motions. When the three months were up, Kepler returned to Graz, but by October he was again knocking on Tycho's door. The Dane gladly took him in, and Kepler resumed work on Mars. A year later, Tycho took ill and died. Within forty-eight hours Kepler was appointed his successor.

Kepler's 'warfare' with Mars, the god of war, was to last for years. His campaign had, he said, a threefold foundation: the Sun-centred vision of Copernicus, the incomparable observational records of Tycho Brahe, and the magnetical philosophy of the Englishman William Gilbert (1544–1603). In 1600, Gilbert, a prominent London physician, had published an experimental treatise, *On the Magnet*, in which he argued that the Earth itself was one vast spherical magnet. The resulting force – which Gilbert used to explain the fall of bodies to the surface of the Earth, the northerly orientation of compasses, and much else – provided Kepler with a suggestion as to how to envisage the Sun's influence on the planets. The Sun, Kepler believed, was rotating and sending out an influence that was pushing the planets round, the nearer planets being influenced more than the distant ones. Since the planetary orbits are not perfect circles with the Sun at their centres, Kepler had also to provide a force to vary the distances of the planets from the Sun. Adapting a concept from magnetism, he supposed that the Sun also exerted on each planet an attraction over one part of its orbit and a repulsion over the remainder.

This physical intuition was to prove decisive. For example, at a crucial moment it led Kepler to realize that he must refer planetary orbits to the real, physical Sun; until then, like Copernicus before him, he had referred planetary orbits instead to the geometric centre of the Earth's orbit. More generally, it caused Kepler to focus his attention on the actual planetary orbit that resulted from the solar forces, whereas his predecessors had been preoccupied with the individual components of the geometrical machinery – epicycles and so forth – that generated the orbit. Accordingly, Kepler would be satisfied with nothing less than a single model that generated the motions of a planet in both longitude and latitude; by contrast, even

Copernicus had been content to work with one geometrical model for the motion in longitude and another (incompatible) one for the motion in latitude.

Yet his physical intuition could also prove a stumbling block, most notably when, after analysing the movements of Mars, he returned to the problem of the exact shape of the orbit. For a time he excluded ellipses from consideration even though their geometry had been fully understood since the time of Apollonius in Antiquity: an ellipse had two axes of symmetry, whereas dynamical considerations suggested that the orbit would have only one such axis. Even today it can strike one as bizarre that a planetary orbit – an ellipse with the Sun at one focus – has geometrical symmetry about the minor axis, whereas dynamically the Sun is to one side of this minor axis and has no physical counterpart at the other, 'empty' focus.

Equally decisive for Kepler's investigations was the legacy of Tycho's observations, even though Kepler's Copernican stance caused him endless diplomatic problems in his dealings with Tycho's heirs. The sheer volume of the observations was to be of the greatest importance to Kepler's assault on Mars. For example, at one stage Kepler realized that any Earth-based observations involved measuring the position of Mars from a moving platform, of whose location in space he was unsure. But he was able to circumvent this problem by using Tycho's ten observations of Mars at 'opposition' (when the Earth lies directly between the Sun and Mars), for then the Earth-based observer sees Mars in exactly the same direction as would a hypothetical observer on the motionless Sun. And when he later needed to reverse the procedure and 'observe' the Earth from a determined position in the orbit of Mars, he found Tycho had left enough observations to make this possible.

As luck would have it, Tycho's observations were also of just the right accuracy to enable Kepler to discover his laws of planetary motion. On the one hand, they were accurate enough to force him to consider non-circular orbits. Kepler had been able to generate the motions of Mars in longitude with a model consisting of a single circle; but when he investigated whether the circle would account for the planet's motion in latitude, he found errors of as much as eight minutes of arc. With any of Tycho's predecessors, such errors would have been acceptable; but Tycho's accuracy was better than eight minutes, and the circle that had promised to be an ideal solution had to be abandoned.

On the other hand, in our post-Newtonian world we know that planets do not move *exactly* in ellipses with the Sun at a focus. Even without the presence of other planets (whose pulls in fact cause disturbances or 'perturbations' of the orbit), a planet would have, at the focus of its orbit, not the Sun itself, but the centre of gravity of the Sun-planet system. Had Tycho's observations been of even greater accuracy, they might have ruled out, not only the circle, but ellipses as well.

Kepler's 'first law', that a planetary orbit is an ellipse with the Sun at one focus, appeared in the book in which he described his warfare with Mars. The law was a strikingly elegant and simple solution to the problem of orbits, and it broke the spell of circularity that had dominated astronomy for two millennia.

His 'second law' (which in fact preceded the first) was, however, confused and confusing. Kepler proposed first a variant of the Ptolemaic 'equant' (see pages 42–3), a position from which the speed of the planet appeared uniform. This position was later to be recognized as the 'empty' focus. He next took the speed v to be inversely proportional to the distance r of the planet from the Sun ($v \propto 1/r$). Finally, he arrived at the rule that we today know as the second law, whereby the line from the Sun to the planet traces out equal areas in equal times. It would be far from clear to other astronomers which (if any) of the three versions was correct and which were approximations, for observations could scarcely distinguish between them. Dynamically, however, the differences were profound, and the situation would become clear only after 1687, when Isaac Newton incorporated motion under the area rule within a unified theory of dynamics. In the meantime, while the elliptical shape of the orbits themselves would be accepted by many, the crucially important question of the ever-changing speed of the planet was left in confusion, and many preferred to work with some form of equant.

Kepler's *New Astronomy* eventually appeared in 1609, a vast book of forbidding aspect, conceptually suspect both in its mathematics and in the underlying physics. The revolution it embodied was proclaimed in the full title: *New Astronomy Based upon Causes, or Celestial Physics, Treated by Means of Commentaries on the Motions of Star Mars*. Kepler had found astronomy a branch of geometry; he had converted it into a branch of physics.

The Harmony of the World

In *New Astronomy* Kepler had investigated how an individual planet behaves as it orbits the Sun, but his long-standing ambition to comprehend the overall structure of God's creation called for knowledge of the relationship of one planetary orbit to another. This was one of the topics he explored in *The Harmony of the World*. The study of cosmic harmony went back to Pythagoras (see page 29), who alone of

According to Kepler, 'The heavenly motions are nothing but a continual music of several voices, which can be comprehended by the intellect, not the ear.' In *The Harmony of the World*, he claimed to set down the notes generated by the individual planets as they moved around the Sun, varying their speeds and distances. Where necessary, he transposed the notes to fit onto the staffs; the notes of Saturn, for example, he thought to be much lower than those of Jupiter.

mortals had been privileged to hear the music of the heavenly spheres. Kepler looked for harmony in the arithmetic ratios to be found in various aspects of geometry and of astronomy, and in Book V he investigated such matters as the speeding up and slowing down of the planets in their orbits, from which he believed he could derive the actual notes of the celestial music. He also looked for arithmetical patterns that might reveal a relationship between the periods and the sizes of the planetary orbits; and he somehow hit upon the fact that the square of the period of a planet is in a fixed ratio to the cube of the radius of its orbit (Kepler's 'third law'). Copernicus had been delighted to find that the further a planet is from the Sun, the longer it takes to complete an orbit; Kepler now knew the arithmetical rule that ensured that this was so.

When *The Harmony of the World* appeared in 1619, Kepler was in process of making his insights and discoveries available to a wider audience, in an *Epitome of Copernican Astronomy*. Because of the disruption to his work caused by his mother's trial for witchcraft, and by the war raging around him, Kepler's *Epitome* was published piecemeal, in 1618, 1620, and 1621. It was in fact to be his largest publication. Cast in the form of question and answer, it covered the whole range of his astronomical thinking. Book IV, on 'Celestial Physics', appeared in 1620 and was conceptually the most important. Its subject matter ranged from the use of regular solids by the divine geometer, to the physical forces operating to produce the orbits not only of the planets but also of the Moon. Copernicus's name appears in the title of the *Epitome*; but nothing better illustrates the revolution brought about by Kepler's creation of 'celestial physics', than a comparison of the forces of the *Epitome* with the geometrical epicycles and deferents of Copernicus's *De revolutionibus*. And though the *Epitome* does not of course list 'Kepler's laws' in the bald manner of a modern textbook, in this work they were at last made accessible to a wider readership.

However, Kepler's astronomy had yet to face the traditional test: could it provide the foundation for planetary tables of improved accuracy? Tycho's own involvement in astronomy had been triggered by his dissatisfaction with the *Prutenic Tables* based on Copernican models; and in 1601, when Tycho presented Kepler to Rudolf II, the emperor had commissioned Kepler to work alongside Tycho on the new planetary tables that Tycho planned. They were to be called the *Rudolphine Tables*.

The tables appeared at last in 1627, and their accuracy was strikingly demonstrated four years later. On 7 November 1631, when Kepler had been dead a year, the French astronomer Pierre Gassendi became the first observer in history to see Mercury crossing the face of the Sun, in fulfilment of a prediction by Kepler. Kepler's tables were in error by only one-third of the solar diameter, whereas even the Copernican tables they had replaced were in error by thirty times that amount. If Kepler's tables were so accurate, surely the planetary laws on which they were based were worthy of serious consideration.

Kepler's Temple of Astronomy

In 1601 the emperor Rudolf II commissioned Kepler to join with Tycho Brahe in calculating new tables of planetary positions, to be named the *Rudolphine Tables*. These appeared only in 1627, when Tycho and Rudolph were both long since dead, and Kepler himself had only three years to live. As the traditional goal of planetary theory was the production of accurate tables, the *Rudolphine Tables* were in this sense the culmination of Kepler's life work, and he furnished them with a suitably prestigious frontispiece.

In the Temple of Astronomy are Kepler's predecessors. At the extreme left and right are the great astronomers of Antiquity, Hipparchus and Ptolemy. In the centre are Copernicus, who inspired Kepler to place the Sun at the centre of the planetary system, and Tycho, whose accurate observations provided the evidential foundation to Kepler's theorizing; Tycho is pointing to the ceiling where there is a representation of his planetary system. One of the legendary Chaldean astronomers stands behind them.

At the very top of the temple the imperial eagle drops coins to symbolize the emperor's role in providing the financial support. Around the dome we see six goddesses who represent the various aspects of Kepler's work in astronomy (others, invisible to us, are mentioned in the introductory poem). In 1604 Kepler had published *The Optical Part of Astronomy*, in which he discussed such topics as refraction; then, in 1611, in response to Galileo's invention of the telescope, he published *Dioptrics*, in which he used a geometrical analysis to propose a new combination of lens that was to become standard in astronomical telescopes. Accordingly, Physica, the figure on the extreme left, holds a globe casting a shadow, and Optica, next to her, a telescope. The next goddess symbolizes logarithms, which had been announced and described by John Napier in books published in 1614 and 1619; Kepler published his own *Thousand Logarithms* in 1624. The remaining goddesses have a

diagram of an ellipse; an unequal arm balance symbolizing the law of areas; and magnets to symbolize the solar forces that Kepler believed controlled the planetary orbits.

At the base of the temple we see in one panel Kepler himself, labouring by candlelight, with the titles of four of his books; and in another, a map of Hven, the Danish island where Tycho had his observatories.

GALILEO DEPLOYS THE TELESCOPE IN SUPPORT OF COPERNICUS

Copernicus had signally failed to promote his own vision of the cosmos: the insights in the brief cosmological Book I of *De revolutionibus* were obscured by the sheer scale of the mathematics that dominated the rest of the volume. Not surprisingly, at the end of the sixteenth century publicly convinced Copernicans were still rare; and as yet they did not include the man who would prove his most effective advocate, Galileo Galilei (1564–1642).

Galileo went to school in Florence and then, in 1581, became an unenthusiastic student of medicine at the University of Pisa. After four years he returned to Florence where he privately studied (and taught) mathematics. In 1589 he became professor of mathematics at Pisa, moving in 1592 to a similar but better-paid post at Padua. It was at Padua that he used the daily and annual motion of the Earth to formulate a possible explanation of the puzzling phenomena of the tides; and when Kepler in 1597 sent him a copy of *Cosmographic Mystery*, Galileo hinted at this theory in his reply. But as yet his Copernicanism lacked conviction, and Kepler's appeal for moral support went unanswered. Galileo's telescopic discoveries in 1609–10 removed any remaining doubts he may have had.

Galileo Galilei

Galileo, depicted here in his early forties, was born in Pisa on 15 February 1564, the son of a musician. He studied medicine at the University of Pisa, but soon turned to mathematics, becoming professor first at Pisa and then, from 1592, at Padua, where he taught for eighteen years. In 1610 he was able to give up teaching when his discoveries with the newly invented telescope earned him the prestigious post of mathematician and philosopher to the Grand Duke of Tuscany, a post that he held until his death in January 1642. These discoveries made him a militant Copernican, and in 1614 this gave his enemies the opportunity to attack him, by fomenting a religious scandal. The issue arose again in 1632 following the publication of his *Dialogue on the Two Great World Systems*, but in more serious form; and Galileo eventually found himself under house arrest for his remaining years.

Extending the Human Senses

Aside from exceptional events such as a new star, a typical Renaissance astronomer looked out on very much the same universe as had his predecessor in Antiquity. If the Renaissance astronomer was blessed with a better grasp of astronomy, this was chiefly because he had been in a position to study printed copies of the books written by his predecessors and contemporaries, or the observational records they had compiled: he had had the advantage of reading more, rather than seeing more.

All this was about to change. From now on, astronomers of every generation would have an overwhelming advantage over their predecessors however brilliant, simply because improved instrumentation would allow them to view objects hitherto unseen and unknown, and therefore unstudied. They would then be able to say of their predecessors, as Galileo did of his, 'If they had seen what we see, they would have judged as we judge.'

In the summer of 1609, when Galileo was in Venice, news arrived that in Holland a device consisting of a tube with two pieces of curved glass was being used to make distant objects seem near. Curved glass, like a curved mirror, was well-known as leading to distortion, the very opposite of the truth. It was likely that still greater distortion would result from the use of two pieces of curved glass in combination, and so Galileo took the precaution of verifying the rumour. Only then did he attempt to construct such an instrument for himself.

The task presented little difficulty to one with Galileo's gift for mechanical improvisation, and by August he was again in Venice, where he demonstrated to the authorities a telescope of 8× magnification, 'to the infinite amazement of all'. He was

Two of Galileo's telescopes, now in a museum in Florence. In March 1611 Galileo went to Rome to publicize his invention, and a banquet was given in his honour by Prince Federico Cesi. Before dinner, guests used a telescope to read an inscription a mile away, and after dinner they viewed Jupiter and its satellites.

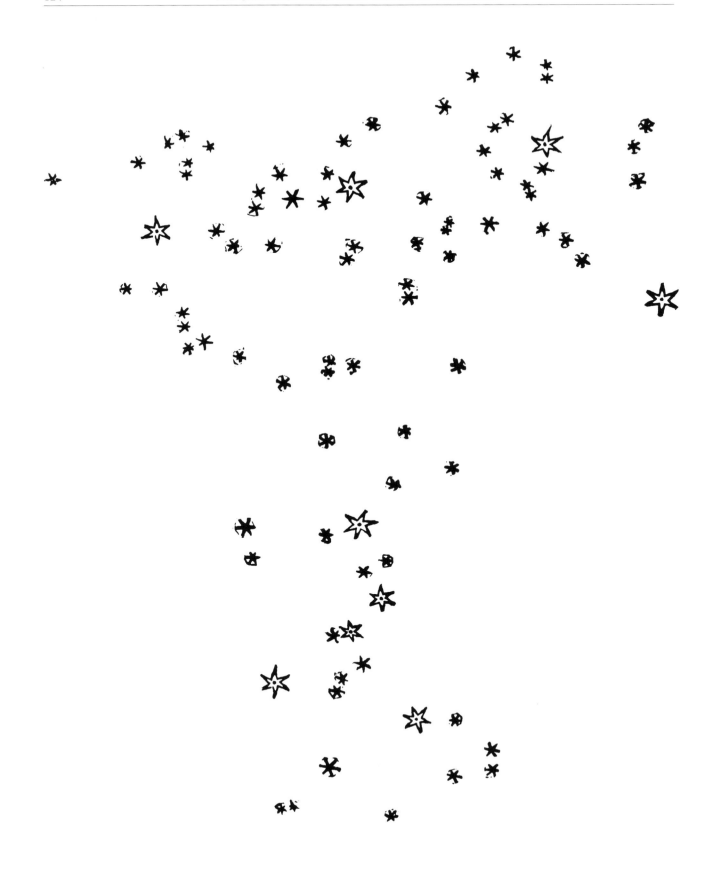

offered a big increase in his Paduan salary. However, the following year the Grand Duke of Tuscany, Cosimo II de' Medici, made him a still better career offer; and so in September 1610 he returned to Florence, this time with a life appointment as mathematician and philosopher to the Duke.

News from the Stars

By the time 1609 drew to a close, Galileo had improved his magnification to some 20×. When he used the instrument to look at the stars, he saw many that had been invisible to the unaided eye – stars that had remained hidden since the Creation, awaiting their discovery by Galileo. The mysterious Milky Way he was able to resolve into myriads of tiny stars, so confirming the speculation reported by Aristotle two millennia before. He also found that whereas the planets appeared proportionately enlarged as expected, the same was not true of stars. This was welcome news for Copernicans. Tycho (see page 111) had estimated that to explain away his failure to detect annual parallax, Copernicans would have to banish the stars to at least 700 times the distance of Saturn; and to appear disc-shaped at such a distance the stars would have to be incredibly big. Now, however, their disc-shaped appearance was exposed as illusory.

His most striking discovery concerned the planet Jupiter. When he first examined it on 7 January 1610, he found the planet in the midst of three little stars ranged – curiously – in a straight line. Jupiter was then moving in westwards ('retrograde') motion, and Galileo therefore expected that the following night Jupiter would be west of the supposed stars; but in fact it was to the east. The next night was cloudy, but on 10 January he found the planet to the west of two stars, with the third star nowhere to be seen. By 13 January the number of stars had increased to four; and by 15 January Galileo had realized that the supposed stars must in fact be satellites, moons circling around Jupiter and carried along by the planet as it orbited the Sun.

This was more good news for the Copernicans. Even in the basic model of the Copernican system presented in Book I of *De revolutionibus* (see page 96), there had been one serious anomaly: the Earth, while in other respects an ordinary planet,

Opposite: Galileo's sketch in the *Starry Messenger* of the stars in Orion's belt (the three large stars across the top of the figure) and sword. Most of the smaller stars in this drawing were being seen for the first time since the Creation. Curiously, he does not depict the great nebula. This may be because he had seen it but could not 'resolve' it into stars (as he did the Milky Way), and so decided to postpone mention of it until some future date, when he hoped to have telescopes powerful enough to show the component stars.

Observations of the moons of Jupiter, from the *Starry Messenger*. The upper sketch shows Jupiter on 7 January 1610, in the midst of three little stars. The planet was then moving west (to the right). In the middle sketch we see the supposed stars on 8 January, with Jupiter to the east (left) of them, quite contrary to what Galileo expected. The lower sketch shows the configuration as it was on 10 January. Galileo devoted nearly half his little book to observations of these satellites.

Ori.   *   * ◯   *   Occ.

Ori.   ◯ * * *   Occ.

Ori.   * * ◯   Occ.

The Copernican system in outline as presented by Galileo in his *Dialogue on the Two Great World Systems* (1632). It is similar to Copernicus's figure (see page 96), with the important difference that the Earth is no longer alone in carrying a satellite. Galileo makes no attempt to represent the two mysterious appendages of Saturn, which disappeared two years after their discovery in 1610, only to reappear a few months later.

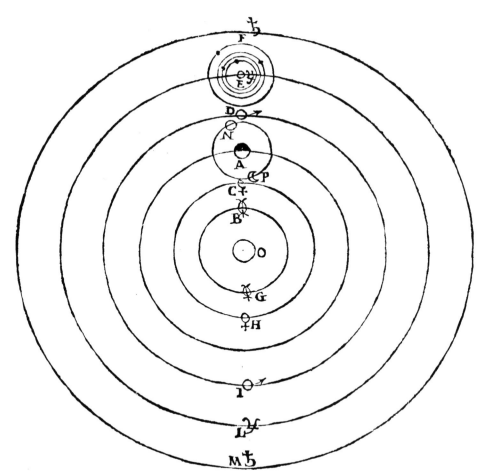

was unique in carrying a satellite Moon around the Sun. The telescope had now revealed that another planet was carrying no fewer than four moons.

Though welcome to Copernicans, the discovery was far from welcome to those who had opted for the system of Tycho, or for another of the compromises then in circulation. In these systems, the Sun orbited the central Earth, and Jupiter orbited the Sun. But now the hierarchy had, implausibly, to be extended a stage further, to accommodate these moons that were orbiting Jupiter (see page 131).

Our own Moon must have been among the objects in the sky that Galileo had examined with his earlier instrument, but now it was time for a serious examination. For all its mottled appearance, the Moon was in the Aristotelian heavens and therefore perfect; yet Galileo's telescope showed its surface to be irregular, with mountains like those on Earth – mountains whose reality he was to bring home to his readers with his estimates of their actual heights, so that the reader might imagine attempting to climb them.

Galileo moved rapidly into print, any lingering doubts about the truth of the Sun-centred system banished by what he had seen. He was anxious to exploit his discoveries and so further his career. His little *Starry Messenger*, written and published in a matter of weeks, proclaimed his astonishing news.

Many were incredulous: peering through a tube with two pieces of curved glass was no way to penetrate the secrets of the cosmos. But decisive support came a year later, when four Jesuit astronomers in Rome signed a statement confirming his discoveries. It would in any case be only a matter of time before telescopes became widely available, and then Galileo's claims could be verified by anyone who had a mind to.

Drawings of the Moon made by Galileo in the winter of 1609/10, possibly between 30 November and 18 December 1609. He concluded that 'the lunar surface is not polished, smooth and perfectly spherical' as befitted a celestial body in the Aristotelian cosmos, 'but on the contrary is uneven, rough, and covered everywhere with cavities and protuberances, just like the surface of the Earth'.

Right: Drawing of sunspots on 10 June 1612, from Galileo's *Letters on Sunspots*. Under suitable conditions, sunspots can be seen with the naked eye. The first telescopic observation of a sunspot was made in England by Thomas Harriot (c. 1560–1621) in late 1610. The following spring, about the time that Galileo was displaying sunspots to observers in Rome, the Frisian astronomers David Fabricius and his son Johann discovered them and made regular observations. In June Johann completed a tract in which he argued that the spots were on the Sun, which was therefore rotating.

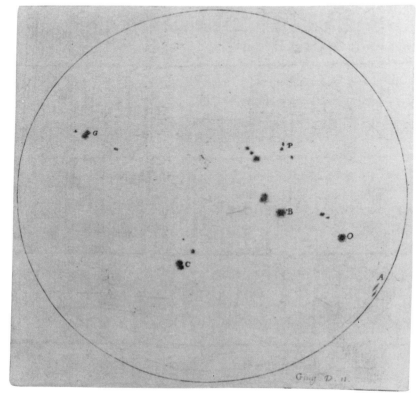

Left: Galileo's drawing of sunspots on 11 June 1612. In 1612 Galileo – who as yet had published nothing on the subject – received from Germany a copy of another tract on sunspots whose author proved to be the German Jesuit, Christoph Scheiner (1573–1650). Scheiner, who had first observed sunspots in March 1611, believed that the spots were satellites of the Sun, rather than being on the body of the Sun itself. The resulting dispute – which was enlivened by the underlying controversy over priority – led Galileo to write his *Letters on Sunspots*, in which, like Fabricius, he argued that the spots were on the rotating Sun.

In the months ahead, Galileo was to make three further discoveries. The Sun, symbol of perfection in traditional cosmology, he found to be 'spotty and impure' (though others would argue that the supposed spots were in fact satellites of the Sun). Saturn, it seemed, had mysterious appendages, which subsequently disappeared and then reappeared; these would puzzle observers until the Dutch physicist Christiaan Huygens (1629–95) hit upon the extraordinary explanation in 1656, long after Galileo's death: the planet was 'surrounded by a thin flat ring which does not touch him anywhere'. Of more immediate importance, Venus went through a complete sequence of Moon-like phases, appearing at times with a circular disc like the full moon, and at others having a thin crescent like the new moon.

This was totally incompatible with the Ptolemaic geometry for Venus. Ptolemy's model (see page 47) implied that Venus was always in between the Earth and the Sun. As a result, the hemisphere of Venus illuminated by the Sun was always facing away from the observer on Earth, who therefore would never see the planet with the circular disc of a full moon.

Unfortunately for Galileo's work as a propagandist for Copernicus, the phases of Venus tell us only about the *relative* motions of Earth, Sun, and Venus, not which of them is absolutely at rest. And since the relative motions were the same in the Sun-centred Copernican and the Earth-centred Tychonic systems, the observed phases of Venus could not chose between them. Galileo had no patience with such logical niceties. To him the Tychonic system was a blatant compromise, and

Christoph Scheiner and a Jesuit colleague measuring sunspot positions with a Keplerian telescope (see page 142). In 1611 Scheiner's sceptical religious superiors had thought his discovery of sunspots so implausible that its announcement would bring ridicule on the Society of Jesus, and at first he was required to publish anonymously. Unlike Galileo, Scheiner persevered for many years in his sunspot studies, which culminated in his massive *Rosa ursina*, published between 1626 and 1630. On the wall in the picture hangs an astrolabe, a relic of a bygone age.

Inuisione Refractoria composita.

Opposite: Tycho Brahe's planetary system, in a 1660 version that incorporates Galileo's discovery of four moons of Jupiter. In the centre is the Earth, with the Moon shown to its right. Above the Earth is the Sun, encircled by the orbits of Mercury and Venus, and, further out, those of Mars, Jupiter, and Saturn. An alternative position for the Sun and its attendant satellites is shown in the lower half of the figure. Seated at the bottom right is Tycho, measuring a globe.

not to be taken seriously; Ptolemy was wrong, and so Copernicus must be right. Others were less dogmatic in their judgement.

The Copernican Propagandist

In both the *Starry Messenger* and his *Letters on Sunspots* (1613), Galileo was careful not to make his support for Copernicus appear too strident. Patronage was as important to astronomers and mathematicians outside the universities as it was to painters and poets; and Galileo was already taking a calculated risk by christening the moons of Jupiter the 'Medicean Stars', for what if the moons proved illusory? But in private Galileo made no secret of his Copernicanism; and if he had the gift of making friends, so too he had the gift of making enemies. In the tense theological climate of the post-Reformation era, his enemies saw good prospects in accusing him of denying truths of Scripture, by holding the Sun to be at rest.

The matter became public in December 1614 when he was denounced from a Florence pulpit. A Scripture reading that day had been from the Book of Joshua. Joshua commanded the Sun to stand still – which implied that it would otherwise be moving. 'Ye men of Galilee', the preacher supposedly took as a punning text from Acts of the Apostles, 'why stand you gazing up to heaven?'

Galileo responded by venturing into theology. In an open *Letter to the Grand Duchess Christina* (1615), he recalled the traditional Catholic teaching, that the scriptural authors wrote in the language of the common man and that allowance must be made for this; and he argued that the purpose of Scripture was to teach men how to go to heaven, not how the heavens go. But his enemies continued their agitation, until at last the Holy Office in Rome became involved. The key figure was the Jesuit Cardinal Robert Bellarmine (1542–1621), a saintly and scholarly man prepared to reinterpret Scripture in the light of new discoveries – but only when the discoveries were established beyond question. In Copernicus's *De revolutionibus*, Bellarmine found from the introduction (which was in fact the unauthorized insertion of the proof-reader, see page 92), that even the Polish canon had – or so it seemed – taken the traditional view: the job of the astronomer was to predict accurately, not to propose truths of cosmology.

The upshot was that in 1616 the various branches of the Holy Office were informed that *De revolutionibus* was to be suspended until amended to ensure it conformed to the traditional role of astronomical texts. Bellarmine notified Galileo privately, that he might no longer believe that the Copernican system was true, or defend this belief.

In 1623 Galileo's Roman fortunes were transformed with the election to the papacy of his friend and supporter, Maffeo Barberini. Memories of the 1616 censure of Copernicanism faded, and after an encouraging series of audiences with the newly elected pope, Galileo finally judged it safe to go ahead with the book on cosmology he had long planned. His *Dialogue on the Two Great World Systems*, published in 1632, was a discussion of the relative merits of Ptolemy and Copernicus. But the title was seriously misleading: by that time the Ptolemaic system had been largely abandoned by believers in a central Earth, and astronomers who could not accept the Sun-centred system of Copernicus – the great majority – were opting for the Tychonic or one of the other Earth-centred compromises on offer.

Written in the vernacular, the *Dialogue* gave a brilliant presentation of the advantages of the Copernican cosmology, and of the telescopic evidence in its favour. So far so good; but its author had to recognize that his readers' sensations were the opposite of what they might have expected as passengers on a spinning sphere hurtling in orbit about the Sun. And many contemporary thinkers agreed with Aristotle: an arrow fired vertically in the air, by falling to ground at the very place from which it had been fired, demonstrated that the ground had not moved while the arrow was in flight.

Galileo's response involved a reappraisal of the very concept of motion. For Aristotle, a natural object, whether living or not, revealed its nature by how it 'moved' (that is, changed – change of place, or 'local motion', being but one of several kinds of motion). Motion, in this general sense, was therefore central to his natural philosophy, which in the thirteenth century had gained added strength from its assimilation into a Christian framework.

For Aristotle, local motions, like all motions, required a cause, and so called for an explanation; rest, on the other hand, did not. Galileo set out to present an

The frontispiece to the massive compendium of astronomical knowledge, *Almagestum novum* (1651), by the Jesuit astronomer Giambattista Riccioli (1598–1671). Riccioli was one of a number of Tycho's contemporaries and successors to propose planetary systems in which the Earth (in some cases spinning daily) was at the centre and the Sun orbited the Earth, carrying some or all of the lesser planets with it. In Riccioli's system, Mercury, Venus, and Mars were satellites of the Sun while the Moon, the Sun, Jupiter, and Saturn orbited the Earth. Urania weighs this system against that of Copernicus, and finds Riccioli's the winner. Ptolemy's system lies discarded, while its author, reclining on the ground, expresses his admiration of the corrections made to it. On the left is Argus Panoptes ('all-seeing'), who in Greek mythology had eyes all over his body and so symbolized the starry heavens. The various phrases are from the Vulgate translation of the psalms, while at the top, as at Belshazzar's feast in the Book of Daniel, a disembodied hand writes 'Number, measure, weight'.

alternative view of local motion. According to him, it was not the motion itself but change of motion – acceleration – that called for explanation. Steady motion – of which rest was simply a special case – was a state, and to continue in the same state generated no sensation of movement. This was why Earth-dwellers were oblivious of the speed of the Earth as it orbited the Sun.

The *Dialogue* is presented as a discussion between three friends: Salviati, who speaks for Galileo; Sagredo, the man of good sense who therefore tends to end up agreeing with Salviati; and Simplicio, the Aristotelian. The real Simplicius had been a sixth-century commentator on Aristotle. But the name also had unflattering overtones, and Galileo made the mistake of putting into the mouth of Simplicio the pope's own view: that, in the last resort, we have to accept that God could – and indeed may – have brought about the observed effects 'in many ways unthinkable to our minds'. Galileo had intended to call his book, *On the Ebbing and Flowing of the Sea*, thus focusing attention on his conviction that the tides provided clear proof of the Earth's motion. He had indeed altered the title at the pope's insistence; but the argument from tides survived in the text, and even though it was dressed up as being merely hypothetical, Galileo's remarks to this effect were unconvincing.

No wonder the pope was displeased, though this does not seem enough to explain why Rome took the extreme step of summoning Galileo and charging him with disobeying the order of 1616. In the end, 'plea-bargaining' took place – Galileo was not of the stuff of which martyrs are made. He abjured his Copernicanism in a scene that was to do incalculable harm to the subsequent reputation of the Catholic Church, even though his 'house arrest' meant a life of comfort, first with his friend the Archbishop of Siena and later in a villa near Florence. Even the symbolic weekly recitation of the penitential psalms was said for him by his daughter.

Had Galileo realized the significance of Kepler's achievement – and possessed the patience to master his writings – how much stronger would have been the arguments at his disposal! But although Kepler had enthusiastically supported Galileo's telescopic claims in *Starry Messenger* by publishing a *Dissertation with the Starry Messenger*, and although we have a letter from a friend to Galileo, written in 1612 and speaking of Kepler's ellipses as common knowledge, Galileo never appreciated the intellectual weapons that Kepler had made available to Copernicans.

Galileo no doubt recoiled from the theological geometry and mystical harmonics that permeated Kepler's thinking, and he would certainly have been repelled by the endless pages of calculation in *New Astronomy*. Another factor was Galileo's lifelong inability to escape from the lure of circles. For him, it was not a ball on a plane that would roll and roll indefinitely (as we have been taught to imagine), but a ball on a smooth horizontal surface; and a horizontal surface was spherical, curving with the curvature of the Earth.

Newton was to synthesize the very different achievements of these two great contemporaries. But first, to make this possible, conceptual abstraction and clarification were needed. The Galileian cosmos still retained privileged places, such as the Earth's centre round which balls could roll, or the Sun round which the planets orbited. These last vestiges of spatial inequality had to be outlawed, and a universe conceived whose space was infinite and perfectly uniform – the universe of Cartesian physics which, as its author explained, was nothing but geometry.

DESCARTES AND THE GEOMETERS' UNIVERSE

René Descartes (1596–1650) was composing his major works at much the same time as Galileo, but he was a generation younger, and the contrast in style is striking. In the medieval universities, Aristotle had been simply 'the Philosopher', and even in Galileo's time, the Philosopher's influence was still dominant. Galileo accordingly saw him as the arch-enemy, and could rarely write a page without making some criticism of him. Descartes, a man of independent means who lived his life outside the university system, saw the battle against Aristotle as already won, and seldom so much as mentions him; to Descartes, the central task was not destruction but reconstruction – what to put in place of the rejected reasoning of Aristotle. His own ruthless methodology led him to an extreme position: that the universe was infinite and uniform throughout, its space completely filled with undifferentiated matter moving under the laws of impact.

Descartes was born at La Haye, between Tours and Poitiers, and was fortunate to study at the Jesuit college of La Flèche. The Jesuits saw (and see) no contradiction in being both a priest and an astronomer or mathematician, and the students at La Flèche were introduced to Galileo's telescopic discoveries within a few months of their first announcement.

The Quest for Certainty

Descartes's teachers instilled in him a love of mathematics, and still more so a love of the certainty that mathematics made possible. After leaving school, in the early 1620s, Descartes pondered the problem of the method to be used in reasoning if one was to attain a similar certainty outside of mathematics; and he gradually came to the conclusion that the method of the geometers was itself the key. But how to apply this method to the study of the natural world? The difficulty was that geometrical reasoning began from established truths, whereas in everyday life people took for true what may in fact have been mistaken assumptions that they had adopted in childhood and had unthinkingly accepted ever since.

How to get rid of these errors? Descartes's solution – or so he claimed – was to attempt to doubt any and all alleged truths, because those that were immune to such drastic treatment must be genuine. His doubting knew no bounds: in particular, he asked himself whether the sensory world, of whose existence he had hitherto felt so sure, might not be as illusory as a dream.

But in the midst of all this Descartes found something whose truth he could not doubt: his own existence, for how could he think about such a question without existing in order to do the thinking? Furthermore, he could conceive of a perfect being (God); and such a concept of perfection could not originate in Descartes's own imperfect mind but must come from outside, from such a being. Furthermore, this perfect being would not play tricks on Descartes, by equipping him with an intellect that was essentially and irredeemably flawed.

The Cartesian Universe

Confident therefore of the power of human reason rightly used, Descartes began to develop an insight into the universe, based on the 'clear and distinct' ideas in his mind – those ideas that were still in the condition in which they had come to him from God. They included mathematical concepts of space and motion, and Descartes tells us that the ancient geometers had understood these concepts with the same admirable insight they had displayed in their method of reasoning. The *space* of the geometers – totally undifferentiated and extending infinitely in all directions – was the space of the real world. The *motion* of the geometers, likewise, was the motion of the real world. The Aristotelian concept of motion Descartes dismissed out of hand as incomprehensible (though it had been comprehended well enough for two millennia!). Motion, he tells us, was now revealed as simply movement from one location to another, as when a point moved to define a curve or a curve to define a surface.

René Descartes

René Descartes was born at La Haye, Touraine, France, on 31 March 1596, the son of a prominent father. He studied at the Jesuit college of La Flèche, and graduated in law from the University of Poitiers. For a number of years his life took no settled direction: he served as a volunteer in more than one army, though a meeting in 1618 with the Dutch physicist Isaac Beeckman revived his fascination with the mathematical sciences. One evening the following year he had a succession of dreams that convinced him that mathematics held the key to true knowledge. Eventually, in 1628, Cardinal de Bérulle urged on him his duty to devote himself to developing this insight, and he retired to Holland in order to do so. He lived there until 1649, when he allowed himself to be tempted to Stockholm by an invitation to be philosopher to Queen Christina. He succumbed to the rigours of the Stockholm winter on 11 February 1650.

Descartes also analysed his concept of matter, and he compared it with his concept of space. Any given piece of matter could have been of any colour, blue, or green, or indeed colourless: colour therefore was not part of the actual concept of matter. Nor was taste, or smell, or any other quality: an isolated piece of matter, as such, had shape and size and location, nothing more. But the same was true of an isolated region of space. And so Descartes drew the astonishing conclusion that – aside from the crucial consideration that matter could move *in* space – matter and space were identical.

Important consequences followed. Since there was but one kind of space, there was but one kind of matter; space did not vary in density, so neither could matter; and space without matter – a vacuum – was conceptually absurd. The result was that, had it not been for the motions in the universe, there would have been no difference whatsoever between one region and another: the entire universe would have been totally uniform.

That there were differences was the result of motion. To understand what Descartes had in mind, it may be helpful to imagine a vast tank of motionless water into which a spherical lump of ice is placed. The lump of ice is made to spin; and as it spins it melts. Suppose that the spinning then continues within this spherical space, occupied by water that was formerly ice. We now have a tank wholly of water (symbolizing Descartes's undifferentiated matter), within which, however, a sphere can be identified because it is distinguished from its surroundings *by its motions*.

In the Aristotelian cosmos, individual bodies existed, and moved in accordance with their natures: a body's movements reflected its nature. It was very different in the Cartesian universe; there, bodies were defined – one might almost say, created – by motion. What more could be said about this motion? God was above all changeless, and this characteristic was reflected in his creation, by what we today might term 'laws of conservation': space/matter was conserved, and (more interestingly) motion was conserved – both the total amount of motion in the universe (speed times amount of matter summed throughout the universe) and, in tendency at least, the motion of individual bodies.

The motion of a body tended to be conserved as it was at that moment, for God lived in the eternal present. The body was then moving with a certain speed in a certain direction, and so it was this motion – straight-line motion – that tended to be conserved.

But in practice this could not happen, for there was no space without matter, and so no body was isolated: nothing could move without compatible movements in the matter surrounding it, both in front and behind. As a result, motions that tended in principle to be straight-line, in practice took the form of a circulation of matter, in whirlpools or 'vortices'.

The solar system was one such vortex, mostly made up of matter whose motions made it invisible. It was this invisible matter (rather than the discredited solar magnetism of Kepler, of whose achievements Descartes was seemingly unaware)

Cartesian vortices and the theory of comets

In a typical region of the Cartesian universe (below left) the Sun, at S in the middle of the figure, is simply an ordinary star, surrounded by its whirlpool of matter, or 'vortex', in which its planets circulate. Space without matter being impossible, the solar vortex shares a boundary with neighbouring vortices, each of which has a star at the centre. The pattern of vortices continues endlessly in all directions, and no place in the universe is in any way special.

Descartes imagined that the vortex around a degenerate star might collapse and that the star would then pass into another vortex. Depending on the dynamical situation, the star would become either a planet and a permanent member of its new vortex, or a comet, in which case it would pass from this vortex to the next. The figure below right shows the collapsed star at N passing through the solar vortex (centre S) and continuing onwards through one vortex after another. Comets seen in the solar system were therefore once-only visitors. The return in 1758 of Halley's Comet, as predicted on Newtonian principles, was striking disproof of Descartes's theory of comets, though by then believers in the Cartesian cosmology were few.

that carried around the visible planets. Meanwhile, by a kind of centrifuge effect, little pieces of agitated matter that the human eye perceived as luminous were being forced to the centre of the vortex, where they collected to form the Sun. The same happened in other large vortices, each of which had a sun (that is, star) at its centre: the Sun, therefore, was a typical star.

A 1574 woodcut of the great
clock of the cathedral at
Strasbourg. The complexity of
such clockwork made it seem
that the universe itself could
be understood as an extreme
example of such machinery,
and God was thought of as the
great clockmaker. At the same
time, the clear and intelligible
explanations a clockmaker
could give of his handiwork
made explanations in terms of
matter in motion the paradigm
of good science.

This conception Descartes believed he had derived from his insights into the immutable nature of God the Creator. Indeed, in his first exposition of his position – withheld from publication on news of the condemnation of Galileo – he made a significant point, that this immutability of God must inevitably be reflected in any world that he created, or indeed could create. This meant that when Descartes showed that a feature of our world was implied by the immutability of God, this feature didn't simply happen to be true of our world, it had to be true. Things could not be otherwise, for God had to be faithful to his own nature.

However, there was a limit to how far Descartes could reason on this basis, not only in practice but even in theory. He explained that when God created the infinite matter/space that constitutes our world, he imposed motion upon it and left it to develop in accordance with the laws of motion. But God was at liberty to impose any one of many patterns of motion, and it was not possible to infer from first principles which of these patterns he had in fact selected; and hence it was not possible to infer any of the detailed consequences that had followed from his particular choice. Similarly, features of the world about us are localized in space and time – for example, we happen to inhabit a planetary system which currently has a given number of planets; but this could have been otherwise, and so to learn about our particular system we have to use observation.

Despite its author's attempt to arrive at certainty, therefore, the Cartesian programme in science depended upon observation and experiment, with all their attendant uncertainties. And since Descartes himself had set out at length the metaphysical foundation for his cosmology, his immediate disciples felt able to skip most of this metaphysics when writing their textbooks. To persuade their readers, they relied instead on the intuitive appeal of using matter in motion to explain the universe. This approach to nature was known as the 'mechanical philosophy', and in Descartes's *Principles of Philosophy* it received its most radical expression.

The mechanical philosophy had been inherited from the atomists of Antiquity. It was especially plausible in a period when machinery (of which the elaborate clock of the cathedral at Strasbourg was the outstanding example) offered a persuasive model of the universe – persuasive because clockwork illustrated how immensely complex were the effects that could result from mechanisms, the individual components of which nevertheless had explanations that were reassuringly clear and intelligible.

Descartes was no astronomer; but so radical was his world picture that whereas all previous cosmologists had looked back (approvingly or otherwise) to Aristotle, now all would look back to Descartes. Following the publication of his *Principles of Philosophy* in 1644, it was increasingly accepted that the Sun was one of innumerable stars in homogeneous and boundless space, and that the planets circulated about the Sun in paths resulting from rectilinear inertia (whereby an unimpeded body moves steadily in a straight line) modified by mechanical forces of impact. Newton was to

The transition from finite to infinite universe

When Copernicus argued that the starry heavens were at rest, rather than spinning daily, he removed the major obstacle to locating the stars at unlimited distances from Earth. But even innovators of the calibre of Galileo and Kepler continued to believe that the universe was finite, and it took the radical approach of Descartes to persuade astronomers that the stars were scattered throughout an infinite universe.

An interesting transitional figure is the English mathematician Thomas Digges (c. 1546–95). In his *Perfit Description of the Caelestiall Orbes*, which he added to the 1576 edition of his father Leonard's

Prognostication Euerlasting, he included a diagram (opposite) of the Copernican system (which he credits to the ancient Pythagoreans in order to add authority to the conception). Outside the sphere ('orbe') of Saturn he has a sphere of fixed stars which 'infinitely up extendeth hit self in altitude sphericallye...' But Digges was still in the grip of medieval cosmology, which located the angelic beings immediately beyond the stars. And so, in his infinite extension of the universe, the outer space is, just as in the Middle Ages, 'the very court of coelestiall angelles' and 'the habitacle for the elect'.

spend his early maturity under the spell of Descartes; and when at last he abandoned the Cartesian plenum for an almost-empty universe bound together by a mysterious 'attraction', he was to scandalize many of his contemporaries.

ASTRONOMY TRANSFORMED

In the century that separated the publication in 1543 of Copernicus's *De revolutionibus* from the publication in 1644 of Descartes's *Principles of Philosophy*, both the science of astronomy and the universe that astronomers studied were transformed. For all its novelties, Copernicus's book was solidly in the tradition of the astronomy of circles. For him, the task of the astronomer was what it had always been, to devise geometrical models that replicated the movements of the planets. Only a few of his contemporaries appreciated that, in carrying out this task, Copernicus had found himself forced to modify a fundamental aspect of the closed, hierarchical cosmos of Aristotle: its opposition between the central earth and the encompassing heavens.

Within decades, astronomers had learned the fundamental importance of repeated and accurate observations, made with instruments designed and refined to meet the goal of precision. They had discovered how to devise apparatus to extend the human senses, and to see celestial bodies hidden since the Creation. And they had extended their goals, to include the study not only of how celestial bodies moved but why – the investigation of the forces to which the planetary movements were responses.

In the process the universe itself had changed, and changed rapidly. Descartes's *Principles* marked a final break with the Aristotelian cosmos with its natural places, substituting instead the totally uniform and undifferentiated space of the

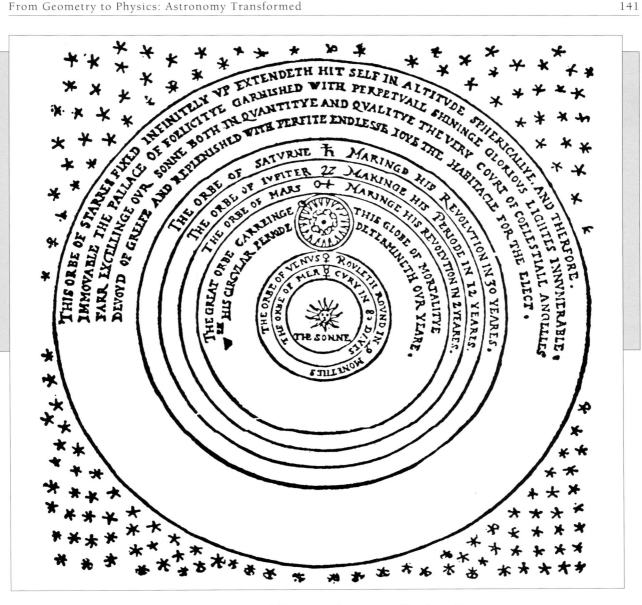

geometers, in which a totally uniform and undifferentiated matter endlessly redistributed itself.

'My physics is nothing but geometry', Descartes wrote in a letter. But the mathematical tools needed to study the Cartesian universe had not yet been invented, and so he and his followers were forced to use words as substitutes for symbols. In the short term this was an advantage: the literate but innumerate gentility who frequented the Paris *salons* could understand Cartesian physics, as could the undergraduates of Cambridge where Cartesianism became all the rage. But although the Cartesians could explain the present state of the celestial bodies plausibly enough, they could rarely forecast how they would behave in the future. The predictive power of quantitative Newtonian physics would in time prove irresistible, and then Descartes would follow Aristotle into history.

When, in 1609, Galileo turned a telescope of his own making to the heavens, he was making imaginative use of an instrument that had been known for a decade or so as an optical device of possible military importance. Galileo's instruments had convex 'objective glasses' (which brought the light to a focus) and concave eyepieces, and magnified up to twenty times. The lenses were available to him because they were manufactured for spectacles, but for telescopic work their quality was poor, and consequently the resolution of his instrument was primitive.

Galileo was the most celebrated telescopic observer at this time, but was not the only one. In England Thomas Harriot (c. 1560–1621) was similarly engaged, but being in the service of the Earl of Northumberland, was less free, and had less need, to publish his results. After Galileo's *Starry Messenger* appeared, however, a telescope became a desirable tool for any participant in the cosmological debate, so that the demand for the instrument grew, along with interest in improving its performance.

Kepler responded to Galileo's initiative by designing a different construction of telescope, one with a convex eyepiece, and his scheme was published in 1611. His instrument had a larger field of view and could be used to project an image onto a screen, useful when observing sunspots (see page 129). A feature that became important later in the century was that a Keplerian telescope could be fitted with an eyepiece micrometer (see page 164) or be adapted as a telescopic sight on an instrument for measurement. The image was inverted, but this was of little matter in astronomy, where the construction – to be known as the 'astronomical telescope' – became standard.

Its resolution, however, was limited by a problem that Kepler himself had indicated: lenses with spherical curvature – the only type that could be made at the time – did not bring all parallel rays of light to a unique focus. In other words, what became known as 'spherical aberration' resulted in a blurred image. All attempts to grind and polish aspherical lenses in the seventeenth century failed. In any case, there was a second problem, 'chromatic aberration', which was not clearly distinguished from spherical: even if a lens could be formed capable

One of the telescopes of enormous length (and with only the skeleton of a tube, to reduce weight and the effects of wind) used by Johannes Hevelius to limit the effects of chromatic and spherical aberration.

The first true lunar map, by Michel Florent Van Langren, published in 1645. Since sunrise or sunset on a particular lunar peak was a time signal that could be seen on Earth over a wide area, Van Langren hoped that if navigators had the necessary tables they could use such an event to compare their local time with a standard time and so determine their longitude.

of focusing monochromatic rays, the different refractive characteristics of differently coloured rays would continue to deny the astronomer a sharp image.

The only practical measures that could be taken were to use lenses of slight curvatures, that is, of long focal lengths, and to introduce apertures that restricted the incoming light to their central portions. Long focal lengths meant long telescopes, and by the mid-century a good instrument for astronomy would be 30 feet or more in length, requiring masts and systems of pulleys to mount and manage it. The leading telescopic observer of the century, Johannes Hevelius (1611–87), had telescopes of extraordinary lengths, up to 150 feet, but an instrument of such a size was quite unmanageable. Christiaan Huygens (1629–93) even resorted to mountings where there was no physical connection between objective and eyepiece, other than a line operated by the observer.

An aerial telescope of Christiaan Huygens, where only a line connects the objective (the lens onto which the light falls) and the eyepiece.

CHAPTER 6   *Newton and Newtonianism*

We can all too easily imagine Kepler's three 'laws' of planetary motion to have been straightforward generalizations from observational data, and therefore trustworthy and uncontroversial. The truth is very different.

What we know as the first law, that planets move in ellipses with the Sun at a focus, was proposed in 1609 in Kepler's *New Astronomy* after complex and confused calculations guided by a very dubious dynamics. It became more widely known when repeated a few years later, clearly and simply, in his *Epitome of Copernican Astronomy*. The law was an astronomer's dream come true: each planetary orbit was described by one single geometrical curve, of a very familiar type. It was no wonder that Kepler's ellipses proved attractive to mid-century astronomers.

Yet we who live after Newton know that the law is not strictly true: planets are 'perturbed' (disturbed) out of pure elliptical orbits by the pulls of the other planets. Mid-seventeenth-century observations were not sufficiently accurate to reveal these complications; but neither were they accurate enough to choose between an ellipse and various ovals that approximated closely to an ellipse.

The second law, again proposed in *New Astronomy* and repeated in the *Epitome*, states that the line from the Sun to a planet sweeps out equal areas in equal times. Not only do perturbations again make this law no more than an approximation, but it was completely beyond the competence of mid-seventeenth-century astronomers to verify such a sophisticated mathematical relationship.

Alternative formulations were, however, to hand. Because the planetary orbits are nearly circular, numerous 'equant' theories (involving locations from which the planet's motion would *appear* uniform) could be devised that were observationally indistinguishable from the law as we know it. These had the great merit of being familiar in form and mathematically tractable.

Nor were these the only candidates in the field. The area rule in fact implies that the speed v of the planet is inversely proportional to the distance r of the planet from the Sun ($v \propto 1/r$) on just two occasions in each orbit, namely, when the planet is at one or other end of the major axis of the ellipse. However, if the ellipse is nearly circular, then the speed is approximately inversely proportional to the solar distance throughout its orbit; and it was this relationship that some astronomers took to be the true law, to which the area law was itself an approximation.

The third law, announced in 1619 in the *The Harmony of the World*, declared that the square of the period of each planet's orbit was in a fixed ratio to the cube of its average distance from the Sun. One might have expected this law to be easily verified, for in dealing with distances and periods astronomers were on familiar ground. But again there were complications: Mercury, and even Venus, being close to the Sun, are difficult to observe; perturbations have their effects; and, as Newton was to show, the law itself needs modification.

THE GENESIS OF NEWTON'S *PRINCIPIA*

Nevertheless, whatever their limitations, Kepler's 'laws' gave astronomers some grip on how the planets moved. How they moved, but not why: the laws were silent on the forces that caused these movements. Kepler believed that a planet was lazy and continued its onward motion only because the rotating Sun somehow pushed it round – that otherwise the planet would immediately come to a halt. Descartes, on the other hand, taught that a planet would – in principle – continue to move in a straight line at uniform speed, unless subjected to outside influence. His universe, however, was completely filled with matter, and so the planets in it were permanently subject to such influences. In practice, therefore, his planets were carried round in the solar vortex, jostled on all sides by matter whose pressure bent the planetary orbits from straight lines into nearly circular paths.

The Magnetical Philosophy

However, the Fellows of the London-based Royal Society (founded in 1660) were heirs to a quite different tradition, one that stemmed from William Gilbert in 1600, who had argued that the Earth is a huge magnet (see page 117). Gilbert was a physician, and typical of the practical men who frequented Gresham College, recently founded by the London merchant Sir Thomas Gresham. Gresham College was an institution of higher learning quite unlike the universities of Oxford and Cambridge. At Gresham the professors lectured mainly in English, and approached their subjects in the manner appropriate for audiences that included instrument makers, seafarers, physicians, and others of practical bent.

From the frontispiece of the 1628 edition of William Gilbert's *On the Magnet*. Gilbert argued that the magnetic properties of the Earth paralleled those of a spherical lodestone (which he termed a 'little Earth', or *terrella*). He investigated the difference between magnetic and true north, and that between the 'dip' of a compass needle and the horizontal. He hoped that if navigators assembled sufficient data on these quantities, the results might provide a substitute for longitude, whose determination was so difficult.

Their basic tool was to be a *terrella* mounted on a float, seen here on the left.

Robert Hooke

Hooke was born on 18 July 1635 on the Isle of Wight, England, the son of a minister. He became a pupil at Westminster School in London, and in 1653 he entered Christ Church, Oxford, where he worked his way through college variously as a chorister and as a servitor.

John Wilkins was among those now making the university the focus of scientific life in England, and within the Oxford circle Hooke became a close friend of the young Christopher Wren, and assistant to Robert Boyle in his work on the air pump. By the time of the restoration of the monarchy in 1660 many of the Oxford circle had moved to London, where the Royal Society was founded that November.

In 1662 Hooke was made the Society's Curator of Experiments. This soon became a salaried appointment carrying lodgings in Gresham College; in 1665 he also became Gresham Professor of Geometry. The Great Fire of London in 1666 led to his appointment as one of the surveyors who were to supervise the rebuilding, though in reality he played a major role as an architect, second only to Wren.

His multifarious duties encouraged his innate tendency to throw out ideas without following them up, and he never fulfilled his immense scientific potential. For a few years from 1677 he was (unsuccessfully) the Secretary of the Royal Society, with responsibility for *Philosophical Transactions*, and it was in this capacity that he initiated in 1679 the correspondence with Newton that set Newton on the path that led to the *Principia*. Hooke died in his rooms in Gresham College on 3 March 1703.

Throughout the early and mid-seventeenth century, the Gresham circle maintained its own tradition of experimental science in which the Gilbertian 'magnetical philosophy' held an honoured place. In the late 1640s, in the aftermath of the English Civil War that was to result in the execution of King Charles in 1649, the group was weakened as the victors imposed some of its members on Royalist Oxford, where they found themselves in key positions. But the tradition survived and, with the approach of the restoration of the monarchy in 1660, Gresham College became once more the focus of scientific life in the capital. Indeed, the informal meeting in November of that year that led to the foundation of the Royal Society took place after an astronomy lecture at Gresham College.

A leading Greshamite of the mid-century was John Wilkins (1614–72). In 1640 Wilkins published a second edition of his *Discovery of a World in the Moone*, and in it he discussed the theoretical possibility of a journey to the Moon. In his view such a possibility existed; for just as Gilbert had believed a spherical magnet to be surrounded by a strictly finite sphere of attraction, so Wilkins thought that the space traveller would be able to escape the magnetic influence of the Earth, once he was perhaps twenty miles above the surface. However, as Wilkins reflected further on this sphere of influence, he decided that it was unreasonable to expect there to be a sharply-defined boundary beyond which gravity had zero effect: 'it is probable, that this magneticall vigor dos remit of its degrees proportionally to its distance from the earth, which is the cause of it.'

Robert Hooke was a militant advocate of the value of incorporating telescopic sights into astronomical instruments for measurement. This quadrant of radius 10 feet, which he designed for the Greenwich Observatory, is an early example of the use of such a sight. Installed in 1676, the instrument proved a failure – partly because, to save graduating the entire 90° scale, Hooke had supplied a moveable graduated 5° arc, which the observer had to clamp in a suitable position.

Robert Hooke and the Early Royal Society

In 1648 Wilkins was imposed on Wadham College, Oxford, as its head, and it was at Oxford that two of the brilliant minds of the next generation came under his influence: Christopher Wren (1632–1723) and Robert Hooke (1635–1703). Wren is remembered as the architect who, in collaboration with Hooke, rebuilt London after the Great Fire of 1666; but he was professor of astronomy, first at Gresham from 1657 to 1661, and then at Oxford.

How the planets moved in their orbits, what paths were followed by comets, and how the magnetic pull of the Earth diminished with increasing distance, were problems repeatedly discussed by the two friends and their circle of acquaintances. In 1662 and again in 1664, Hooke attempted to confirm by experiment that the pull of the Earth varied with height. Perched high up in Westminster Abbey or in old St Paul's Cathedral, he measured the weights of bodies, first when alongside him and then when suspended near to ground level – but of course without managing to detect any change.

In 1666, to help illustrate his developing ideas concerning the interaction of the Sun with the Earth and Moon, Hooke showed to the Royal Society a pendulum whose string was divided near its end into two strings, one with a large weight (representing the Earth) and the other with a small one (the Moon). These two weights were set rotating about each other, together forming a system which was then rotated as a whole about the 'Sun'.

Thereafter Hooke's progress in understanding was rapid. In 1674 he took the opportunity to summarize the position he had reached, in one of the most remarkable passages in the history of astronomy ever published. It was in the form of three 'Suppositions' appended to his *Attempt to Prove the Motion of the Earth*:

> First, That all Coelestial Bodies whatsoever, have an attraction or gravitating power towards their own Centers, whereby they attract not only their own parts, and keep them from flying from them, as we may observe the Earth to do, but that they do also attract all the other Coelestial Bodies that are within the sphere of their activity...

Hooke held that the Earth both attracted and was attracted by, not only the Sun and Moon, but all the other planets as well – and that this interplanetary force was the same as the gravity that held the parts of the Earth together. However, his mention of 'the sphere of their activity' shows that he did not think of these attractive powers as universal.

> The second supposition is this, That all bodies whatsoever that are put into a direct and simple motion, will so

continue to move forward in a streight line, till they are by some other effectual powers deflected and bent into a Motion, describing a Circle, Ellipsis, or some other more compounded Curve Line.

This was the first correct published statement of the dynamics of motion in an orbit.

> The third supposition is, That these attractive powers are so much the more powerful in operating, by how much the nearer the body wrought upon is to their own Centers.

The attractive powers, Hooke held, diminished with increasing distance. But did they vary inversely with the distance itself ($f \propto 1/r$), or with the distance squared ($f \propto 1/r^2$), or what? Hooke could not say; but he regarded the answer as relatively unimportant, merely one of the loose ends:

> Now what these several degrees are I have not yet experimentally verified... He that understands the nature of the Circular Pendulum and Circular Motion, will easily understand the whole ground of this Principle.

Of the possible laws, the obvious candidate was that the forces varied inversely with the distance squared (the 'inverse-square' law). After all, the brightness of the Sun or a planet reduces in just this way. But there was another reason for suspecting the inverse-square law, which had to do with the attempts of the Cartesians to analyse the forces at work in orbital motion. When a slinger began swinging the stone around him, he felt the stone pulling forcefully away from him as he stood braced against it. Prior to the release of the stone, the circular path in which the stone travelled at the end of the rope seemed the result of a balance between the

The observatory of the wealthy Danzig brewer, Johannes Hevelius (1611–87). The platform spans three adjoining houses, and carries numerous instruments, including a long-focus refractor slung from a pole, and a portable version of the sextant shown on page 205. The doors of a separate turret (extreme right) open to reveal a quadrant. The observatory, the finest of its day, was destroyed by fire in 1679.

Christiaan Huygens

Christiaan Huygens was born in The Hague on 14 April 1629, the son of a distinguished father who numbered Descartes among his friends. He studied jurisprudence at Leiden and Breda, before devoting himself to physics and astronomy. In 1655, with his brother, he learned to grind telescopic lens, and that winter he discovered Titan, the largest moon of Saturn, and recognized that a detached, flat ring would explain the planet's mysterious appendages. Soon afterwards he invented the pendulum clock.

In 1661 Huygens visited London and addressed the Royal Society on the laws of collision. Five years later he became a founder-member of the French Academy of Sciences, and he lived in Paris until 1681, when he returned to The Hague, dying there on 8 June 1695.

Right: Huygens's diagram showing how the ever-changing appearances of Saturn's appendages could be explained by supposing the planet to be surrounded by a ring. Interestingly, the idea occurred to him in 1655–56, when the ring was edge-on and invisible.

restraining pull of the slinger directly towards the centre and the outwards pull of the stone directly away from the centre. (Today we can see that this is a misleading analysis, the result of working with a rotating frame of reference: the analysis involved consideration of the pulls, inwards and outwards, in the direction of the rope – which was rotating all the while.) In 1673, in a book by the Dutch physicist Christiaan Huygens (1629–95), the outwards pull of the shot was shown to be proportional to the speed squared divided by the radius (v^2/r).

Now Kepler's third law stated that the square of the period of a planet was in a fixed ratio (say, k) to the cube of the radius of its orbit. Suppose the planets had all been travelling in circles with constant speeds, rather than in nearly circular ellipses with varying speeds. Then since the period was equal to the circumference $(2\pi r)$ divided by the speed v, the law would have stated that for each planet, $(2\pi r/v)^2 = kr^3$; that is, v^2/r – Huygens's pull – would be in a fixed ratio to $1/r^2$.

The inverse-square law was therefore the obvious candidate. However, the real planetary orbits were not circular with constant speed, but elliptical with varying speed; and so the crucial question was: would an inverse-square law of attraction to the Sun result in elliptical orbits? In January 1684, in discussion with Wren and Edmond Halley (c. 1656–1742) at the Royal Society, Hooke claimed to have a demonstration that this was so, though he would not reveal it. Wren, who knew Hooke well enough to take the claim with a pinch of salt, offered a book prize to him or to anyone else who actually produced a proof within two months. The prize went unclaimed.

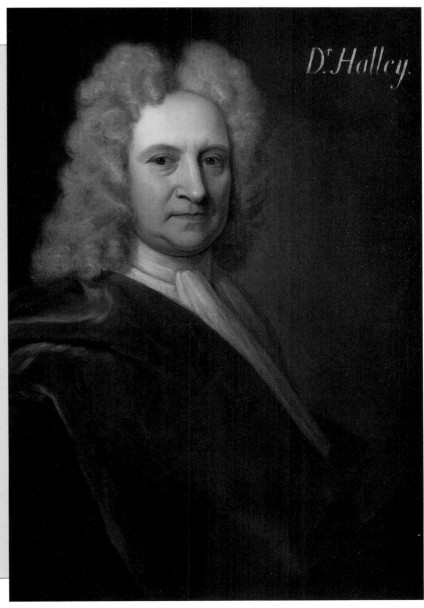

Edmond Halley

Halley was born about 1656, the son of a prosperous London merchant, and was educated first privately, then at St Paul's School, London, and afterwards at Queen's College, Oxford, which he entered in 1673. In 1676, and despite his youth, he left Oxford to sail to the island of St Helena off the west coast of Africa, to observe the southern stars, being the first astronomer to do so (see page 251). In 1678 he was elected to the Royal Society, and the following year he visited Johannes Hevelius in Danzig, to defuse the clash between Hevelius and Hooke over whether it was any longer acceptable to measure celestial positions without employing telescopic sights. In the mid-1680s Halley's diplomacy was fully stretched as he coaxed Newton into writing the *Principia*, while Hooke made matters more difficult by demanding printed acknowledgement of the ideas he believed he had contributed; although the *Principia* is dedicated to the Royal Society, Halley paid for the publication himself.

In 1696 Halley became Deputy Comptroller of the Mint at Chester, and then in 1704 he was elected Savilian Professor of Geometry at Oxford. In 1720 he succeeded Flamsteed as Astronomer Royal, a post he held until his death on 14 January 1742.

Later that year, Halley visited Isaac Newton (1642–1727) in Cambridge, where he was professor, and asked him what would be the orbit of a planet moving under an inverse-square law of attraction to the Sun. Newton gave Halley the answer he had hardly dared hope for: the orbit would be an ellipse.

The reflecting telescope

In early astronomical telescopes, the light passed first of all through a glass lens, the objective glass. This bent or refracted the rays so that they came to a focus, where they could be inspected through the eyepiece.

The practical difficulty in giving the objective glass of a refractor the ideal geometrical shape, and the excessive tube lengths required to circumvent this problem (see page 142), in 1663 led the Scots mathematician James Gregory (1638–75) to propose an alternative design. In it the light passed down the tube to a concave mirror at the bottom, which reflected it back up the tube. A second, small concave mirror, centrally placed, reflected the light back again down the tube. The light then passed through a hole in the primary mirror, and so entered the eyepiece.

Gregory took the design to a London firm of opticians, but their rudimentary trial did no more than confirm the feasibility of the design.

A similar design, but with a convex secondary mirror, was proposed in 1672 by a Frenchman called Cassegrain, of whom almost nothing is known. A century later it was shown that the convex secondary mirror helped to avoid the 'spherical aberration' introduced by departures of Gregory's concave mirrors from their ideal shapes, and in modern telescopes the Cassegrain design is widely used.

Meanwhile, Isaac Newton in Cambridge had made a discovery of fundamental importance (see pages 261–2). It had long been believed that white light was simple and that colours were 'modifications' of white light, but Newton showed that – on the contrary – white light was compounded of the colours, each of which was refracted by a given objective lens through a slightly different angle. The blurring caused by this 'chromatic aberration' was, he thought, inherent in refracting telescopes. The mirror that Gregory used in place of a lens avoided this problem; but unfortunately Gregory's design called for a second curved mirror, and the primary mirror had to be pierced.

In 1668, therefore, Newton built himself a little reflector to a design that avoided these drawbacks. A mirror at the foot of the tube again reflected the incoming light, but now a small, flat mirror inclined at 45° to the axis of the tube reflected the light sideways to where the eyepiece was located.

The three forms of reflector invented in the 1660s and 1670s. In the Newtonian design (below), it was not necessary to pierce the primary mirror, and the secondary mirror was flat rather than curved. This simplified the task of the builder, and so this form became popular in the eighteenth and nineteenth centuries, especially for major instruments with massive mirrors.

In 1671 the Royal Society heard that Isaac Newton had designed and built a reflecting telescope, and they asked to see it. It arrived at the end of the year, and caused an immediate sensation. This is a copy of the drawing that the Secretary of the Society sent to Christiaan Huygens in Paris. The two crowns are the same object, an ornament on a weathercock 300 feet away, as seen (left) through Newton's reflector and (right) through a refractor 25 inches long.

This reflector became known only to a handful of Newton's Cambridge acquaintances, but in 1671 he made a similar instrument, which he presented to the Royal Society. Neither of these reflectors survives, but parts of a third instrument, which he made in the winter of 1671/72, are believed to be incorporated in a reflector that was presented in 1766 to the Royal Society.

Because it called for only one curved mirror, the Newtonian design was widely used in the eighteenth and nineteenth centuries, especially for the study of faint objects for which a mirror with large 'light-gathering power' was required.

Newton was overly pessimistic in thinking the problem of chromatic aberration insoluble. In 1729 a London barrister, Chester Moor Hall (1703–71), devised an 'achromatic' lens by combining two glasses of different refractive properties, a concave of flint and a convex of crown glass. But Hall did not pursue the idea commercially, and this allowed John Dollond (1706–61), a leading London instrument maker, to revive the idea in a paper to the Royal Society in 1758, and to patent it. Thereafter 'Dollond achromatics' (see page 202) became much sought-after among observatories and amateurs alike.

Opposite: An engraving of
1703 embodying the Cartesian
conception of the solar vortex.
The Moon is carried around
the Earth (numbered 4) in the
terrestrial vortex, while the
terrestrial vortex as a whole is
carried round in the solar
vortex. Not surprisingly,
Newton found it difficult to
form a clear conception of the
dynamics of the Moon's motion
in such a situation. Jupiter
(numbered 6) is shown with
the four moons discovered in
1610 by Galileo, while Saturn
(numbered 7) has the five
moons discovered by
Christiaan Huygens and
G.D. Cassini between 1655
and 1684. The solar vortex is
(somewhat ineptly) shown
surrounded on all sides by
similar systems, each
consisting of planets in
orbit about a central star.

The Dynamics of Elliptical Orbits

The Cambridge that Newton had entered in 1661 still respected the primacy of
Aristotle in its statutory teaching, but the semi-independence of college tutors had
permitted the unofficial introduction of Cartesian physics, and with it the mechanical philosophy – the doctrine that matter in motion is the explanation of all things
and that one body can affect another only by direct contact.

In the universe of Descartes, who thought the very idea of a vacuum to be a self-
contradiction, the Earth and the other planets were carried round in the huge
vortex that was the solar system, while the Earth in turn was the centre of a subord-
inate vortex that carried the Moon. The motion of the Moon therefore involved
both the solar and the terrestrial vortices, and this presented the investigator with
an ill-defined situation.

By comparison, the analysis of the abstract notion of motion in a circle was
straightforward, and by the mid-1660s the young Newton had independently
derived for the outwards pull the same formula as Huygens. But thereafter his
progress was slow and uncertain. One obstacle lay in the variants of Kepler's second
law in circulation: although observationally these variants were almost indistinguishable from each other, conceptually the differences were profound. Newton
himself, between 1665 and 1679, worked with a number of equant versions of
the law. Eventually, however, he came across a published statement of the law in
its area form.

In 1679, Newton was still a Cartesian in planetary dynamics, still struggling to
make sense of the complex interactions within the solar vortex, and still confused
over inwards and outwards pulls. That November he received a letter from Hooke,
with whom he had earlier clashed over the nature of light. Hooke was now Secretary
of the Royal Society and his purpose in writing to Newton was to involve this
temperamental genius in the Society's activities. In his letter he invited Newton to
consider the consequences of 'compounding the celestiall motions of the planets of
a direct motion by the tangent [inertial motion] and an attractive motion towards
the centrall body'. Where Newton had thought of orbital motion as a tussle between
opposing forces, Hooke was treating it as one in which a body that tended to
continue in a straight line was instead pulled into a curved path.

Newton was intrigued by what was to him a purely academic challenge, and he
responded with an analysis of where a body dropped from a height above the Earth
would reach the surface – and of how the body would continue to move, if the Earth
was imagined as offering no resistance, and if the Earth's pull was constant throughout. In his analysis Newton made a slip, which Hooke was quick to point out; and
nothing concentrated Newton's mind more than being proved wrong. Yet Hooke
soon tired of pursuing the consequences of this constant pull that was little more
than a mathematician's game, for he was very much the physicist: 'my supposition
is that the Attraction always is in a duplicate proportion to the Distance from the
Center Reciprocall', in other words, the inverse-square law.

By this time, however, all Newton's defence mechanisms were fully deployed, and he withdrew into his shell. But, unknown to Hooke, he solved the problem Hooke had set him: the imagined planet would move in an ellipse with the central body at a focus, just as the real planets did according to Kepler's first law.

Universal Gravitation

Newton indeed had gone further. Provided Hooke's approach to the analysis of planetary orbits was valid, he now had reason to think that an inverse-square pull exerted by the Sun would result in planetary orbits that obeyed both the first and the second of Kepler's laws, and doubtless the third as well.

Hooke believed that attraction operated between the Sun, the planets, and sometimes – but not invariably – the comets. Indeed, in his recent *Cometa* he had suggested that the matter forming comets could lose its attracting power, owing perhaps to a 'jumbling' of its internal parts. Cometary orbits had long presented mathematicians with a challenge, for to derive an orbit in three dimensions from observations made from Earth was very difficult – so difficult that Newton was later to struggle with the problem for months while composing the *Principia*. However, comets were seen to follow paths across the sky that were roughly arcs of 'great circles' (circles centred on the human observer), and the general opinion was that these arcs were projections onto the heavenly sphere of orbits that were nearly straight-line. It was thought that when a comet appeared, heading for the solar system, it approached along a straight line, accelerating as it came; then, after a small swerve as it passed the Sun, it continued straight on and out of the solar system, decelerating as it went.

John Flamsteed (1646–1719), however, the recently appointed Astronomer Royal at Greenwich, was one of several astronomers who took a different view. A comet had appeared in November 1680 heading towards the Sun, and another comet was first seen on 10 December heading away from the Sun in pretty much the opposite direction. Flamsteed believed that these were not two distinct comets, but one and the same. To explain the reversal of direction, he argued that as the comet had been moving inwards in response to the Sun's attraction, it had been thrown off course by the solar vortex. Eventually it had found itself in a new magnetical relationship to the Sun, and been repelled. As a result, it had been driven back before it ever reached the Sun.

Flamsteed went out of his way to inform Newton of his theory and to provide him with the observational data on which it was based. In the exchange that followed, Newton floated the alternative suggestion that the comet had 'fetched a compass about the Sun' – that it had passed around the back of the Sun. He may well have been wondering whether the comet was attracted by the Sun in accordance with the inverse-square law; should this prove to be the case, cometary orbits would be no different in principle from those of the planets. Eventually, however, he concluded that the evidence told otherwise.

We know little else of how Newton's ideas developed between the exchange with Hooke in the winter of 1679/80 and the arrival of Halley at his door in 1684 – a visit that revived his interest in the dynamics of planetary orbits. Halley asked him 'what he thought the Curve would be that would be described by the Planets supposing the force of attraction towards the Sun to be reciprocal to the square of their distance from it'. Newton revealed to the young and sympathetic Halley what he had concealed from the combative Hooke: the curve would be an ellipse.

Halley was 'struck with joy and amazement'. When he could speak, which we may suppose was not immediately, he begged to see the calculation 'without any further delay'. Newton was unable to lay hands on it, but he promised to send it to him.

The Writing of the Principia

What arrived was a draft treatise of only nine pages, much of it devoted to bodies moving in empty space. One of the 'hypotheses' is a statement of the law of inertia: 'Every body, unless impeded by something extrinsic, by its innate force alone proceeds uniformly into infinity, along a straight line.' It is expressed in terms of infinite straight lines, lines that had been inconceivable in closed, finite universes such as those of Kepler and Galileo.

Newton proved – or believed he had proved – a series of propositions relating to Kepler's laws or to generalizations of them. If a body moved under a pull to a 'centre', it would obey the area law. If the pull was inverse-square, then the orbit would be conical (possibly, but not necessarily, elliptical); and conversely. If bodies moved in elliptical orbits and the pull was directed towards their foci, then the orbits would obey Kepler's third law; and conversely.

These abstract results in mechanics could be applied to the solar system, raising to the status of true laws what had until now been viewed as Keplerian generalizations of questionable reliability: 'The major planets orbit, therefore, in ellipses having a focus at the centre of the Sun [Kepler's first law], and with their radii drawn to the Sun describe areas proportional to the times [the second law], exactly as Kepler supposed.'

The focus of a planetary orbit, we note, was 'at the centre of the Sun'. Newton was still thinking of the Sun as doing the pulling and the planet as being pulled. He had yet to match the insight that Hooke had expressed, however imperfectly, in the first of his 'suppositions' of 1674: that the attraction between celestial bodies is mutual.

This is a little surprising, in that his draft stated that Kepler's third law applied also to the moons of Jupiter discovered by Galileo, and to the moons of Saturn (Huygens had come across Titan in 1655, and in Paris Gian Domenico Cassini had recently discovered four more). These moons were therefore subject to the inverse-square pull of the parent planet. If Saturn pulled Titan, why should it not also pull the Sun?

By the time Newton composed a revised draft a few weeks later, he had taken the crucial step: celestial bodies did pull each other. He was now committed to the path that would lead him to universal attraction. Conceptually, confusion was giving place to clarity on a scale and at a pace unique in the history of the physical sciences. But even Newton was aghast at the implications for the unfortunate investigator. With so many celestial bodies, each pulling all the others, how could anyone hope to cope with the mathematics involved: 'But to consider simultaneously all these causes of motion and to define these motions by exact laws admitting of easy calculation exceeds, if I am not mistaken, the force of any human mind.'

The transformation that had recently taken place in Newton's world picture could hardly have been more dramatic. When the 1680s began, his world was crammed full of matter, a confused Cartesian plenum in which bodies endlessly impacted on their immediate neighbours. By the middle of the decade his world was almost empty; in it, isolated bodies were diverted from their onward paths by the attractive pulls of other isolated bodies, reaching out to them across seemingly empty space.

Halley had no trouble in realizing the revolutionary significance of all this, and he began to coax Newton towards publication. He knew that Newton was what was termed a 'nice' person, one who had to be handled with great delicacy. Newton was also an inveterate drafter and polisher, an author who rarely considered his work to be fit for publication in its present state. But he became increasingly obsessed with the task in hand, and the proposed treatise grew and grew.

One prize within his grasp was a theory of comets – even before writing the first draft he had revised his opinion of the comet of 1680 and now believed it had indeed 'fetched a compass' about the Sun. Comets, he was convinced, obeyed the inverse-square law. Their paths were as lawlike as those of the planets, and after the mathematical problems had been overcome it would be possible 'to ascertain... whether the same comet returns with some frequency to us'. Newton's investigations were to result in a Proposition in the *Principia* declaring that comets moved in conic sections – possibly, but not necessarily, ellipses – with the Sun at a focus (Kepler's first law, generalized), and that the line to the Sun swept out equal areas in equal times (the second law).

In his first 'supposition' of 1674, Hooke had identified the pull that the Earth exerted on other celestial bodies at great distances with the pull it exerted on falling stones a few yards away. The same idea now occurred to Newton. Could it be that the inverse-square law of attraction applied to matter large and small? In particular, did the Earth exert the same attractive force on a stone as on the Moon?

Before Newton could attempt an answer to this question, his faith in the law – like that of his future readers – had to survive a test. Could one seriously maintain that two equal lumps of matter at the same distance from a stone (or the Moon), one lump deep in the bowels of the Earth and the other lying on the surface, would attract the stone (or the Moon) with exactly equal force? Surely in the case of the lump inside the Earth, the screen provided by the surrounding rocks must

Isaac Newton

Newton was born at Woolsthorpe in Lincolnshire, England, on 25 December 1642 (in the unreformed calendar). He entered Trinity College, Cambridge in 1661, but was undistinguished as a student. Because of the plague, the university was effectively closed for parts of 1665 and 1666. This coincided with a period of exceptional productivity in Newton's intellectual development, and so it was while working alone and mostly at home that Newton laid the foundations for his future achievements in optics, mathematics, and dynamics.

In 1667 Newton became a Fellow of Trinity, and two years later he succeeded Isaac Barrow as Lucasian Professor of Mathematics in the university. After the publication of his *Principia* in 1687, Newton – who was as devoted a student of alchemy and biblical chronology as of 'science' – began to look for work outside academic life, and in 1696 he was appointed Warden (and later Master) of the Mint in London. He supervised the great recoinage that was then just beginning, and he was to control the Mint for the rest of his life. He finally resigned his Cambridge appointments in 1701. In 1703 he was elected President of the Royal Society, which he ruled with a rod of iron until his death on 20 March 1727.

Newton's explanation of 'variation'

In his *Principia*, Newton was able to explain the 'variation' in the Moon's movement discovered by Tycho. To a first approximation the centre of gravity of the Earth–Moon system moved around the Sun in an ellipse, while the Earth and Moon moved in ellipses around this centre of gravity, the Earth in a smaller ellipse and the Moon in a larger one, each opposite the other. But the Sun interfered with this simple picture: it pulled the two bodies with different pulls if their distances from the Sun were different, and it pulled them in different directions when their positions formed an angle at the Sun. These different effects had to be combined with the Earth's pull on the Moon.

Twice a month, around new moon and full moon, the effect was to subtract from the Earth's pull on the Moon, and twice a month, around the quarters, it was to add to the Earth's pull. As the Moon passed from the quarters to new or full moon, the Moon's linear speed was increased; as it passed from new and full moons to the quarters its linear speed was diminished. As a result of these fluctuations, the Moon's angular position underwent deviations having extremes four times each month, in the octants – just as Tycho had found.

As the Earth–Moon system orbited the Sun each year in its elliptical path, it of course varied in its distance from the Sun. When the system was near the Sun, the Sun's force was more powerful than usual and this tended to diminish the Earth's influence on the Moon, while when the system was further from the Sun the effect was reversed. This caused the further, annual irregularity that Tycho had noticed.

reduce its pull? So strong did this objection seem to some of Newton's later disciples that they declared that, since attraction depended on distance alone, and intervening obstacles made no difference whatsoever, attraction must be a law imposed directly by God.

Granted, then, that attraction varied precisely with the inverse-square of the distance, Newton had to overcome a mathematical hurdle before he could make the Moon test. The total pull of the Earth on a falling stone was the combination of innumerable individual pulls, some over distances of a few yards, others over distances of thousands of miles. How to sum up the combined effects of all these pulls, operating in different directions and over such disparate distances? Newton succeeded in proving a very remarkable theorem, that in a uniform spherical body (as the Earth approximately is), all such pulls are together equivalent to the pull of the combined matter imagined as concentrated at the very centre of the sphere: the falling stone was, in effect, exactly one Earth's radius away from the attractive pull of the entire Earth.

This done, Newton was able to compare the Earth's pull on the stone (at a distance of one Earth radius) with the pull on the Moon (at some sixty times the distance). The pull on the stone was the cause of its acceleration in free fall; the pull on the Moon caused it too to 'fall', out of straight-line motion and into an orbit around the Earth. His calculations showed that the pulls were indeed in about the ratio of $1 : 60^2$. It was the final blow to Aristotle's dichotomy between terrestrial and celestial: Kepler's planets and the falling stones studied by Galileo were moving under the same law of force.

As Newton's treatise grew, so too did the number of observed facts for which attraction now emerged as the cause. For example, he gave the first satisfactory explanation of the tides, as resulting from differences in how the solid Earth and the flowing water of the seas were pulled by the Moon and the Sun. And he pointed out that, as a result of its spinning, the Earth would bulge slightly at the equator and be slightly flattened at the poles; because of this non-sphericity, the pulls of the Moon and Sun would cause the Earth to wobble like a child's top, resulting in the precession of the equinoxes that had been discovered by Hipparchus in Antiquity (see page 40).

Because the Moon is so near to the Earth-based observer, its complex movements can be studied in unique detail; and because the Moon's mass is significantly large in comparison with that of its parent planet (unlike, say, those of the moons of Jupiter), its dynamical interaction with the Earth is exceptionally complicated. Indeed, in the eighteenth century the Moon was to become a testing ground for Newtonian dynamics: could the law of attraction explain, not only qualitatively but quantitatively, the many departures of the Moon from a simple orbit?

Tycho, for example, had found that, when other factors had been taken into account, there was a residual variation in the speed of the Moon's movement around the sky (that is, in longitude), such that the resulting displacements reached alternating positive and negative maxima in the four 'octants' (the positions that lie midway between one of the quarters and either full or new moon). He had also found that the Moon's position in longitude displayed a further annual irregularity, amounting to as much as 11 minutes of arc.

Newton's qualitative derivations of these and other lunar irregularities are to be found in Book I of the *Principia*. But his attempts to obtain quantitative results in Book III and elsewhere were only partially successful. The *Principia* was indeed the end of one era in the history of astronomy, but it was also the beginning of another. It set the agenda for 'celestial mechanicians' in the decades to come.

The law of attraction gave Newton other insights into the planetary system. From the inward accelerations of the moons of those planets known to be endowed with them – Earth, Jupiter, and Saturn – Newton could calculate their causes, the masses of the parent planets. The Earth, it transpired, was small compared to Jupiter and Saturn. There was reason to think that the three remaining planets were likewise small (see page 186), in which case the major example of attractive inter-action between planets would be the pull between the massive Jupiter and Saturn. Astronomers were indeed having difficulty getting Keplerian-style theories to work for Jupiter and Saturn over the long term: there was trouble here, though its exact nature would not be clarified for nearly another century.

Newton believed that the foresight of Providence in constructing a stable, clock-like universe was demonstrated by the layout of the solar system: the planetary orbits were in planes that were nearly parallel, all the planets moved in the same direction, and the two massive (and potentially disruptive) planets had been banished to the outer recesses. Yet even this degree of planning (he believed) would not ward off forever the collapse of the solar system. To prevent this disaster, Providence must have arranged to intervene from time to time, to repair the disruption resulting from the perturbations. To those able to read the Book of Nature, this would be one illustration of how much God cared for his creation. We shall meet another when we come to discuss Newton's model of the universe of stars (see page 221).

Meanwhile, as Newton's treatise took shape in response to Halley's anxious encouragement, it occurred to Hooke that important elements of the conceptual jigsaw had been announced earlier in his own writings, as indeed they had; and he asked to have this recognized in the forthcoming publication.

It fell to Halley to break the news to Newton. Newton allowed himself few favourites and Hooke was not one of them. In his view Hooke was a poor mathematician and a worse philosopher of science, with no appreciation of the difference between unsupported speculations and a fully-articulated mathematical theory in which hypotheses are used to explain a wide range of phenomena. The long-suffering Halley poured copious oil on the troubled waters, and the crisis passed. When Newton, however, came to redraft the section he had intended to call *The System of the World*, he chose instead to replace it with material that was pitched mathematically at a level well beyond anything that Hooke could write, or even understand.

THE IMPACT OF THE *PRINCIPIA*

Newton's *Principia* appeared in 1687. The title – in unabbreviated form, *Mathematical Principles of Natural Philosophy* – was a clear challenge to the generally accepted view of nature, as embodied in Descartes's systematic textbook, *Principles of Philosophy*. The two books were themselves in sharp contrast. Where Descartes had offered the ambiguity of language, Newton offered the precision of mathematics; where Descartes had been content to offer explanation, Newton's geometry opened the way to quantitative prediction.

But Descartes's world view had the great merit of intelligibility: in the Cartesian universe one lump of matter could affect another only if they were in direct contact, like the cogs of a clock. Descartes had taught natural philosophers to reject 'tendencies', 'appetites', 'attractions', and the like, as words coined to disguise ignorance. To Cartesians, therefore, Newton's 'attraction' was retrograde beyond belief. Newton himself several times tried to identify a physical cause for gravity, but with little success. He could only insist that the force must be real because its consequences were evident. It was a subtle methodological point that not all would accept.

Another problem for Newtonians stemmed from the antiquated style of geometry that Newton had chosen to employ. The mathematics of the *Principia* made

ENTRETIENS,
SUR LA PLURALITÉ DES MONDES.

Cartesian physics became very popular in fashionable French society. Here, in *Conversations on the Plurality of Worlds,* first published in 1686, Bernard de Fontenelle, as author, talks on successive evenings with 'the Marquise de G.' He instructs her on the systems of Ptolemy, Copernicus, and Tycho Brahe, explains the planets, their satellites, and the stars, and discusses whether these other 'worlds' might be inhabited.

its argument incomprehensible to all but a very few. In 1697 Newton's deputy in the Cambridge professorship, William Whiston (1667–1752), was approached by a young man with a moral dilemma: his college tutor, a Cartesian 'zealot', had asked him to prepare a new Latin edition of the classic classroom text of Cartesian physics, the *Treatise of Physics* of Jacques Rohault. Was it morally acceptable to perpetuate in this way a physics that was false? Whiston replied to the effect that as Newton's work was at present unintelligible, and as Cartesian physics was a big improvement on the Aristotelianism that had preceded it, the student – Samuel Clarke (1675–1729) – might go ahead with clear conscience.

In the event, in the later editions of his translation Clarke eased his scruples by using editorial footnotes to wage a running battle with the Cartesian argument of the main text. He did not hesitate to dismiss Rohault's teaching with 'Hoc falsum est', and to offer in extended notes the alternative, Newtonian explanation. And so, in this roundabout way, students became familiar with the basics of Newtonianism.

To his followers, Newton was unique among mortals, the one man in history privileged to reveal to the rest of the human race the fundamental truths about the universe: in the couplet by the English poet Alexander Pope,

> Nature and nature's laws lay hid in night.
> God said, Let Newton be!, and all was light.

Measuring with the telescope

The telescope allowed observers to see new things, and to see familiar things enlarged. By contrast, the astronomer's traditional instruments were for the measurement of angles. It was not immediately obvious that the two could be combined.

In the telescope as designed by Kepler, the image was brought to a focus and examined with the eyepiece, which served as a microscope. About 1640, an English amateur astronomer named William Gascoigne

The principle of the wire micrometer. To measure the angle separating two stars, the observer aligns the micrometer so that the fixed wire *A* passes through the two stars. He then adjusts the moveable wires *B* and *C* so that each passes through one of the stars. The amount of the adjustment required is then a measure of the angle separating the stars.

found that a spider had spun a web in the focal plane of his telescope, and this web appeared superimposed on the astronomical image. He realized that cross-hairs could be inserted in this same plane, to locate the exact centre of the field of view and so enable the telescope to be aligned accurately on the target object. Alternatively, a measuring device or 'micrometer' could be similarly placed, and used to measure either the width of an object such as a planet, or the angle between two nearby objects, such as two mountains on the Moon.

Unfortunately Gascoigne was killed in 1644 in a battle in the Civil War, but his techniques were applied in Oxford in the 1650s, notably by Christopher Wren, who used micrometers in a survey of the Moon. In 1663 Wren also demonstrated to the Royal Society an astronomical instrument that incorporated telescopic sights.

Meanwhile Christiaan Huygens announced his (independent) discovery of a form of eyepiece micrometer in 1659 in his *Systema Saturnium*. Other forms of micrometer were developed in Paris in the 1660s, by Adrien Auzout, Pierre Petit, and Jean Picard. Soon, the transformation of the telescope into an instrument of measurement was complete.

Once the Newtonian framework was accepted, the agenda was clear: to demonstrate mathematically that *all* the observed movements of the celestial bodies were the result of attraction.

But even among those prepared to countenance attraction as a valid concept in physics, a choice between the two philosophies of nature was by no means clear, nor was it evident that a compromise was ruled out. A direct choice was the more difficult in that Cartesianism did not lend itself to quantitative predictions, so that issues in which experiment might decide between contradictory forecasts were few. One such, however, concerned the shape of the Earth.

The incorporation of a telescope into traditional measuring instruments, to improve the resolution of the unaided eye, was slower in coming. Johannes Hevelius defended his refusal to use telescopic sights when measuring the positions of astronomical bodies, by claiming that his traditional instrumentation was supplying an accuracy never before achieved, of only a few seconds of arc. Robert Hooke, however, had little patience with what he saw as a perverse refusal to accept a technical advance, resulting in a waste of time and resources. A public squabble ensued, with Hevelius refusing to budge; but by the time of his death he was almost alone in denying the value of telescopic sights.

Hevelius using his azimuthal quadrant, which he termed *exquisitissimus*, 'most exquisite'. He rotated the quadrant until it was aligned in the desired direction. He then aimed the sight rule at the star, using his right hand to position it with the aid of counterbalanced weights, and read the angle from the scale. The bronze figures on the left helped the balance the instrument.

Fig. **H**

Newtonians versus Cartesians on the Shape of the Earth

Since Antiquity, every educated person had known that the Earth was roughly spherical (see pages 30–1): Christopher Columbus understood perfectly well that the Earth had no edge for his ship to fall off, and the approximate size of the Earth had been known since the work of Eratosthenes in the third century BC (see page 39). But no one person would be able to determine with precision the shape of the

The south front of Paris Observatory. Construction began in 1667, and though the building was completed in 1672, the interior was not finished until 1683. On the ground a telescope is slung from a pole. To the right is a wooden tower some 130 feet high. This had originally been a water-tower at Marly, near Versailles; it was re-erected at the Paris Observatory in 1685, to furnish a stable platform for the objectives of the immensely long aerial telescopes favoured by G.D. Cassini and his colleagues (see page 143). A ladder was provided, and a balcony added to ensure the safety of the assistants.

Veüe et perspectiue de l'Obseruato

Earth; this called for an organization with sufficient funds to attract the necessary talent and to meet the considerable expenses involved.

Such an organization had come into being with the formation of the Paris Academy of Sciences in 1666 by Jean Baptiste Colbert, the finance minister of Louis XIV. One immediate result was the foundation of the Paris Observatory, and the recruitment of astronomers from all over Europe, to be led by the Italian Gian

scituée au bout du fauxbourg St Jacques a Paris
fait par Aueline auec Priuilege du Roy

Louis XIV visits the Paris Observatory. The observatory was one result of the proposal, in 1665, from a number of French scientists, for the foundation of a broadly based academy of sciences. It was intended that the observatory building should provide meeting rooms and laboratories for the academicians as a whole. However, its location proved inconveniently remote from the centre of Paris, and even from the start, the activities carried out in the observatory were purely astronomical. In this visit by the king on 1 May 1682, we see around the walls vestiges of the observatory's wider function.

Centre: The map of the Moon, 22 inches in diameter, compiled at Paris in 1679 by G.D. Cassini. In its preparation, Cassini and his assistants produced some sixty drawings of all or part of the lunar surface.

Domenico Cassini (1625–1712, the founder of a dynasty of astronomers and who is therefore known as Cassini I).

Perhaps the most eminent recruit to the Academy, acquired at great expense, was Christiaan Huygens from Holland. Another was the Frenchman Adrien Auzout (1622–91), who had played a central role in the development of the wire micrometer; this enabled the astronomer to measure a tiny angle by adjusting two hairs or wires in the field of view until they coincided with the angle to be measured (see page 164). He in turn promoted the recruitment of Jean Picard (1620–82), who brought to the problem of the size of the Earth the skills of a dedicated observer.

Huygens had assumed that the time a pendulum of given length took to swing was the same everywhere on Earth, in which case the length of a pendulum that beat so that each swing took one second was a natural unit; and in 1661 he had proposed to the Royal Society that this length be taken as a universal standard. However, different investigators began to obtain significantly different lengths. Matters came to a head after the Academy sent Jean Richer in 1671 to the island of Cayenne off the Atlantic coast of South America, where he was to make observations, especially of the close approach of Mars to Earth the following year. He took with him a seconds pendulum; or, more exactly, a pendulum that in Paris had beaten seconds. But in Cayenne Richer found that it would beat seconds only if shortened by a tenth of an inch.

Newton in his *Principia* offered an explanation: the Earth bulged at the equator. Huygens, working in the Cartesian tradition, came to a similar conclusion, but differed as to the amount of the bulge.

A lengthy debate ensued as to who was right. Eventually the Academy decided to use its financial muscle to settle the matter, by funding expeditions to measure the length of a degree of latitude in two very different locations: Lapland and Peru. The conclusion was that the Earth did indeed bulge at the equator; but by exactly how much was still uncertain. Other measurements were made, but still there was no definite conclusion.

P.L.M. de Maupertuis (1698–1759) led the Academy of Sciences' expedition to Lapland, charged with measuring a degree of latitude. He set out in 1737 and returned the following year, satisfied that he had proved the Earth was flattened at the poles. In this engraving he is shown dressed in Lap costume and compressing the Earth. Above the reindeer pulling a sledge is a poem praising his achievement.

The foundations of astronomy

Eighteenth- and nineteenth-century observatories saw it their duty to record the positions of the stars and other celestial objects with ever increasing precision and completeness. The equivalent of terrestrial longitude is known as right ascension (RA), and is measured from the point on the celestial equator where the Sun crosses the equator at the spring equinox. As the Earth rotates once a day, RA can be expressed in degrees or in time (with one hour equivalent to 15°), and the fundamental measurement is of the moment when the celestial object crosses (or 'transits') the meridian of the observatory. The first precision-mounted instrument dedicated to the measurement of transits was built by Ole Römer after he returned from Paris in 1681 to take charge of the Copenhagen Observatory. The location of an observatory's

Ole Römer at his transit instrument in Copenhagen, 1684. The telescopic sight was accurately mounted in the meridian. To measure the position of a star in right ascension, Römer noted the time at the exact moment when the star crossed the meridian.

Similar instrumentation, serving the same purpose, a century and a half later. The transit room of the observatory of the English amateur, William Henry Smyth, in 1832.

This attempt to decide between Newton and Descartes had ended in confusion. But the issue was soon to be settled, among astronomers and the general public alike, by the appearance of a comet in 1759. This would lead to the apotheosis of Newton, for the comet had been forecast two generations earlier by Edmond Halley: a Newtonian had predicted the unpredictable.

transit instrument came to define the longitude of the observatory, and since 1884 the transit instrument at Greenwich has defined the world's zero of longitude.

To measure the altitude of the celestial body at the moment of transit, and so obtain its declination (the celestial equivalent of latitude), observatories followed the example of Tycho Brahe and used quadrants or other sectors of a circle mounted against a wall precisely aligned north–south – but of course with telescopic sights. In the nineteenth century, leading English makers such as Edward Troughton developed a mural circle, which (like its predecessors) was to be used in conjunction with the transit instrument.

This, however, was expensive, not least in manpower, since two observers were required. Continental makers responded to demand by building transit instruments that incorporated circles of sufficient accuracy to render a mural arc (or circle) unnecessary, and English makers were soon forced to follow suit.

Below: The mural circle by Edward Troughton installed at Greenwich in 1812. It complemented a transit instrument by measuring the altitudes of stars as they crossed the meridian.

Above: The transit circle installed at Greenwich in 1850. It combined in a single instrument the functions of a transit instrument and a mural circle, with corresponding economies in manpower. In 1880 Greenwich Mean Time, based on the meridian passing through this instrument, became legal time in Great Britain, and in 1884 a conference in Washington adopted the same line as the world's prime meridian.

The Return of Halley's Comet

Newton had shown in the *Principia* that comets were lawlike and moved through the solar system in conic sections, though their paths were very different from the nearly-circular ellipses of the orbits of the planets. So elongated were the cometary orbits that near the Sun they approximated to a parabola.

However, unless its velocity was sufficiently large, the departing comet would be unable to escape from the domination of the Sun. Instead, it would leave along an elliptical orbit, destined one day to be hauled back again to the solar system – though on a path that was modified, compared to that of its previous visit, if it had then passed close to a major planet. But even with such modifications, the comet's orbit would have similar characteristics on each appearance, and these appearances would be separated in time by similar intervals. Did historical records provide any examples of this?

Halley undertook the necessary enquiry, and quickly found a prime candidate in the comet of 1682: its orbit was retrograde (that is, opposite in sense to the orbits of the planets), and so had been the orbits of the comets of 1531 and 1607. Further enquiry showed that the orbits had much else in common, and in 1695 Halley told Newton he felt sure these were reappearances of the same comet.

The problem was that the intervals between these appearances were similar but not identical. Halley realized that this was because the comet's orbit had been modified as the comet had experienced the pull of one or more planets, and he

Halley's 1705 table of the features of the orbits of 24 comets. If Newton was right and comets were subject to the inverse-square law of attraction, then a comet that rounded the Sun would be in an orbit that was elliptical, parabolic, or hyperbolic; and if it was elliptical, the comet would return. In 1695 Halley noted the resemblance between the orbits of the comets of 1531, 1607, and 1682, and after studying them further concluded they were reappearances of the same comet (Halley's Comet).

Cometæ Anni.	Nodus Ascend.	Inclin. Orbitæ.	Perihelion in Orbe.	Perihelion in Ecliptica	Latitudo Perihelii	Distantia Perihelia à Sole.	Log. dist. Perihelii à Sole.	Temp. æquat. Perihelii Londini.
	gr. ' "	gr. ' "	gr. ' "	gr. ' "	gr. ' "			die. h. '
1337	♊ 24. 21. 0	32. 11. 0	♉ 7. 59. 0	♉ 12. 45. 15	22. 40. 30 B	40666	9. 609236	Junii 2. 6. 25
1472	♑ 11. 46. 20	5. 20. 0	♉ 15. 33. 30	♉ 15. 40. 20	4. 25. 50 A	54273	9. 734584	Feb. 28. 22. 23
1531	♉ 19. 25. 0	17. 56. 0	♒ 1. 39. 0	♒ 0. 48. 15	17. 3. 05 B	56700	9. 753583	Aug. 24. 21. 18½
1532	♊ 20. 27. 0	32. 36. 0	♋ 21 7. 0	♋ 16. 59. 40	15. 57. 00 B	50910	9. 706803	Oct. 19. 22. 12
1556	♍ 25. 42. 0	32. 6. 30	♑ 8. 50. 0	♑ 11. 6. 0	31. 10. 20 B	46390	9. 666424	Apr. 21. 20. 3
1577	♈ 25. 52. 0	74. 32. 45	♌ 9. 22. 0	♍ 7. 53. 00	69. 35. 20 A	18342	9. 263447	Octo. 26. 18. 45
1580	♈ 18. 57. 20	64. 40. 0	♋ 19. 5. 50	♋ 19. 17. 10	64. 40. 0 B	59628	9. 775450	Nov. 28. 15. 00
1585	♉ 7. 42. 30	6. 4. 0	♈ 8. 51. 0	♈ 8. 59. 10	2. 55. 25 A	109358	0. 038850	Sept. 27. 19. 20
1590	♍ 15. 30. 40	29. 40. 40	♏ 6. 54. 30	♏ 2. 55. 50	22. 45. 50 A	57661	9. 760882	Jan. 29. 3. 45
1596	♒ 12. 12. 30	55. 12. 0	♏ 18. 16. 0	♏ 22. 44. 35	54. 44. 30 B	51293	9. 710058	Julii 31. 19. 55
1607	♉ 20. 21. 0	17. 2. 0	♒ 2. 16. 0	♒ 1. 29. 40	16. 10. 5 B	58680	9. 768490	Oct. 16. 3. 50
1618	♊ 16. 1. 0	37. 34. 0	♈ 2. 14. 0	♈ 6. 10. 00	35. 50. 0 A	37975	9. 579498	Oct. 29. 12. 23
1652	♊ 28. 10. 0	79. 28. 0	♈ 28. 18. 40	♊ 10. 41. 35	58. 14. 0 A	84750	9. 928140	Nov. 2. 15. 40
1661	♊ 22. 30. 30	32. 35. 50	♋ 25. 58. 40	♋ 21. 37. 30	17. 17. 0 B	44851	9. 651772	Jan. 16. 23. 41
1664	♊ 21. 14. 0	21. 18. 30	♌ 10. 41. 25	♌ 8. 40. 35	16. 1. 50 A	102575½	0. 011044	Nov. 24. 11. 52
1665	♏ 18. 02. 0	76. 05. 0	♊ 11. 54. 30	♉ 24. 6. 35	23. 8. 0 B	10649	9. 027309	Apr. 14. 5. 15½
1672	♑ 27. 30. 30	83. 22. 10	♉ 16. 59. 30	♋ 9. 26. 00	69. 27. 40 B	69739	9. 843470	Feb. 20. 8. 37
1677	♏ 26. 49. 10	79. 03. 15	♌ 17. 37. 5	♋ 16. 21. 05	75. 44. 10 B	28059	9. 448072	Apr. 26. 00. 37½
1680	♑ 2. 2. 0	60. 56. 0	♎ 22. 39. 30	♐ 27. 26. 50	8. 11. 10 A	00612⅓	7. 787106	Dec. 8. 00. 6
1682	♓ 21. 16. 30	17. 56. 0	♒ 2. 52. 45	♒ 2. 0. 30	16. 59. 20 B	58328	9. 765877	Sept. 4. 07. 39
1683	♍ 23. 23. 0	83. 11. 0	♊ 25. 29. 30	♋ 10. 36. 55	82. 52. 00 B	56020	9. 748343	Julii 3. 2. 50
1684	♎ 28. 15. 0	65. 48. 40	♏ 28. 52. 0	♐ 15. 15. 25	26. 35. 20 A	96015	9. 982339	Maii 29. 10. 16
1686	♓ 20. 34. 40	31. 21. 40	♊ 17. 00. 30	♊ 16. 24. 00	31. 17. 35 B	32500	9. 511883	Sept. 6. 14. 33
1698	♎ 27. 44. 15	11. 46. 0	♑ 00. 51. 15	♑ 0. 47. 20	0. 38. 10 A	69129	9. 839660	Oct. 8. 16. 57

The map used to detect Halley's Comet as it returned. In the race to detect the comet, predicted to return in the winter of 1758/59, Joseph-Nicolas Delisle judged from past records that it might be seen telescopically a month before it rounded the Sun (at 'perihelion'). He calculated its position at ten-day intervals, on the supposition that it would be visible either thirty-five days or twenty-five days before perihelion, and plotted these positions on this map. He then instructed Charles Messier to observe along arcs joining corresponding dates. On 21 January 1759, Messier noticed what proved to be the returning comet.

predicted the comet would return 'about the end of the year 1758, or the beginning of the next'.

In the popular mind comets had always been portents of disaster, while even to astronomers the nature of comets and their role in the cosmic order were still shrouded in mystery. As 1758 approached, therefore, interest mounted. Some feared that the returning comet might collide with the Earth and put an end to the human race.

In his prediction, Halley had taken into account the acceleration the comet had experienced from the pull of Jupiter as it approached its 1682 passage round the Sun, but he had not allowed for the opposite consequences of the pull of Jupiter as the comet left the solar system. To remedy this called for detailed and laborious calculations, and in Paris, in June 1757, Alexis-Claude Clairaut embarked on the task with the help of two associates. They first analysed the orbit of the comet as it left the solar system after the 1531 encounter, and on this basis 'predicted' its return in 1607 which they compared with what actually occurred. They then analysed the comet's orbit as it left in 1607, and similarly 'predicted' its return in 1682. Learning from these quasi-experiments, they similarly predicted the comet's forthcoming return: it would swing round the Sun in mid-April 1759, give or take a month.

A 'backstaff' made in London in 1736. Navigators who wished to determine their latitude by using the cross-staff (see page 81) to measure the altitude of the Sun above the horizon needed to look directly at the Sun, and to keep one end of the cross bar aligned with the horizontal as the ship moved. The backstaff improved the situation. The navigator stood with his back to the Sun and held the instrument vertically. He allowed the shadow of the shadow vane (on the right-hand arc) to fall on the slit of the horizon vane (far right). At the same time he viewed the horizon through the pin-hole on the sight vane (left) and the horizon vane. He did this while the Sun crossed the meridian at noon, and the meridian altitude of the Sun was obtained by adding the angles indicated on the two arcs.

It was a good forecast. The returning comet was first seen on Christmas Day 1758 by a farmer living near Dresden, while the first professional to observe it was the Parisian comet-hunter, Charles Messier (1730–1817), four weeks later. On 13 March the comet rounded the Sun.

The characteristics of the comet's orbit were very similar to those of the comets of 1531, 1607, and 1682. Clearly, the comets were one and the same: Halley had been proved correct, and Newtonianism had won its most public triumph.

The Newtonian Programme

While these very public trials of Newtonian attraction were taking place, a tiny band of the mathematical elite were working at their desks, exploring the consequences of the attractive pulls between the Sun, the planets, and their moons. To implement this Newtonian programme, Providence supplied an extraordinary cluster of talent in the middle decades of the eighteenth century: Jean le Rond d'Alembert (1717–83), who took his name from the Paris church where he had been found abandoned, and who later devoted his life to mathematics while living in Paris on an annuity from his natural father; the Swiss Leonhard Euler (1707–83), lured by Catherine the Great to St Petersburg, and then by Frederick the Great to Berlin, and who finally returned to St Petersburg where he died; the precocious Parisian Alexis-Claude Clairaut (1713–65), whose life-long association with the Paris Academy of Sciences began when he was only twelve; and Joseph Louis Lagrange (1736–1813) of Turin, who was persuaded first to Berlin and then to Paris. The last decades of the eighteenth century and the early decades of the nineteenth would be dominated by Pierre Simon de Laplace (1749–1827), who survived the French Revolution and ended his life as a marquis.

Often in competition with each other, these mathematicians tackled one problem of 'physical astronomy' after another, developing the necessary mathematical tools in the process, as Newton's cumbersome geometrical approach to trigonometry was replaced by techniques involving infinite series that could be treated in an abstract manner. Even so, investigators were forced to proceed by approximation, and in handling infinite series they often had to make a judgement as to which terms in a series were significant and which might safely be neglected.

One example of the difficulties this entailed is given by Clairaut's investigation of the motion of the lunar apogee (the point in the Moon's orbit where it is furthest from the Earth). In 1747 he announced that attraction accounted for only half of the observed motion (a conclusion independently reached by Euler and d'Alembert), and he suggested adding a fourth-power term to the inverse-square law. But a year later, having carried the approximation further, he found that in fact the law as stated by Newton did account for the whole of the observed motion.

The Moon was of special interest, not only because observational data concerning the many irregularities in its movements were exceptionally complete, but because there were hopes that a mastery of these irregularities would allow mathematicians to solve the urgent problem that faced every mariner who sailed far from land: how to determine his longitude. The 'finding' of longitude was also of great importance to geographers and astronomers: geographers needed it in order to prepare accurate maps, and astronomers in order to correlate observations made at different observatories.

The Finding of Longitude

Latitude north of the equator is, by definition, equal to the altitude of the celestial North Pole, the point in the sky about which the stars revolve, and so on land the determination of latitude presented no problem. Even at sea, the Renaissance mariner could derive his latitude from measurements of the altitude of the Pole Star or of the Sun at noon (see pages 84–5).

Longitude – the angle by which a town, or ship, is east or west of a standard location such as Greenwich – was altogether more difficult to determine. Differences in local times correspond to differences in longitude, one hour being equivalent to 1/24 of a circle or 15°, and to determine local time in a given place by the Sun or stars was straightforward – for example, if the Sun was due south, it was midday. But to determine the difference in local time between two places, it was necessary to determine their local times *simultaneously*. How was this to be done?

In Antiquity, Hipparchus had pointed out that an eclipse of the Moon was in effect a time-signal that would be simultaneously visible from wherever on Earth the Moon could be seen; and on occasion down the ages, at each of two places on land, a lunar eclipse had been observed and the local times recorded, and from the difference in the times the corresponding difference in longitude was later calculated. For example, an eclipse was due to take place in 1631, and Henry

Using an eclipse to measure longitudinal difference. Observers located at different longitudes see an eclipse of the Moon simultaneously, but at different local times, each hour of difference in local time corresponding to a difference of 15° in longitude. In this 1585 figure, the Sun is lower right, and the Moon (upper left) is in the shadow of the Earth (in the centre). The observer drawn vertically sees this eclipse at the time shown on the outer scale, while the other sees it at the time shown on the inner scale, namely two hours later. Their difference in longitude is therefore 30°.

Gellibrand (1597–1636), professor of astronomy at Gresham College in London, arranged with a sea captain setting out to find the Northwest Passage, that both men would observe the eclipse and note the times when it occurred. In an appendix to *The Strange and Dangerous Voyage of Captaine Thomas James*, published in 1633, Gellibrand concluded that the difference in longitude between London and Charlton Island in James Bay, Canada, was 79° 30′ (a value some 15′ in error). But eclipses of the Moon were much too rare to be of use to navigators on the high seas, and even observers on land found the moment of eclipse poorly defined.

There were periods in the sixteenth and seventeenth centuries when it seemed that the mariner equipped with suitable charts might be able to use the difference between magnetic north and true north, or between the angle of dip of the magnetic needle and the horizontal, as a co-ordinate taking the place of longitude (see page 145); but these hopes came to nothing. Seafarers meanwhile continued to rely on 'dead reckoning' – in other words, on 'guestimates'.

The public were constantly reminded of the perils of navigation out of sight of land, as regular toll was taken of warships and merchantmen alike. For Britain the worst such disaster occurred in September 1707, when a fleet en route from Gibraltar, and sailing east in the belief that it was in the English Channel, had four of its ships wrecked on the Scilly Isles with the loss of two thousand men.

In 1713 William Whiston and a schoolteacher named Humphry Ditton announced they had a solution to the problem of longitude at sea which they would disclose if suitably rewarded. The method was fantastic – it involved mooring ships at fixed positions across the oceans – but their claim stimulated the British Parliament to establish a Board of Longitude, with authority to award the huge prize of £10,000 to whoever produced a practical method of determining longitude at sea to one degree, or twice the amount for twice the accuracy.

There were two methods seriously in contention. The first was a question of technology: how to develop precision clockwork capable of withstanding the motion of a ship buffeted by a gale. This would provide the navigator with a standard time, which he could compare with his local time. Christiaan Huygens had devised a pendulum-driven clock that he hoped would keep a regular beat even in a rough sea; but tests on voyages between France and the Mediterranean in 1668–70, and later as far as the Cape of Good Hope, showed that the problem had not yet been solved.

The other strongly-backed candidate in the race was of considerable scientific sophistication: the method of lunar distances. The Moon travels rapidly, moving every hour among the background stars by the equivalent of its own diameter. Just as the hour hand of a conventional clock tells the time as it moves among the numbers engraved on the clockface, so – it was hoped – the Moon might be used to tell the time as it moved among the stars.

To read the time by the lunar clock, the mariner would require three things: (i) an accurate and convenient instrument for measuring the angle between the

Moon in its current position and a suitable reference star, (ii) an accurate catalogue of star positions, and (iii) accurate tables of the Moon's motion. All three were lacking at the end of the seventeenth century. No suitable instrument existed; those available, like the cross-staff (see page 81), required the user to look at both objects simultaneously, and it was difficult to do this at sea with any accuracy. As to star positions, Tycho Brahe's catalogue of 777 stars had been compiled before the invention of the telescope, while Johannes Hevelius had spurned the use of telescopic sights when compiling his catalogue of over 1500 stars that was published posthumously in 1690. Yet it was the lack of reliable tables of the Moon's motion that posed the greatest problem. Even after Newton had revealed the dynamics underlying the Moon's complex manoeuvres, the errors resulting from his lunar theory as set out in the 1713 edition of the *Principia* still came to several minutes of arc, and this alone could easily have resulted in an uncertainty of a hundred miles in a ship's position.

Longitude and the moons of Jupiter

In 1610 Galileo discovered four moons of Jupiter, and two years later he observed one of them eclipsed by the parent planet. He realized that such eclipses were much better defined than those of the Moon, and much more frequent. They might therefore be used as time-signals for the measurement of longitude. Admittedly they were not frequent enough to answer the needs of mariners; but tables that set out the changing configuration of Jupiter's moons would provide a useful, though much less precise, alternative.

In 1598 the King of Spain, whose ships were sailing the oceans of the world, had offered a handsome reward to whoever could 'find longitude'. Galileo, never slow to turn his discoveries to his financial advantage, entered into negotiations over the use of Jupiter's moons for this purpose. But from the deck of a rolling ship it was rarely possible to see Jupiter through a telescope, let alone its moons; and despite attempts to rig for the navigator a swinging platform that would reduce the movement, the practical difficulties proved too great for the method to be adopted at sea.

For use on land, where accuracy was the main criterion, the Jupiter method called for reliable tables of eclipses of the moons; but it was not until 1668, through the work of Cassini I, that such tables became available. French astronomers in particular were quick to exploit the new technique. For example, before the observations made in the previous century by Tycho Brahe on the Baltic island of Hven could be correlated with observations in progress at Paris, the difference in longitude between the two sites had to be determined with the greatest possible care. In 1671 Jean Picard was entrusted with the task by the Paris Academy of Sciences and, in company with the Danes Erasmus Bartholin and Ole Römer, he spent eight months on Hven making a series of observations of eclipses of the first satellite of Jupiter. Meanwhile, Cassini did the same in Paris, and on Picard's return, a comparison of their data gave the required difference in longitude.

Pages for January 1668 from G. D. Cassini's tables of phenomena associated with the moons of Jupiter. Top: The appearances at 7 p.m. each day. Right: The times (in hours and minutes) of eclipses, etc.

On two of the three counts, matters improved early in the eighteenth century. The invention in 1731 of a double-reflection quadrant, the ancestor of today's sextant, provided a method of measuring angles between heavenly bodies that was both accurate and suitable for use at sea. The positions of 3,000 stars were to be

Greenwich Observatory

In 1673 King Charles II of England set up a committee to investigate persistent claims that the behaviour of the magnetic compass could be used to supply a co-ordinate that could be used in place of longitude. While the committee was considering the question, a French visitor to the court claimed to have a method of finding longitude that depended on the movements of the Moon. To the surprise of the king, it emerged that even if such a method were feasible in principle, astronomical data of the necessary accuracy had not yet been assembled.

It lay within the power of the king to remedy this lack. Early in 1675 a royal warrant was issued that began: 'Whereas, we have appointed our trusty and well-beloved John Flamsteed, master of arts, our astronomical observator, forthwith to apply himself with the most exact care and diligence to the rectifying the tables of the motions of the heavens, and the places of the fixed stars, so as to find out the so much-desired longitude of places for the perfecting the art of navigation.'

For this task Flamsteed would need a permanent observatory, and at the suggestion of Christopher Wren the site chosen was that of a castle on a hill (right) above Greenwich. Construction began the same year, Flamsteed moved in on 10 July 1676, and on 19 September observing began.

The observatory remained at Greenwich, headed by successive Astronomers Royal, until the end of the Second World War, when war damage, interference from the lights and the smog of London, and the generally run-down state of the institution compelled an exodus from the capital.

found in the 'British Catalogue' of John Flamsteed published posthumously in 1725 – half-a-century earlier King Charles II had appointed Flamsteed Astronomer Royal, and established him in Greenwich Observatory, expressly in order to supply the needs of navigators.

Christopher Wren designed this Great Room of Greenwich Observatory 'a little for pomp'. The clocks to the left of the door were of the pendulum type invented by Christiaan Huygens in 1657, and had been specially commissioned. Their pendulums, which were suspended behind the panelling, were no less than 13 feet long, and beat every two seconds, over a very narrow arc. The telescope was used for observations, such as eclipses, that had to be timed, and the ladder supporting it could be moved from window to window. The quadrant was for keeping check on the clocks.

It was now up to the mathematicians to get a firm grip on the motion of the Moon and to use this to calculate tables of lunar positions. Around the middle of the century both the Paris Academy of Sciences and the St Petersburg Academy focused attention on questions of lunar theory by means of challenge prizes; and it was by using equations developed in this connection by Leonhard Euler, that the Göttingen professor Tobias Mayer (1723–62) – an astronomer of practical bent – produced tables of the Moon which he sent to London in late 1754. Trials of the tables were interrupted by the Seven Years War, and Mayer prepared improved tables shortly before his death. These eventually earned his widow £3,000 from the British Parliament (and a surprised Euler £300), and they enabled the Astronomer Royal, Nevil Maskelyne (1732–1811), to publish in 1766 the first of the annual volumes of *The Nautical Almanac*, designed to make it convenient for navigators to determine longitude by the method of lunar distances.

But meanwhile advances in technology were offering a more straightforward alternative. In 1735, the English clockmaker John Harrison (1693–1776) produced his first marine timepiece (known today as H1), and this was taken in trials to Lisbon and back the following year. The Board of Longitude then granted Harrison £250 towards an improved clock; and things continued in this way, with successive improvements and successive grants, for nearly thirty years.

PROSPECTUS INTRA CAMERAM STELLATAM.

Shown left is Harrison's first marine timepiece, 'H1'. It is 3 feet in height, width and depth, and because of the amount of brass used in its construction it weighs 72 pounds. The balls at the ends of the two balances swing together and apart in unison, even when the clock is tilted by a ship's movement. The dials show hours, minutes, and seconds, and also the days of the month. Below is Harrison's prize-winning H4, in which he pioneered the use of jewelled bearings. It is a mere 5$^{1/4}$ inches in diameter.

In 1764 Harrison and H4 sailed to Barbados in a warship. H4 did all that was asked of it, but Parliament decided that Harrison should have £10,000 then, and the second £10,000 when copies had been made and tested. Finally, in 1773, the aged Harrison was awarded the bulk of the second £10,000. A copy of H4 was taken by Captain James Cook on his voyage to the tropics and the Antarctic (1772–75), and 'our never failing guide, the Watch' performed triumphantly. As soon as suitable chronometers could be constructed, they became the preferred solution to the problem of longitude. Astronomical observatories were established alongside major ports, and astronomers found a new role, making accurate

The design of Captain Robert Wauchope for the world's first timeball, tested at Portsmouth, England, in 1829. Time by the ship's chronometer was crucial to the determination of longitude at sea, and so a navigator would take the opportunity to monitor the chronometer's performance whenever his ship was in port. Under Wauchope's proposal, at exactly the agreed hour, an astronomical observer on shore would order the lower ball to be dropped, and the watching navigators could then check their chronometers. A timeball was established at Greenwich in 1833, and time signals of one form or another were soon in use at ports through-out the world.

observations of noon-time and immediately (or at 1 p.m.) dropping timeballs as a signal by which the ships' chronometers could be checked.

On land, however, the roughness of the roads tended to disturb the running of chronometers as they were transported from one site to another. Yet pairs of obser-vatories were increasingly anxious to establish their differences in longitude in order to correlate their observations. Sometimes artificial means – the firing of rockets from an intermediate hilltop – were used to generate a signal to be seen simultaneously from two different places. Greenwich and Paris, for example, were linked in this way in 1825 by a succession of such signals, the astronomers being supplied with large parties of troops for the purpose. But before long the introduction of the electric telegraph solved the problem of longitude on land, just as the chronometer had solved the problem at sea.

The Past and Future of the Solar System

By the mid-eighteenth century, 'celestial mechanics' based on Newtonian principles had registered many successes. In particular, the new solar and lunar tables took account of perturbations, and were far more accurate than the tables they replaced. But two puzzling anomalies remained: an apparent acceleration of Jupiter and deceleration of Saturn, evident since the time of Tycho Brahe, and an apparent acceleration of the Moon, shown by Halley to have been going on since Antiquity. These trends had profound implications: if they continued indefinitely, with Jupiter spiralling into the Sun, Saturn receding, and the Moon falling into the Earth, the solar system was doomed to change drastically, and perhaps to perish.

In the work of Euler (who was one of those prepared to accept the possibility of such a catastrophe), Lagrange, and Laplace, a distinction came to be made between two kinds of variations, the 'periodic' and the 'secular'. Periodic variations were seen as pendulum-like oscillations, in longitude, latitude, and distance from the central body, that were restored in a relatively short time. The secular variations were long-term, and Euler at first conceived them as operating always in one direction. They affected the shape and orientation of the ellipse in which the body orbited: its eccentricity, the position of the axis of the ellipse, the inclination of the orbital plane to the ecliptic (the plane of the Earth's orbit), the position of the line of the nodes (where the ellipse cut the ecliptic), and perhaps even the average distance to the Sun.

In 1772 Laplace, who was then only twenty-three years old, proved that, to a high degree of approximation, the attractive pulls between two planets could not cause permanent, one-directional change in the average Sun–planet distance. In the light of Kepler's third law (as corrected by Newton), this implied that these attrac-tive pulls could not cause a one-directional change in the periods of the planets. Laplace drew the conclusion that the apparent acceleration of Jupiter and decelera-tion of Saturn could not be due to their mutual gravitational interaction – he thought it likely that these effects were due to interactions with comets.

Transits of Venus

Edmond Halley pointed out in 1716 that on the rare occasions when Venus crossed the face of the Sun, astronomers had an opportunity to make accurate measurements of the astronomical unit (the mean distance of the Earth from the Sun, which is fundamental to measurements within the solar system and beyond). The next such 'transit' would occur in 1761, and there would be another in 1769.

The method required comparison of observations from widely separated locations, and astronomers of many nations organized expeditions for each of the transits. The least fortunate was the French astronomer Guillaume Le Gentil, who set out in 1760 for Pondicherry in French India. The British captured the settlement before the transit took place and his view of the transit from shipboard was of no scientific value. He decided to remain in the region and await the transit of 1769, only to be frustrated by cloud. He returned to France to find himself presumed dead, his estate divided, and his post filled.

The overall results were very disappointing, partly because of the 'black-drop' effect encountered by observers and the difficulty this created in deciding when the planet had entered the disk of the Sun.

Left: In a transit of Venus (shown as a black circle against the curved outline of the Sun), the planet appears for a time to be joined to the limb of the Sun. As a result of this 'black-drop' effect, the moment of internal contact is poorly defined.

Right: This portable astronomical quadrant was made specially for an expedition mounted by the Royal Society to observe the 1769 transit of Venus. There are two telescopic sights, one fixed in the horizontal and the other pivoted, and the instrument is levelled by means of the plumb-bob (which was suspended in water to dampen its motion). The elliptical ring (top left) reflected light, to illuminate the cross-wires by which the moveable sight was centred on the object observed.

Pierre-Simon de Laplace

Laplace was born on 28 March 1749 at Beaumont-en-Auge in Lower Normandy, and educated by Benedictines. He began studies for the priesthood at the University of Caen, but a teacher of mathematics recommended him to d'Alembert, and a mathematical essay of Laplace's so impressed d'Alembert that he arranged for Laplace to become professor of mathematics at the Ecole Militaire. Laplace lived in Paris until his death on 5 March 1827, occupying a succession of official posts. He managed to survive and prosper through the French Revolution, the Napoleonic era, and the return of the Bourbons, by whom he was made a marquis. His semi-popular *Exposition du système du monde*, published in 1796, made celestial mechanics available to a wide readership, while his five-volume *Mécanique céleste* (1799–1825) synthesized the achievement of the Continental mathematicians – not least Laplace himself – who had investigated the solar system on Newtonian principles.

In 1774 Lagrange showed that the secular variations in the inclinations of the planetary orbits to the ecliptic and in the position of the line of nodes were, to a first approximation, oscillatory and periodic, with periods measured in thousands of years. On reading Lagrange's demonstration, Laplace immediately applied the same type of analysis to other aspects of the planetary orbits. It now appeared that all long-term changes in the planetary orbits resulting from their mutual attractive pulls were oscillatory and periodic.

In 1785, Laplace discovered the cause of the long-continued acceleration of Jupiter and deceleration of Saturn: the variation was not one-directional but periodic, with a period of about 900 years. It differed from the still longer-term secular changes because it depended on the positional relationship between the two interacting planets (Jupiter and Saturn) and the Sun. To find this effect, Laplace had to pursue the series approximations to the order of the third power of the eccentricities and inclinations – an undertaking that called for finesse but also involved a great deal of sheer calculative drudgery.

In 1787 Laplace identified a cause for the secular acceleration of the Moon. It was a secondary effect, arising from the secular reduction currently taking place in the eccentricity of the elliptic orbit of the Earth, which led to a net reduction in the Sun's action on the Moon. This effect would be reversed when, in its long-term oscillation, the eccentricity stopped diminishing and began increasing again. (In the 1850s, it would be found that Laplace's explanation accounted for only about half the Moon's apparent acceleration; the other half came to be attributed to the slowing down of the Earth's rotation, caused by the friction of the Moon-induced tides in shallow seas.)

These discoveries led to a picture of a solar system whose internal motions and geometrical parameters were subject only to minor oscillations about their average values. Laplace in fact believed that he had proved the solar system to be a stable, self-regulatory system, similar in this respect to the self-regulation evident in living Nature (though, as we shall see on page 197, in the late twentieth century it would become clear that Laplace's insight here was only part of the truth).

Laplace described this picture of a stable solar system in a brilliant work of popularization, his *Exposition du système du monde*, which appeared in 1796. But he did something else as well: he attempted to explain how such a system could have come about. There could have been, in the first beginnings of the solar system, a giant nebula or vortex whirling about the Sun, and the planets and their satellites could have condensed out of this whirling matter as the myriad particles attracted each other. This supposition would account for the fact that all the planets and then-known satellites circulated about the Sun in the same direction, from west to east, and in nearly the same plane. Granted this explanation, the nicely balanced oscillations in the motions of the solar system would be the result of spatial relationships that had survived from the system's chaotic origins.

The *Exposition du système du monde* was translated into many languages, and widely read. Laplace's 'nebular hypothesis' of the origins of the solar system harmonized well with William Herschel's contemporary theorizing about the origins of stars and of systems of stars, which we discuss in the next chapter, and helped prepare the way for Darwin's promulgation in 1859 of his theory of organic evolution. Today the nebular hypothesis is the accepted account of the beginnings of the solar system, and one of the main preoccupations of solar-system theorists is to understand how the solar system has evolved, by means of tidal and other gravitational interactions.

Laplace's magisterial five-volume synthesis, *Traité de mécanique céleste* (1799–1825) – especially in the English translation by the American amateur Nathaniel Bowditch (1773–1838), which contained many helpful explanatory notes – became the bedside reading of astronomers and celestial mechanicians, describing the methods and setting the problems for future research.

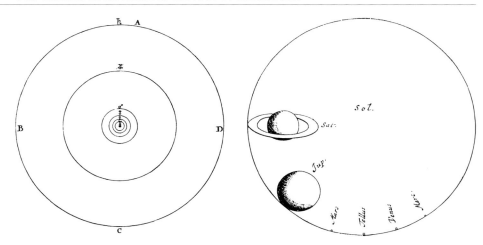

Left: A diagram of the solar system, from David Gregory's *Astronomiae elementa* (1702). The orbits of Mercury, Venus, Earth, and Mars are clustered at the centre, and there is then what seems a disproportionately large gap between the orbits of Mars and Jupiter. Right: The Sun ('Sol') and the planets drawn to scale, from Christiaan Huygens's *Cosmotheoros* (1698). Jupiter and Saturn are huge by comparison with Mercury, Venus, Earth ('Tellus'), and Mars.

'MISSING' PLANETS

Early on in his attempt to discover what had motivated God the geometer in his choice of the layout of the solar system, Johannes Kepler had been disturbed at what he saw as the disproportionately large gap between the fourth planet, Mars, and the fifth, Jupiter; and he had toyed with the idea that the gap was occupied by an undiscovered planet. A century later, Newton saw the gap as evidence of how Providence had contrived to limit the disruption caused to the structure of the solar system by the attraction of the two massive planets, Jupiter and Saturn, by banishing them to the outer recesses of the system (see page 162).

Other, more prosaic minds offered physical explanations of how the gap might have been generated. The Alsatian polymath, J. H. Lambert (1728–77), thought that Jupiter and Saturn may have 'plundered' planets that were once in the gap, while the eccentric English amateur Thomas Wright of Durham (1711–86), in an unpublished speculation, wondered if a planet that once existed in the gap had been destroyed by collision with a comet.

Bode's 'Law'

The suggestion that the gap contained (or had contained) a 'missing' planet was encouraged by an arithmetical relationship that emerged during the eighteenth century. In his *Astronomiae elementa* of 1702, the Oxford professor David Gregory (1659–1708) noted that the radii of the planetary orbits were roughly proportional to the numbers 4, 7, 10, 15, 52, 95. Christian Wolff, a German philosophical popularizer, republished these figures in a work that came to the attention of Johann Daniel Titius (1729–96), professor of physics at Wittenberg University. In 1766 Titius produced a German translation of the *Contemplation de la nature* by the distinguished French naturalist Charles Bonnet; and into Bonnet's text he interpolated a paragraph in which, by altering Gregory's 15 into 16 and 95 into 100, he made the numbers equal respectively to 4, 4+3, 4+6, 4+12, 4+48 and 4+96.

None of the known planets corresponded to the missing term in the sequence, 4+24. Bonnet commented: 'But should the Lord Architect have left this space empty?

Never!' His own method of filling the gap was absurd – undiscovered satellites of Mars – but in 1772 a second edition of his translation came to the eyes of a young German astronomer, Johann Elert Bode (1747–1826), who was then putting the finishing touches to a new edition of his highly successful introduction to astronomy. Bode dismissed the nonsense about the satellites of Mars, but he was fascinated by the relationship and agreed about the gap: 'Can one believe that the Creator of the Universe has left this position empty? Certainly not!' Bode became convinced that a primary planet lay undiscovered in the Mars–Jupiter gap, at some 4 + 24 units from the Sun (the Sun–Earth distance being 4 + 6 units).

The Discovery of Uranus

Inclusion in Bode's book guaranteed the relationship wide publicity. And in 1781, any remaining doubts that Bode entertained vanished. That March William Herschel (1738–1822), a Hanoverian-born organist living in the English spa resort of Bath, became the first person in history to discover a planet, which – against Herschel's wishes – was eventually named Uranus.

An organist discovers a new planet

When he came across Uranus, William Herschel had not been looking for a planet – indeed, as a self-taught amateur astronomer, he was unaware of speculations about 'missing' planets. His interests lay in the stars, not the solar system, and at the time he was using a home-made reflecting telescope to familiarize himself with the brighter stars, which he was examining one by one. On 13 March 1781, his systematic 'review' of the sky led him to the constellation Gemini. There he came across an object that professionals had earlier mistaken for a star. But so good was Herschel's telescope, and so skilled was he as an observer, that he was able instantly to recognize the anomalous nature of the object. It was, he noted in his journal, 'a curious either Nebulous Star or perhaps a Comet'.

If it belonged to the solar system, it might well be moving perceptibly against the background stars. Herschel therefore returned to the object four days later, and found that his suspicions had been justified, for it had already moved. Since it was not one of the known planets, he assumed it must be a comet. A friend with scientific contacts informed the Astronomer Royal, Nevil Maskelyne, and the astronomy professor at Oxford, Thomas Hornsby. Though observing with professionally made instruments, neither could see any unusual object in that region of the sky, and it was some time before they could identify Herschel's 'comet'. It proved instead to be a planet, the first to be discovered since the dawn of history.

Herschel's 7-foot reflector, with which he discovered Uranus. It was of Newtonian design (see page 152), with the mirror at the base of the tube, and the eyepiece at the top and to one side.

A page from William Herschel's observing book for March 1781 (crossed out to show his sister Caroline had made a fair copy). For Tuesday 13th he notes that 'in the quartile near ζ [Zeta] Tauri the lowest of the two is a curious either Nebulous Star or perhaps a Comet'. He examined the object again four days later, and found that it had moved and so was close to Earth. Professional astronomers could see nothing unusual about the objects in that region of the sky.

Uranus – astonishingly – was found to have an orbit that corresponded well to the next term in the sequence, 4 + 192 = 196. The court astronomer at Gotha, Baron Franz Xaver von Zach (1754–1832), now became a convinced believer in the arithmetical relationship, and in 1787 he undertook a search for a planet in the Mars–Jupiter gap – but without success. In 1799 Zach visited a number of German colleagues, and from their discussions the idea of an organized attack on the problem began to evolve. On 21 September 1800 Zach met with five other astronomers in Lilienthal, home of the prominent amateur J. H. Schröter (1745–1816). They decided to enlist the co-operation of leading observers throughout Europe, to form a team of twenty-four 'celestial police', each of whom would be assigned a share of the zodiac, with the duty to keep a lookout for strangers at large in his district.

The Discovery of the Asteroids

Their plans were overtaken by events. One of the co-opted members was to have been Giuseppe Piazzi (1746–1826) of Palermo, the southernmost of the European observatories. In the 1780s, by dint of residing in London and breathing down the neck of the renowned instrument-maker, Jesse Ramsden, while work progressed, Piazzi had coaxed Ramsden into completing a superb 5-foot vertical circle of unique design (see page 214). The turn of the century found Piazzi using this circle to assemble a star catalogue of greater accuracy than any of its predecessors.

Jesse Ramsden, greatest of the eighteenth-century instrument makers. The accuracy with which an observer could measure the position of a star depended crucially on the precision with which the scale on his instrument was marked or 'divided'. Ramsden's success was based on his dividing engine, seen here in the foreground. Behind is the circle with which Giuseppe Piazzi discovered Ceres.

On New Year's Day, 1801, unaware of the task for which he had been 'volunteered' by the Lilienthal astronomers, Piazzi was at work as usual on his catalogue. He measured the position of no. 87 of N.-L. de Lacaille's catalogue of zodiacal stars, and took the opportunity of measuring a (supposed) star of about the eighth magnitude that preceded it. His careful method of working involved him in remeasuring his positions on a subsequent night; and when he did this, he found that the eighth-magnitude 'star' appeared to have moved, a movement that he confirmed on the nights that followed. It was therefore no star, but a member of the solar system.

The object, named Ceres by Piazzi in honour of the patron goddess of Sicily, turned out to lie at about the distance predicted by Bode's 'law', and at first there seemed no reason to doubt that it was, like Uranus, a major planet. But Herschel found to his surprise that, even with his large telescope, he could scarcely detect a planetary disk. It was, he thought, even smaller than our Moon. Worse still, in March Olbers found another moving body, which he named Pallas. Herschel measured its diameter also, and thought this to be less than 111 miles. Clearly, Pallas was no planet. As a descriptive term for this new species of celestial body, Herschel proposed 'asteroid'.

Ceres lost and found

Throughout January 1801 Piazzi kept watch on his newly-discovered member of the solar system whenever weather conditions allowed, and on the 24th he wrote letters to Bode and others to announce his discovery. In his letter to Bode he claimed no more than the finding of a comet; but to his friend Barnaba Oriani of Milan he admitted a suspicion that 'it might be something better than a comet'.

By mid-February the object was too close to the Sun to be seen, and so Piazzi began to investigate its orbit on the basis of the twenty-four observations he had been able to make. He tried to fit parabolas, first to one triad of observations and then to another, because a parabola was known to approximate well to the elongated orbit of a comet when the comet was near the Sun. But no parabola came near to accounting for all the measured positions. Piazzi then tried circles, and found that radii of around 2.7 times the Earth–Sun distance – close to the $(4+24)/(4+6)$ implied by the unassigned term in Bode's 'law' – gave promising results. The object, it would seem, must be a planet, in an elliptical (but nearly circular) orbit. The problem was, how to determine its orbit with enough precision for astronomers to recover it when it emerged from the glare of the Sun – for it had been only briefly observed, while moving over a very short arc.

Illness prevented Piazzi from making any further progress with the mathematical analysis of his observations, and so in April he sent his data to Oriani in Milan, to Bode in Berlin, and to J.-J. L. de Lalande in Paris, thereby handing over the problem to the astronomical community of Europe. By good fortune a brilliant new mathematical talent had emerged in the person of Carl Friedrich Gauss (1777–1855); and by November, Gauss had devised a method that allowed him to calculate the characteristics of the orbit. On the basis of this information, Zach began to search for the lost planet; and on the last night of the year he found it, in just the position forecast by Gauss.

Asteroids by the hundred

For a long time after the discovery of Vesta in 1807, no further asteroids were identified, and astronomers soon wearied of the search. It was revived by the German K.L. Hencke, an ex-postmaster. Hencke was a man of extraordinary dedication, for he began work in 1830 and laboured for fifteen years before enjoying his first success. He found another asteroid in 1847, and at this leading observers began to take up the hunt.

By 1891, more than 300 asteroids had been discovered, and the pace of discovery then greatly increased with the application of photography. Max Wolf at Heidelberg would photograph a large star field with an exposure of several hours, the camera care-fully tracking the rotation of the sky. On the resulting photograph, the stars appeared as point-like images, but the asteroids betrayed themselves by the small streaks caused by their movements during the time the photograph was exposed.

Today great numbers of asteroids are known. Some two thousand of them have well-determined orbits and revolve around the Sun as part of the 'main belt', in the gap between Mars and Jupiter. They are tiny, only a handful having a diameter of more than a few tens of miles, and the combined mass of all the asteroids is thought to be only a fraction of that of the Moon.

To rescue the 'law', Olbers suggested that the two asteroids were fragments of a full-sized planet that had once occupied the gap. If so, other fragments were waiting to be discovered, and so it proved: by 1807 Juno and Vesta had also been found.

If Olbers had been right, and asteroids were indeed the fragments of an exploded planet that had obeyed Bode's 'law', then (initially at least) their orbits would all have intersected, in the place of the explosion and again on the opposite side of the Sun. But as increasing numbers of asteroids became known in the second half of the nineteenth century, it was realized that this was far from being the case. It is now clear that with only a handful of asteroids having diameters in excess of 250 km (150 miles), the combined mass of the asteroids is far less than that of the Moon, let alone of a planet.

The Discovery of Neptune

Herschel had come across the planet Uranus, and Piazzi the asteroid Ceres, quite unexpectedly, when they were engaged in studies of the stars. But in the mid-nineteenth century another planet was to be found, this time by observers searching in a region determined by astronomer-mathematicians who had been armed with nothing more than pen and paper and a knowledge of Newtonian dynamics.

Soon after Uranus was discovered in 1781, Bode found that the planet's position had been recorded by Tobias Mayer in 1756, and by John Flamsteed as long ago as 1690; both had taken it for a star. These additional observations allowed his friend Placidus Fixlmillner and others to determine the characteristics of its elliptical orbit and to calculate tables of its future positions.

The planet, however. soon began to deviate from its predicted orbit. Matters improved in 1790 when the Paris mathematician J.-B.J. Delambre (1749–1822) published tables that seemed to match the observations well enough, but in the 1820s and 1830s the theory of Uranus's orbit was again in trouble.

Various explanations were floated. Some were quickly rejected – that the planet was being impeded by a cosmic fluid, that it had an unseen but massive satellite, that it had been struck by a comet about the time of discovery – but two others were more worthy of consideration: perhaps the law of attraction departed noticeably from the inverse-square at great distances, or perhaps Uranus was being pulled out of its orbit by an outer planet as yet undiscovered.

Modifications in the law of attraction had been considered from time to time in the mid-eighteenth century, but by now the law was firmly established. A consensus therefore emerged, that the anomalous behaviour of Uranus was the result of perturbations by an undiscovered planet.

By 1845 Uranus's movements were under the scrutiny of a proven master of the techniques of Newtonian mechanics, Urbain Jean Joseph Le Verrier (1811–77) of the Paris Observatory. He presented his first paper on the topic to the Paris Academy of Sciences in November of that year, and a copy soon reached the Astronomer Royal at Greenwich, George Biddell Airy (1801–92). The following June Le Verrier presented his second paper. In it he made the assumption that the undiscovered planet occupied the next place in the sequence embodied in Bode's 'law', and after a lengthy analysis he concluded that its current longitude as seen from the Sun must be within a few degrees of 325°.

The chart with which Neptune was discovered. On 23 September 1846, J.G. Galle of Berlin Observatory and a student assistant, H.L. d'Arrest, began to search for the planet predicted by Le Verrier. When Galle found nothing unusual, d'Arrest suggested comparing the star field with a chart. In a drawer they came across the relevant sheet of the Berlin Academy's new Star Atlas, which had not yet been distributed abroad. Near the position predicted by Le Verrier (marked by a square), Galle found an object (marked by a circle) missing from the chart, and this proved to be the planet.

Urban Jean Joseph Le Verrier

Le Verrier was born on 11 March 1811 at St-Lô, Normandy, the son of a local government official. He studied at the Ecole Polytechnique in Paris from 1831 to 1835, and after flirting briefly with a career in chemistry was appointed in 1837 to a junior post in astronomy at the Ecole. He immersed himself in the great French tradition in 'celestial mechanics', and first extended Laplace's work on the stability of the solar system. He then studied the orbit of Mercury, and problems involved in the identification of periodic comets, and in 1845 he was invited by François Arago, director of the Paris Observatory, to tackle the problem of why Uranus was deviating from a normal planetary orbit. His prediction that an unknown planet was pulling Uranus was verified visually in the autumn of the following year.

Le Verrier went on to study in immense detail the other planetary orbits. In 1854 he succeeded Arago as director of the Paris Observatory, but his autocratic rule finally brought about his dismissal in 1870 – though he was reinstated (with reduced powers) in 1873, after the death of his successor. He died at Paris on 23 September 1877.

Unbeknown to Le Verrier, a young Cambridge graduate, John Couch Adams (1819–92), was at work on the same problem. He too had assumed the planet obeyed Bode's 'law', and he had arrived at an approximate solution for the position of the planet by October 1843. Distracted by teaching duties, he did not derive a more precise result until September 1845; the heliocentric longitude of the planet on 1 October he reckoned to be 323°34′.

Armed with a letter of introduction from James Challis (1803–82), the professor of astronomy at Cambridge, Adams called on Airy to present his analysis, but owing to a chapter of accidents failed to speak with him. He did, however, leave him a summary of his results.

The arrival next summer of Le Verrier's paper, with its nearly identical prediction of heliocentric longitude, stirred Airy to action. In his opinion, a hunt for an undiscovered planet was not the business of the publicly-funded Royal Observatory, but he persuaded Challis to make a search at Cambridge, and this Challis eventually did. Unfortunately, Challis did not possess accurate charts of that region of the sky. This meant that the only way he could identify a planet – a temporary visitor to the area – was to re-examine the region at a later date, to see whether any of the 'stars' had moved in the interval. This was a chore that Challis undertook with no sense of urgency.

The delay cost Adams priority, for Le Verrier had meanwhile prevailed on astronomers at Berlin Observatory to make a search, and they were fortunate enough to

Two cartoons by Cham of
Paris. Le Verrier's triumph
in predicting the position
of Neptune was galling to
British astronomers: Adams
had made a similar prediction,
but had been thwarted by the
slow pace of the search at
Cambridge, and by the
unavailability there of the
latest sky chart. Unsurprisingly, the British suggestion
that Adams have a share of the
credit was not well received in
Paris. In the first cartoon, we
see 'M. Adams looking for M.
Le Verrier's planet' in quite the
wrong direction, and in the
second, 'M. Adams discovering
the new planet in the account
by M. Le Verrier'.

(M. Adams cherchant la planète de M. Leverrier.)

(M. Adams découvrant la nouvelle pianète dans le rapport de M. Leverrier)

have the relevant sheet – not yet distributed – of the Berlin Academy's new Star Atlas (see page 192). The search began on 23 September 1846, and that same night the Berlin observers came across a 'star' that was not on the sheet. It was the missing planet.

The discovery of Neptune in 1846 was the ultimate triumph of Newtonian dynamics: two astronomer-mathematicians, sitting at their desks, had calculated from the effects – the deviations of Uranus from its predicted orbit – to the cause, and had pinpointed the whereabouts of the culprit, a major planet whose very existence had until then been unsuspected.

The Failure to Discover Vulcan

In 1846 celestial mechanics appeared to be the queen of the sciences, the most successfully mathematicized, and the most exact in its predictions. Fate, however, had a trick in store.

Like Uranus, Mercury had an unexplained feature in its orbit: its point of nearest approach to the Sun ('perihelion') was advancing in longitude faster than expected, by about half a minute of arc per century.

That such a tiny discrepancy – a degree every hundred centuries or so – should arouse concern is astonishing testimony to the success of mathematicians in explaining the planetary movements on Newtonian principles. Le Verrier naturally considered whether the cause might again be an undiscovered planet, this time orbiting inside Mercury, and in September 1859 he announced the results of his calculations. A body of the same size as Mercury but at half the distance from the Sun would produce just such an advance; so would a similarly placed ring of asteroids.

As luck would have it, earlier that year, in the French town of Orgères, an unknown physician named Lescarbault had seen, crossing the Sun, what he believed to be just such an intra-Mercurial planet. He said nothing about this at the time, but in December he learned of Le Verrier's prediction, and he wrote to him. Le Verrier hurried to Orgères, satisfied himself of the physician's *bona fides*, and named the planet Vulcan. Its period he calculated to be just under twenty days.

In 1876 Vulcan was supposedly seen once more crossing the disk of the Sun; but on further investigation the object proved to be a sunspot. By this time some twenty possible observations of the supposed planet had been assembled, five of which Le Verrier was prepared to accept as authentic. On the basis of these he calculated the planet would probably cross the Sun in March 1877, and again in October 1882; but in the event, nothing was seen.

Searches during solar eclipses revealed nothing, although evidence of the advance of the perihelion of Mercury became ever stronger. By the end of the century Vulcan was accepted as spurious, a 'pseudo-planet', and the advance reverted to being an unsolved problem of celestial mechanics, a rare exception to the seemingly endless stream of successes for Newtonian dynamics. But by a strange stroke of fate, what had been expected to lead to another Newtonian triumph was to prove the very opposite. As we shall see in Chapter 8, Albert Einstein showed in 1915 that his General Theory of Relativity, based on a radical reappraisal of Newton's basic concepts, implied that Mercury's perihelion would advance by almost precisely the observed amount.

The puzzling movement of the planet had at last been explained, but not on Newtonian principles. Pope's couplet in praise of Newton required amendment:

It did not last: the Devil howling 'Ho,
Let Einstein be,' restored the status quo.

CELESTIAL MECHANICS IN THE TWENTIETH CENTURY

But despite the Einsteinian revolution, most work in celestial mechanics – the plotting, for instance, of orbits of artificial satellites – has continued to be based on the Newtonian equations. Such calculations are inevitably no more than approximations, but for many purposes the Newtonian theory gives results that are entirely adequate.

In the accurate measurement of time, however, this is not the case. Until the mid-twentieth century the 'mean' (average) solar day was taken as the unit of time; but recognition of the role of tidal friction in slowing the Earth's rotation – to say nothing of changes in the rotational rate due to changes in the Earth's shape – made this standard obsolete. Atomic clocks now replace the Earth's rotation in supplying a standard of time. However, like all clocks on Earth, atomic clocks exist in an accelerated frame of reference, and are therefore subject to alterations in rate predicted by Einstein's General Theory of Relativity. Celestial mechanicians must therefore undertake to incorporate these relativistic effects in the computer programs whereby they generate predictions of planetary and satellite positions.

In the late twentieth century, celestial mechanics has had to face up to still another challenge, perhaps even more disconcerting than that posed by Einstein. The view of celestial mechanics promulgated by Laplace, and accepted until recently, was that its predictions could be made as precise as the observations. As Laplace put it, a demon that knew for any instant the exact positions and velocities of all the bodies in the universe, and had a mind capable of performing the necessary calculations, would be able to know the whole future and past of the universe. Human astronomy was but a 'pale image' of such a knowledge, but an image it was. If observations gave the positions of the planets to a certain precision, then their future positions could be predicted to similar accuracy.

In the 1890s the French mathematician Henri Poincaré (1854–1912) introduced a new way of thinking about the problems of celestial mechanics. Instead of focusing on problems that appeared to be soluble by the traditional method of approximations by equations with an infinite series of terms, Poincaré used quasi-geometric methods to test the assumption that small changes in the parameters of the equations would result in only small differences in the numbers forming the solution. He found, instead, that for many dynamic systems, small differences in initial conditions could lead to drastic differences in outcome.

With the advent in recent decades of electronic computers of ever increasing power, it has become feasible to explore such systems quantitatively. Take, for instance, a system consisting of a simple pendulum whose point of suspension is set in oscillation; if there is a near match between the pendulum's natural frequency and the frequency of the imposed oscillation, the sensitivity of the behaviour to the actual initial conditions turns out to be indefinitely large. Since there is a limit to the precision with which the initial conditions can be measured, the motion of the system becomes unpredictable – even though in principle it is completely determined. The connection between determinism and predictability that Laplace had assumed is here broken, and we have a situation of the type now known as 'dynamical chaos'.

It has recently been shown that the inner planets Mercury, Venus, Earth, and Mars, as well as the outer planet Pluto, provide instances of dynamical chaos. Certain perturbations of the Earth's motion, for instance, are in 'near resonance' with its annual motion, in close analogy with the case of the simple pendulum mentioned above. From observation we can never know exactly – with unlimited precision – where any planet is; and for the inner planets and Pluto the initial range of uncertainty increases with time by leaps and bounds. Calculations with computers have shown that the initial uncertainties for the Earth increase by a factor of three every 5 million years, so that an initial error of 10 metres produces an error of 1 million kilometres after 100 million years.

One implication of the discovery of 'chaos' among the planets is that earlier attempts to prove the stability of the solar system are fundamentally flawed. Systems subject to dynamical chaos are sensitive to chance perturbations: we live in a less certain world than Laplace imagined.

Celestial mechanicians, their dream of deterministic certainty gone, have nevertheless an exciting prospect for future investigation. Their tasks include exploring the detailed implications of planetary 'chaos', refining predictions with the limitations now revealed, and working out plausible scenarios for the evolution of the solar system and its sub-systems.

CHAPTER 7

The Astronomy of the Universe of Stars

In 1833 John Herschel published *A Treatise on Astronomy*, an introductory text for the interested amateur. Although Herschel was the world authority on the universe that lay beyond the narrow confines of the solar system, he allotted just one chapter to the study of the stars: astronomy was still preoccupied with the Sun's family of planets and comets. Yet nearly two centuries had passed since Descartes had portrayed an infinite universe in which there was no centre and no outside boundary; one in which the Sun was merely our local star, and the stars were distant suns.

During these two centuries, the stars had continued to play their traditional humble role, as reference points in the sky against which to measure the current positions of members of the solar system. The lack of interest in them as bodies free to move in space is not surprising. Because they are so far away, their movements as seen from Earth are minute – so tiny that even at the end of the seventeenth century, not a single star was known to have altered its position since the first catalogues were compiled in Antiquity. The stars were too remote and appeared too static to arouse the curiosity of most astronomers.

Stellar astronomy became important only with the development of astrophysics in the second half of the nineteenth century. Today, the pendulum has swung, and it is the solar system that is of minority interest. As a result, the investigations pioneered by the handful of earlier astronomers – many of them amateurs – who had the courage to tackle the distant stars have proved to be of immense significance for the future development of astronomy.

THE SUN'S NEIGHBOURS IN THE UNIVERSE OF STARS

Depending on their temperament, these men adopted one of two possible approaches. Some bold spirits – often mixing astronomy with theology – sought to understand the cosmos as a whole. Others preferred to proceed step by step, outwards from the solar system to the Sun's nearest neighbours among the stars. It is with work of these latter, more cautious souls, that we begin.

Stars that Change in Brightness

It may seem surprising that astronomers of Antiquity and the Middle Ages believed the stars to be unchanging in brightness as well as in position. There are stars whose light varies regularly, and some of these variations are not difficult to detect with the naked eye – once you know where to look and, more fundamentally, once you know that such variations exist to be observed. These changes escaped detection for so long because Nature so arranged the universe that none of the brightest stars varied sufficiently to force such changes on the attention of early observers.

Opposite: Kepler's chart showing the position of the new star of 1604 and the locations of the nearby planets.

Below: the same bodies in detail. Above and immediately left of the body of Cancer (the Crab) are overlapping circles representing Jupiter and Saturn when the 'great' conjunction of 1603 occurred (17 December in the Gregorian calendar). By 10 October 1604 Jupiter had moved (leftwards) into Ophiucus (the Serpent Holder) and was then close to Mars; Saturn had moved a shorter distance in the same direction. That day, in the midst of the three planets, the new star blazed forth.

17 December 1603: SATURN

JUPITER

10 October 1604: SATURN

JUPITER

MARS

New star

The dramatic new star of 1572 ('Tycho's nova', see pages 100–1) compelled astronomers to recognize that changes in brightness can and do occur. Remarkably, Nature forced home the lesson only a generation later, in 1604, when a second such star ('Kepler's nova') blazed forth.

It appeared at a time of the utmost astrological significance. Jupiter and Saturn were the slowest moving of the known planets in their journeys among the stars, and so their conjunctions – the occasions when Jupiter caught up with and over-took Saturn – were the rarest of all planetary conjunctions. In December 1603, for the first time in eight centuries, one of these conjunctions took place in a zodiacal house associated with the element of fire, the most portentous of the elements. By the autumn of the following year, Jupiter and Saturn had been joined there by the third-slowest planet, Mars. Suddenly, in the midst of the three planets, and on this occasion of the very highest astrological significance, a new star shone out.

Some thought the end of the world had come. Others expected the overthrow of the Turkish kingdom, or the appearance of a great new monarch: 'nova stella, novus rex'. Johannes Kepler, as Imperial Mathematician in Prague, rushed out a short tract

The astrology of the 'great' conjunctions

Jupiter took some twelve years and Saturn thirty to complete one circuit of the zodiac, and so their 'conjunctions', when Jupiter caught up with Saturn, occurred at intervals of some twenty years. Because of their rarity, these conjunctions were thought of as 'great'.

In the twenty years before the next conjunction, Saturn moved 2/3rds the way around the zodiac, in the next twenty years another 2/3rds, and after a further twenty years another 2/3rds. By this time it had completed two full circuits and had returned to near the place of the first conjunction.

The positions of three consecutive conjunctions therefore formed (roughly) the vertices of an equilateral triangle around the sky, and the signs of the zodiac within which they fell would normally be either the first, fifth, and ninth of the twelve signs; or the second, sixth, and tenth; and so on. These groups of three signs were known as 'trigons' (that is, triangles), and each trigon was associated with one of the four elements, the fiery trigon being the most portentous.

The conjunctions in practice drifted from one trigon to the next every two centuries or thereabouts; and so after every eight centuries, the conjunctions returned to their original trigon. It happened that around the birth of Christ, the conjunctions entered the fiery trigon (and today some use this fact in interpreting the Star of Bethlehem); and in 1603/4, after two such cycles of eight centuries, the same profoundly ominous event was happening again.

Kepler's diagram from his *Cosmographic Mystery* (1596) illustrating in idealized form the pattern of 'great' conjunctions (with dates and elements added). Any three consecutive conjunctions fell nearly, but not quite, at the vertices of an equilateral triangle. A sequence of fifteenth- and sixteenth-century conjunctions, all falling within the watery trigon of Cancer, Scorpius, and Pisces, ended with that of 1583 in Pisces (centre right). The 1603 conjunction (bottom left) was not in Scorpius but in Sagittarius, which with Aries and Leo formed the fiery trigon – so initiating a sequence of conjunctions in the fiery trigon.

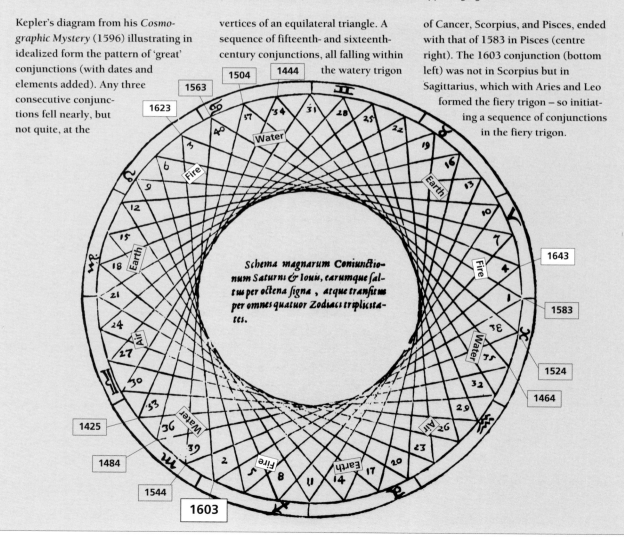

on the nova, and then, two years later, he published *On the New Star*, in which he ranged over all the issues involved, astronomical and astrological.

Tycho, Kepler and their contemporaries were privileged: astronomers have now been waiting nearly four centuries for the appearance in our Galaxy of another 'supernova', as these celestial fireworks are now classified.

There had already been talk of another newcomer among the stars: in 1596 the Frisian astronomer David Fabricius (1564–1617) claimed that a nova had appeared in the constellation of the Whale (Cetus). Then, in 1638, another Frisian observer, Johannes Phocylides Holwarda, noticed a star, also in the Whale, that was missing from the star catalogues of Ptolemy and Tycho. Before long, it disappeared. The pages of Holwarda's account of the discovery had already been printed off, when he was astonished to find the star had reappeared.

Eventually it was realized that Fabricius's nova and Holwarda's nova were one and the same. But the reappearances of '*mira stella*' ('wonderful star') seemed capricious. In 1662, Johannes Hevelius gathered together all the known observations, including those that he had himself accumulated in three years of careful scrutiny. By bad luck, three years was not quite long enough to reveal that, although Mira varied in brightness not only within each cycle but even in the corresponding stages of one cycle and the next, it did observe one significant regularity in its behaviour. It attained a maximum brightness every eleven months; and to this extent at least, Mira is predictable.

It was left to the French priest-astronomer Ismael Boulliau (1605–94) to announce this in a work published in 1667. Boulliau also offered a physical explanation of 'variable stars', one that was to endure into the nineteenth century. The Sun, he reminded readers, was a star that rotated. Furthermore, it had spots on its surface, and these spots varied. A variable star, he believed, likewise rotated, and it too had dark regions on its surface, though these regions were much more extensive than the spots on the Sun. Any regular variation in the star was due to its rotation, and any irregular variation to alterations in the dark regions.

It was an explanation that could account readily – perhaps too readily – for almost any change in the stars. Meanwhile, the hunt was on. Reputations could be made overnight through the discovery of (supposedly) variable stars. In their haste, observers were deceived by changes in seeing conditions, or perhaps by wishful thinking; and sceptics had no way of proving them mistaken. Yet the discoveries – even if genuine – seemed to be leading astronomy nowhere, and by the end of the seventeenth century the whole subject had fallen into disfavour.

Part of the problem lay in the lack of a sufficiently delicate technique for monitoring the apparent brightness of a star. Stars were simply grouped according to the crude classification inherited from Antiquity, whereby the brightest stars were first magnitude and the faintest, sixth. The mid-nineteenth century would see the invention of new instruments to give an objective measure of the brightness of stars, and a new definition of magnitude (see page 297). But before then, in the

A 'Dollond achromatic' (see page 153) of 3¹⁄₂-foot focal length, somewhat larger than the one acquired in 1781 by John Goodricke. This instrument has an aperture of 3³⁄₄ inches. Both the tube and the tripod base are made of mahogany, and the 'finder' (a low-power telescope mounted alongside to help point the instrument in the desired direction) is of brass. 'Slow motion' screws allow the observer to track the object as the sky rotates.

closing years of the eighteenth century, astronomers were at last provided with a simple method of determining whether a star had in fact altered in brightness, when John Herschel's father, William Herschel, at his observatory near Windsor Castle, prepared a series of 'Catalogues of the Comparative Brightness of the Stars'. In these catalogues, each of the major stars in a given constellation was compared with neighbouring stars that had been selected because they were almost equal in brightness to the listed star; any future change in the brightness of a listed star would reveal itself by disturbing these published comparisons.

Herschel's technique of comparative brightnesses was a refinement of a simple method using sequences of stars arranged in order of brightness, that had been developed in the 1780s by two amateur observers in York in the north of England: Edward Pigott (1753–1825), whose father had recently established a private observatory in the city, and his neighbour, the young deaf-mute John Goodricke (1764–86). In 1781 Pigott decided to dedicate himself to the investigation of new and variable stars, and the following summer the two friends re-examined the known variables, notably Mira, but also Algol (Beta Persei), which a century earlier had on two occasions been seen as fourth magnitude instead of the usual second.

On 7 November 1782, Algol was second magnitude as usual, but five nights later an astonished Goodricke found it reduced to fourth magnitude. The following night it was back to second again, a change so rapid as to be utterly without parallel in the literature.

In the weeks that followed, the two friends kept independent watch on the star, and were rewarded on 28 December, when they both saw it start the evening at third or fourth magnitude and then brighten to second magnitude before their very eyes. Pigott at once suspected that 'the alteration of Algol's brightness, was maybe occasioned, by a Planet, of about half his size, revolving round him, and therefore does sometimes eclipse him partially'; and he even calculated possible orbital periods for the planet.

More of these supposed eclipses followed, and the longest orbital period that would account for every one of them grew ever shorter, until by April it had reduced to a mere sixty-nine hours! Pigott generously let Goodricke appear as sole author of the paper to the Royal Society announcing their results; but the young man concentrated on observational facts, and mentioned the eclipse hypothesis merely as an alternative to the usual explanation in terms of dark patches.

Whereas dark patches could explain almost anything, and so could scarcely be either proved or disproved, the eclipse hypothesis made exact predictions: that the brightness at minimum would always be the same, that the length of time occupied by one cycle of variation would always be the same, and that on either side of minimum the light curve would be symmetric. Interestingly, in the years to come the two friends abandoned this correct explanation for Algol. Perhaps they were deceived by changes in seeing conditions into thinking that the predictions had not been fulfilled, or perhaps they were influenced by the impossibility of using the eclipse hypothesis to explain the three other short-period variables they had discovered. But in their brief years of activity, before 1786 when the Pigotts left York and Goodricke met an untimely death through illness, the two friends had enriched the study of variables with a new and wholly unsuspected class, those whose periods were measured in days rather than months or years.

Thereafter the study of variable stars was to stagnate, until the development of astrophysics offered the possibility of insights into the underlying physical processes at work.

Opposite: Johannes Hevelius observing with the help of his second wife, Elisabetha. Significantly, this brass sextant of some 6-foot radius was closely modelled on the instrument developed a century earlier by Tycho Brahe (see page 106). To measure the angle separating two stars, Elisabetha would keep the fixed radius aligned on the first star, while her husband directed the adjustable radius towards the second star, and then read the required angle on the scale.

The Direction in which the Solar System Is Travelling

In the Newtonian universe, the stars were isolated bodies, free to move in any direction; yet in Newton's own mind the stars were effectively as 'fixed' and motionless as ever. The first announcement of the discovery of motions of individual stars (their 'proper' motions) appeared in *Philosophical Transactions* in 1718. In Oxford, Edmond Halley (*c.* 1656–1743) had been comparing contemporary positions of stars with those recorded for the same stars in Antiquity. He found that he could make sense of the latitudes ascribed by Ptolemy to three of the brightest stars only on the assumption that the stars had moved. After all, 'these Stars being the most conspicuous in Heaven, are in all probability the nearest to the Earth, and if they have any particular Motion of their own, it is most likely to be perceived in them'.

One might have expected that thereafter the number of known proper motions would rapidly increase, as other astronomers compared contemporary positions with those recorded by earlier observers. But such was not the case. Changes resulting from proper motions accumulated century by century, and so knowledge of these motions depended on two factors: the accuracy of the positions recorded in the past, and the length of the time interval between past and present observations. Unfortunately, the search for accuracy in astronomy had been inaugurated by Tycho Brahe little more than a century earlier – and that was before the invention of the telescope and the use of telescopic sights, and before there was a proper

The refraction of starlight

The observed position of a celestial body is the direction in which light from this body is travelling at the very moment it reaches the observer. Unless the body in question happens to be directly overhead, this light will have entered the atmosphere obliquely, and in the last few miles of its journey its path will have been bent by refraction.

Astronomers in Antiquity had been aware of this problem. Cleomedes, who lived somewhere around the time of Christ, spoke of seeing the Moon eclipsed when both Sun and Moon were above the horizon – even though an eclipse occurs when Sun, Earth, and Moon are in a straight line. He (correctly) explained this by saying that the light from the Sun and the light from the Moon were both raised during their passage through the Earth's atmosphere. In fact the Sun on the horizon is raised by refraction by an amount roughly equal to

its own size, so that at the moment when an observer on an atmosphere-free Earth would have seen the last rays of the Sun disappear from sight, an observer on the real Earth can still see the whole of the Sun's disc.

The effect of atmospheric refraction on a body anywhere near the horizon is therefore very considerable; but to establish by exactly how much the observed position of a body differs from what the position 'ought' to be is not easy. Tycho Brahe attempted to draw up tables of refraction; but he believed that refraction had different effects on the Sun, the Moon, and other bodies, and he also took refraction to be effectively zero above a certain angle from the horizon – 45° for the Sun, but only 20° for stars. The accuracy of his star catalogue suffered correspondingly.

James Bradley

James Bradley was born at a small town in Gloucestershire, England, in March 1693. His uncle, Rev. James Pound, was an able astronomer who encouraged the boy's interests in that direction. Bradley entered Balliol College, Oxford, in 1711, and in 1719 he took holy orders. He was appointed vicar of a parish in Monmouthshire, but his duties were light and he was able from time to time to collaborate in his uncle's astronomical observations.

Pound introduced Bradley to other astronomers, including Edmond Halley, and in 1718 Bradley was elected to the Royal Society. In 1721 he became Savilian Professor of Astronomy at Oxford, and it was while professor that he collaborated with Samuel Molyneux in the measurements that led to the discovery of the aberration of light.

In 1742 Bradley succeeded Halley as Astronomer Royal. He found serious defects in the Greenwich instruments, but by 1750 he had re-equipped the observatory, and from then until his health deteriorated he devoted himself to a massive programme of observations of unsurpassed accuracy. He died on 13 July 1762 after a lengthy illness.

Opposite: The mural quadrant installed at Greenwich in 1750. In 1724, Bradley's predecessor, Halley, had ordered two 8-foot quadrants, one to face north and the other south, but the available funds proved sufficient for only one. In 1748, Bradley was granted £1,000 for new equipment and this allowed him to order a second, matching quadrant, the chief difference being that it was in brass rather than iron, and so less likely to bend under its own weight.

understanding of the amount by which starlight alters direction as it enters the Earth's atmosphere.

To make matters worse, a large and unforeseen source of error in earlier observations was identified in 1728 by James Bradley (1693–1762), Savilian Professor of Astronomy at Oxford, through his discovery of 'the aberration of light'. This is the name given to the effect whereby the observed position of any given star is constantly altered by the ever-changing velocity of the Earth-based observer as he is carried round the Sun, much as the direction of rainfall on a train window is affected by the movement of the train. A further problem was revealed by Bradley's later discovery that the Earth's axis nods or 'nutates' (mainly as a result of the gravitational pull of the Moon on the non-spherical Earth), and this affects the very co-ordinate system we use to measure the positions of stars.

How it was that Bradley came to discover aberration is a fascinating story that we shall discuss later (see page 213). His achievement meant that future astronomers would know they had to correct their raw observations to allow for these effects, and this ushered in an era of greatly improved accuracy. Bradley himself later took the lead in this, assembling a treasury of stellar positions which later observers would use to determine proper motions. As Astronomer Royal at Greenwich from 1742, he first re-equipped the observatory, and then, from 1750 until his health began to fail, carried out a massive observing programme during

which he carefully recorded all the circumstantial information necessary to amend the raw data and so derive stellar positions of unrivalled accuracy.

Bradley himself did not live to make these 'reductions', but his unreduced observations were published at the end of the century. The German astronomer and mathematician F. W. Bessel (1784–1846) later undertook the necessary reductions, and in 1818 he published a catalogue of over three thousand stellar positions for 1755, a convenient mid-date in Bradley's observing programme. Bessel's volume bore the proud title, *Fundamenta astronomiae*, and its appearance allowed students of proper motion to use the year 1755 as the starting point in time from which to measure future changes in the positions of stars.

Meanwhile, attempts were being made to determine proper motions, however provisional, and to make sense of them. This was no easy task. In 1748 Bradley himself re-emphasized in *Philosophical Transactions* that such apparent motions were relative, and could arise either from movements in the stars themselves, or from the motion of the solar system, or a combination of the two.

Tobias Mayer (1723–62) of Göttingen explained in 1760 how these two causes might be disentangled. A *pattern* of the appropriate kind in the observed movements was to be ascribed to a single cause, the motion of the Earth-based observer through space. The *residual movements*, however, were to be ascribed to motions of the individual stars themselves.

But what kind of pattern would it be that resulted from the motion of the observer as he was carried through space with the solar system? Mayer pointed out that someone walking in the forest saw the trees ahead appear to move

Friedrich Wilhelm Bessel

Bessel was born at Minden, Germany, on 22 July 1784, and as a young man worked in a merchant's counting house in Bremen. To train himself in navigational theory he 'reduced' a set of 200-year-old observations of Halley's Comet by Thomas Harriot, and the impression this made on the astronomer H.W.M. Olbers led in 1806 to Bessel's becoming astronomical assistant to J.H. Schröter at Lilienthal (see page 188).

In 1810, at the age of twenty-five, Bessel became director of the new observatory at Königsberg, a post he held for the rest of his life. His talents were in mathematics as much as in astronomy, and he combined the two disciplines when reducing James Bradley's legacy of accurate observations, in a book published in 1818. Twenty years later he ended centuries of frustration for astronomers with his convincing determination of the distance of a star (61 Cygni), a feat which John Herschel described as 'the greatest and most glorious triumph which practical astronomy has ever witnessed'. He died at Königsberg on 17 March 1864.

aside at his approach. In the same way, if the solar system was moving towards a particular point on the heavenly sphere (the 'solar apex'), the stars would appear to move aside – that is, away from this apex, each star moving along a 'great circle' towards the opposite point of the sky.

Mayer himself could see no such pattern in the (often unreliable) proper motions known to him, and he concluded that the Sun is at rest. 'Perhaps the true and genuine reason for these movements', he wrote, 'will still remain unknown for many centuries'.

Mayer could not have foreseen the inventiveness of William Herschel. In a most uncharacteristic investigation, carried out at his desk and simply using the limited data available to everyone, Herschel managed in 1783 to find indications that the Sun was travelling in the direction of the constellation Hercules.

A sketch by Herschel for his 1783 study of the direction in which the solar system is travelling. He has projected the northern hemisphere onto a plane through the celestial equator, and marked the stars whose proper (individual) motions were known. He has then drawn 'great circles' radiating from Lambda Herculis. If this star lies in the path of the solar system, the proper motions will tend to be along these circles, in the directions away from the star.

In 1818, however, working from the much more accurate positions contained in *Fundamenta astronomiae*, Bessel failed to discern any pattern in the proper motions that might indicate the whereabouts of a solar apex. The patterns supposedly discovered by Herschel and others had, it seemed, been illusory. But as the years passed, the ever-increasing time span permitted true motions to be distinguished from observational errors. In 1837, F. W. A. Argelander (1799–1875), professor of astronomy at Bonn, published the results of a long investigation into no fewer than 390 proper motions, a number large enough for him to divide the motions by size into three groups, which he treated independently. Each of the three groups yielded a solar apex not far from that proposed by Herschel.

Similar analyses by other astronomers quickly followed, with similar results; but all the analyses depended on Bradley's observations, and all were confined to stars visible from Europe. Fortunately, records of positions of southern stars in 1750 were available in the catalogue that had resulted from the expedition of Nicholas-Louis de Lacaille to the Cape of Good Hope (see page 251); and reliable modern positions were available from the new Royal Observatory at the Cape and from the East India Company's observatory on the island of St Helena.

In 1847, Thomas Galloway (1796–1851), a London-based actuary, analysed the resulting proper motions. Although the number of motions at his disposal was small, the apex he obtained from these completely independent data matched the positions recently derived from the northern stars – a striking confirmation of the reality of the phenomenon.

Since then, improvements in the data, and in the sophistication of the mathematics used to analyse them, have permitted regular improvements in the position of the solar apex. But the reality of the solar motion, and its general direction, have not been in doubt.

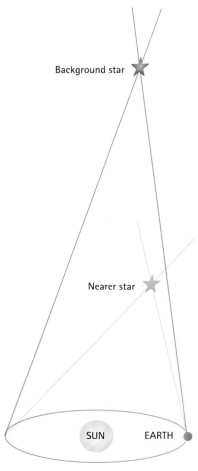

Galileo's method for detecting annual parallax. To do this directly, by tracking the tiny changes in the co-ordinates of the star over the year, was impracticable: atmospheric refraction varied, and the measuring instrument itself would alter shape in response to changes in temperature and humidity. By adopting a background star as a quasi-fixed reference point these problems could be avoided, the positions of the two stars being affected equally by refraction, and other factors.

The Distances that Separate Us from the Nearest Stars

The stars appear to us as points of light, whose positions we measure on the two-dimensional surface of the heavenly sphere. To investigate the distribution of the stars in three-dimensional space, and so discover the structure of the star system to which the Sun belongs, we must somehow determine the third dimension, that of their distances from us.

According to Copernicus in the sixteenth century, the Earth was orbiting the Sun. If so, two measurements of the position of a given star that were separated by six months in time had been made from widely separate locations – the opposite ends of a diameter of the Earth's orbit. His opponents, reasonably enough, demanded to know why in that case the position of the star always appeared the same – why the star did not show 'annual parallax'?

Galileo recognized that no evidence in favour of Copernicus would be more convincing than the detection of one or more examples of annual parallax. Accordingly, in 1632, in his *Dialogue on the Two Great World Systems*, he suggested a way of carrying out such delicate measurements. Suppose that two stars *appeared* to be close together in the sky, but that this came about purely by accident: the two stars happened to lie in almost the same direction from Earth, though in fact one of them was, say, six times further away than the other. As the Earth-based observer was carried round the Sun, the two stars would each appear to him to move in orbits that were exactly similar in shape; but the orbit of the nearer star would be a whole six times bigger in scale than that of the more distant.

If, therefore, he measured the position of the nearer star *relative* to the more distant – treating the distant star as a quasi-fixed reference point provided by a helpful Nature – then the changes he was measuring would in fact be five-sixths of the true parallax of the nearer star. This slight reduction in the quantity to be measured was an acceptable price to pay for the great convenience of measuring small relative angles rather than large absolute ones.

A still greater advantage of this method was that the positions of the two stars would be equally affected by atmospheric refraction (and by aberration and nutation, factors unknown to Galileo), so that by measuring the position of one relative to the other, such unwelcome complications would be avoided entirely. Typically, however, Galileo made no effort to follow up this brilliant suggestion, though in the nineteenth century measurements of annual parallax by reference to background stars would become routine.

Meanwhile, as the seventeenth century wore on, the context of the search for annual parallax altered. As the study of the forces at work in the heavens became an accepted part of astronomy, the advantages of the Copernican hypothesis, whereby the planetary system was centred on the massive Sun rather than on the relatively-small Earth, became ever more persuasive. The search for annual parallax then became less a matter of polemics, and more an investigation into the inverse problem, that of the distances of stars – for the nearer a star, the more it would appear to move. Successful measurement of the annual parallax of a star would yield its distance (in multiples of the radius of the Earth's orbit, or 'astronomical units').

The first to mount a well-considered instrumental attack on this problem was Robert Hooke. He was alert to the dangers arising from the uncertain effects of atmospheric refraction; but as luck would have it, Nature had provided him with a solution to the problem. A bright star, Gamma Draconis, passed directly overhead

The scale of interstellar distances

While astronomers pondered the challenge of how to track tiny angular changes over a twelve-month timescale, Descartes's claim that the Sun is simply the star closest to Earth was seen as offering an approach that might reveal at least the scale of the distance of one star from another.

Let us suppose that the stars (including the Sun) are not merely similar in their physical nature, but virtually identical. Let us further suppose that starlight reaches Earth without being reduced *en route*, by obscuration or some other unwelcome complication. Then, since light falls off with the square of the distance (double the distance of a candle and it appears one-quarter as bright), we can derive the relative distances of the Sun and a given star – say Sirius – if we can somehow contrive to measure their relative brightnesses. If the Sun proves to be a million times brighter than Sirius, for example, then Sirius will be a thousand times further than the Sun (and be 1,000 'astronomical units' from us).

The Dutch physicist Christiaan Huygens (1629–95) attempted to carry out the measurement by putting a screen between himself and the Sun, and making a hole in the screen of such a size that the Sun, viewed through the hole, appeared as bright as Sirius. Measurement of the fraction of the Sun's total surface that could be seen through the hole would then yield the desired result.

Unfortunately the Sun was so bright that it proved impossible to make a measurable hole sufficiently small in size. Huygens reduced the sunlight further by placing a lens in front of the hole, but he could only estimate the effect of this. In the end he concluded that Sirius lay at 27,664 astronomical units.

This result made the scale of the stellar universe astonishingly great. Yet, unknown to Huygens and most of his contemporaries, Newton was working along similar lines and with a more reliable technique, and arriving at a still larger value.

In a book published in 1668, the Scots mathematician James Gregory (1638–75) had suggested a brilliant solution to the problem of comparing the brightness of the Sun with that of Sirius. It involved using a planet as an intermediary: one waited until the planet was equal in brightness to Sirius, and from then on, one disregarded Sirius and focused on the planet instead. The problem thereby reduced to that of comparing the brightness of the Sun – that is, the light of the Sun that reaches us directly – with the brightness of the planet – that is, the light of the Sun that reaches us via the planet.

For the calculations, one needed to know the distances separating members of the solar system; one also had to estimate the fraction of the incoming light that was reflected by the planet; and one had to assume that the planetary regions were free from obscuring matter. Granted all this, the arithmetic was straightforward.

Gregory arrived at a value of 83,190 astronomical units for the distance of Sirius. But he expressly pointed out that he had worked with a scale of the solar system that needed upwards revision, and that with an improved value the result would be much larger. Newton used such a value in 1685 when drafting his *The System of the World*, and he employed Gregory's method to arrive at a staggering one million astronomical units for the distance of Sirius.

Newton had intended *The System of the World* to be part of his *Principia*, which was published in 1687; but he decided instead to replace it with more technical material, and the scale of interstellar distances was one of the topics he omitted in the process. The result was that Newton's insight into the immensity of the distances separating the stars from each other was known only to members of his immediate circle, until his *The System of the World* at last appeared posthumously in 1728. Meantime, Huygens's estimate (published in 1698, and much inferior, in both senses of the word) held the field.

The methods used by Huygens, Gregory, and Newton were based on the questionable hypothesis of the physical uniformity of the stars; and no one could regard their conclusions as a substitute for the measurement of the annual parallax of specific stars. But they had revealed, for the first time, the incomprehensible immensity of the universe of stars.

Hooke's zenith telescope
(below), and Bradley's zenith
sector (opposite), stable
instruments designed to detect
annual parallax through
measurements of the position
of one or more stars
passing overhead
the observer.
Zenith measure-
ments avoided
the complica-
tions caused by
atmospheric
refraction.

his lodgings at Gresham College in London. Being overhead, the star's light entered the atmosphere at right angles, and so would be unaffected by refraction.

But there was another difficulty: his telescope would have to remain motionless and undisturbed throughout the months of observation. Hooke therefore decided to incorporate a purpose-built instrument into the very fabric of his own house.

Telescopic astronomy was still in its infancy, and so it is remarkable to find a research programme so focused that it called for the construction of a special form of telescope, designed to measure the position of one single star, and that only when the star was near the zenith. Unfortunately the brilliant conception was ruined by poor execution: in 1669 Hooke was able to make only four measurements before illness, and an accident to the telescope lens, put an end to his investigation. Never one to undervalue his own achievements, he nevertheless declared himself satisfied that his 'Archimedean Engine that was to move the Earth' (by proving the Copernican hypothesis) had done just that; but few were convinced.

Half a century later Samuel Molyneux (1689–1728), a prosperous amateur observer living near London, decided to make another attempt to measure the parallax of Gamma Draconis. He enlisted the help of James Bradley, and commissioned from the instrument-maker George Graham a telescope designed to measure positions of stars almost directly overhead. Late in 1725 this 'zenith sector' was mounted against a chimney stack within Molyneux's house, so that it could be moved very slightly to either side of the zenith in a north–south direction. As Gamma Draconis passed overhead, the telescope was tilted a little so that the star passed through the centre of the field of view, and the angle of tilt was then measured.

Like Hooke before them, Molyneux and Bradley were measuring the position of Gamma Draconis in only one co-ordinate, the north–south direction. A simple calculation showed that annual parallax would cause this particular star to reach an extreme southerly position on 18 December, around which date its movement from day to day would be vanishingly small. It was therefore with great surprise that on 21 December Bradley found it passing further south than it had a few days earlier. This continued in the weeks that followed, until by March the star was some 20 seconds of arc south of its December position – although annual parallax should by this time have been causing it to move north. The star then stopped, and began to move north, passing through its December position in June and reaching an extreme northerly position in September.

The two friends considered a variety of possible explanations. In particular, could it be that the atmosphere of the Earth did not form a truly spherical envelope, and so even a star in the zenith was affected by refraction?

Bradley decided they needed to bring more stars within the scope of their investigation, and for this he commissioned another zenith sector from Graham, similar in construction but shorter and with a wider field of view. Soon the pattern of movements became clear, but the physical explanation did not occur to Bradley until – according to a likely story – he was on a pleasure boat on the River Thames. He noticed that the weather vane altered direction whenever the boat put about – not of course because the wind then blew from a different quarter, but because the direction in which the weather vane pointed depended not only on the velocity of the wind but also on that of the boat.

Back in the 1670s, the Danish astronomer Ole Römer (1644–1710) had shown that the speed of light, though very great, was finite: eclipses of Jupiter's moons were observed on Earth ahead of schedule if the planet was nearer to Earth than usual, but behind schedule if the planet was unusually distant. Bradley realized that since the speed of light was finite, the position of a star as seen by him depended, by analogy with the weather vane, not only on the velocity of its light, but also on that of the Earth. He had been looking for annual parallax, an effect that results from the Earth-based observer being located at the end of the radius of the Earth's orbit, rather than at the Sun itself. He had found instead 'the aberration of light', caused by the Earth-based observer's velocity, which is tangential to the Earth's orbit. Radius and tangent are at right angles, which was why aberration and annual parallax were out of phase by three months.

The implications of Bradley's discovery were profound. First, he had identified an unsuspected error involved in previous measurements of stellar coordinates, including those in John Flamsteed's great 'British Catalogue' published as recently as 1725. As a result of aberration the apparent position of a star might change by as much as 40 seconds of arc over a six-month period, and Bradley's discovery – in conjunction with his subsequent demonstration (see page 206) that the Earth's axis nutates – ushered in the era of exact positional astronomy.

Second, it was proof – though in a wholly unexpected form – of the motion of the Earth around the Sun.

Third, since all the stars involved were similarly affected, their light must be reaching Earth at the same speed, irrespective of the distance it had travelled – and, as shown by analysis of the timing of eclipses of Jupiter's moons, irrespective of

whether the light was direct or reflected. The speed of light was a constant of nature, and Bradley calculated that light took 8 minutes and 12 seconds to reach Earth from the Sun, within 8 seconds of the modern value.

Fourth, Bradley's failure to measure annual parallax implied that this parallax was too small for detection even with his precision instrumentation: it must be less than one second of arc! A simple calculation showed that the stars under scrutiny must therefore be *at least* some 400,000 times the distance of the Sun.

Bradley's results were read to the Royal Society in January 1729, a few months after Newton's *The System of the World* appeared. In *The System*, Newton had calculated that if – and it was a big 'if' – Sirius was physically similar to the Sun, then comparison of its brightness with that of the Sun implied that it was a million times more distant than the Sun (see page 211).

Two major contributions to the problem of the distances of stars had now been published in quick succession: Newton's argument, based on the hypothesis that the stars were physically uniform, that the nearest (and brightest) stars lay at a distance of a million or so astronomical units; and Bradley's observational proof that certain stars were *at least* 400,000 astronomical units from us. The convergence of these two investigations established once and for all that interstellar distances were to be measured in millions of astronomical units. It also showed that the measurement of annual parallax was a technical challenge of extraordinary delicacy: that of measuring, over an extended period of months, a movement that amounted to no more than the width of a coin several miles away.

Not surprisingly, this led the astronomical community to hesitate long before resuming the attempt to measure parallax. But with each new generation of instrument-makers, the precision of astronomical instruments improved, and by the early nineteenth century astronomers were once more looking for ways to detect these tiny movements.

It was of course crucially important to select for scrutiny the stars that are nearest to Earth, and therefore show the greatest parallax. The obvious criterion for this had been brightness: other things being equal, the brightest stars would be the nearest. But evidence was accumulating to suggest that stars varied enormously in 'luminosity' (the star's 'absolute magnitude'), and astronomers were increasingly coming round to the view that large proper motion was more reliable as a guide to nearness.

Of all the proper motions then known, the largest was, surprisingly, that of a relatively faint (fifth magnitude) star in the constellation of the Swan, 61 Cygni, which was moving across the sky at over 5 seconds of arc per annum. Its motion had been noticed by Giuseppe Piazzi as long ago as 1804, but he was isolated at Palermo, and 'the flying star' did not receive the attention it deserved until 1812, when F. W. Bessel independently published a notice of its large motion. Even so, several attempts to measure its parallax proved fruitless – as did efforts with a variety of other stars.

Opposite: The Palermo circle by Jesse Ramsden, with which Giuseppe Piazzi discovered Ceres (see page 190), catalogued nearly 8,000 stars, and detected the large proper motion of 61 Cygni. While manufacturing the circle in the 1780s, Ramsden twice abandoned it, but eventually completed it in 1789. Azimuth was read on the horizontal scale with the help of a micrometer microscope, and altitude on the vertical scale with the aid of two diametrically opposed microscopes. Ramsden is to be seen on page 189 with the dividing engine he used to engrave the scales (and with the circle in the background).

It was time to regroup and consider strategy. What were the criteria by which astronomers might identify the nearest stars? There were three, according to the German-born Wilhelm Struve (1793–1864) in a paper published in 1837 when he was professor at Dorpat (now Tartu in Estonia): is the star one of the brightest, does it have a rapid proper motion, and, if it happens to be a 'binary star' (not one star but two that are bound together by gravity), do the two component stars seem widely separated in view of the time they take to orbit each other? Struve listed the stars that satisfied each criterion, and drew attention to those that satisfied more than one; and if Struve's paper is checked against a modern list of the nearest stars, it becomes evident that astronomers were now selecting the most suitable objects for their parallax measurements.

Equally important was the quality of instrumentation available. At Dorpat Struve possessed a magnificent refractor of 24-centimetre (9½-inch) aperture, by the German craftsman and theoretician, Joseph Fraunhofer (1787–1826). This instrument was the largest of its kind in the world. At Königsberg in Germany, Bessel had a 16-centimetre (6½-inch) Fraunhofer 'heliometer', the ideal instrument for measuring small angles (see page 219). Both men decided to measure the position of their target star by reference to neighbouring stars that they had good reason to believe were very distant – an adaptation of Galileo's method.

Wilhelm Struve

Friedrich Georg Wilhelm Struve was born at Altona, Germany, on 15 April 1793. His parents sent him out of Germany to avoid conscription, and so it was at Dorpat in Russia that he studied philology. He then received permission to work in the university observatory, and in 1813 was appointed an observer and a teacher of astronomy and mathematics. He was an observer of astonishing industry, especially in the study of double and multiple stars, and he also made notable improvements to Dorpat's instrumentation, including the acquisition in 1824 of a 24-centimetre (9½-inch) refractor by Fraunhofer, the largest such instrument in the world.

His achievements brought him an invitation from the St Petersburg Academy of Sciences to organize an observatory at Pulkovo, of which he was to be director. Struve was given the resources to buy the very best instruments obtainable, and soon after Pulkovo Observatory was opened in 1839, it was recognized as arguably the leading observatory in the world.

Struve retired in 1862, being succeeded by his son Otto (1819–1905), four of whose descendants later became distinguished astronomers. Wilhelm Struve died at Pulkovo on 25 November 1864.

Fig 1

'The Great Dorpat Refractor', installed in 1824. Its equatorial mounting (see page 218), and the unique size and quality of its achromatic object glass, made it ideally suited to the study of double stars. Working at the rate of seven stars a minute when observing, Struve spent two years examining 120,000 stars to see which were double, and ten further years measuring the doubles he had found. In 1837 he was able to publish data on over 3,000 double stars, three-quarters of which were previously unknown. The advantage this instrument gave him led his friendly rival, James South, to take precipitate action to restore the balance, with disastrous consequences (see page 249).

Tracking the heavens: the equatorial mounting

The simplest mounting for a telescope, whether reflector or refractor, is 'altazimuth': such an instrument can be rotated horizontally (in azimuth) and vertically (in altitude). However, the sky spins about the celestial poles, which means that an observer tracking an object with an altazimuth telescope must constantly make adjustments in both directions. By contrast, a telescope mounted 'equatorially', with one axis pointing to the celestial pole, needs adjustment only in one direction.

The advantages of equatorial mountings had been well recognized since the time of Tycho Brahe, but it was in the eighteenth century that construction techniques improved to the point where such mountings became popular, if only for the portable instruments beloved of amateur observers. However, by the early nineteenth century equatorial mountings could be provided for even the largest refractors, such as the Dorpat equatorial (see page 217). Such mountings became standard for refractors and reflectors alike, until the advent of modern computer-controlled guidance systems provided an alternative solution to the problem of the smooth tracking of a celestial body.

The 'Shuckburgh equatorial', made in 1791 by Jesse Ramsden for Sir George Shuckburgh's private observatory in Warwickshire, England. The axis points towards the celestial North Pole, so that to track a star the observer has to move the instrument in only one direction. This was the first large telescope to have an 'equatorial' mounting.

The divided-lens micrometer

If an object glass is cut along a diameter into two semicircles, each semicircle will form a complete image of the object viewed, though at half the previous brightness. If the two halves are displaced sideways with respect to each other, the images are correspondingly displaced.

This can be used to measure small angles, such as the angle between the two stars that make up a double star. At first the observer simply sees the two stars in the normal way; but as the two half-lenses are displaced, there appear two images of each star. The amount of displacement required to make one image of the first star coincide with the other image of the second star is a measure of the angle separating them.

In 1753 the London instrument-maker John Dollond (1706–61) made this the basis of a divided object-glass micrometer, which soon became an accessory sold with many reflectors. Because they were well suited to measurements of the Sun's apparent diameter at different times of the year, such instruments were often termed 'heliometers'.

Early in the nineteenth century, particularly through the work of Joseph Fraunhofer and his successors in business at Munich, the technique was applied to the object glasses of refractors, though it required the coolness of a diamond-cutter to undertake the halving of a large achromatic lens of high quality. It was an outstanding example of a Fraunhofer heliometer that Bessel used in 1837–38 in the first convincing measurement of stellar distance.

A divided-lens micrometer from the late eighteenth century. The scale for measuring the displacement between the half-lenses is 5.5 inches long, and can be read to 0.002 of an inch.

In 1835 Struve selected Vega as the subject for his investigation. The star was exceptionally bright and had a large proper motion, so that it amply fulfilled two of his three criteria of nearness. In 1837 he announced the results of seventeen observations, from which he inferred a parallax of one-eighth of a second of arc (close to the modern value). But he promised to continue his measurements, and in 1840 gave the results of nearly 100 observations, from which he now inferred a parallax that was twice as great. Given the long history of fallacious claims to the measurement of parallax, astronomers remained sceptical.

But meanwhile Bessel was directing his heliometer at 61 Cygni. He began observations in 1834, but was soon distracted by the arrival of Halley's Comet, and it was not until 1837 that he returned to the task. Encouraged by Struve's preliminary results, for over a year he subjected 61 Cygni to intensive scrutiny, commonly repeating his observations an astonishing sixteen times every night, and still more when the 'seeing' was especially good. His mathematical skills enabled him to deal with problems such as refraction, and by the end of 1838, he announced that the parallax of the star was about one-third of a second of arc. What carried conviction was the way in which the pattern of his many observations matched the prediction from theory. John Herschel, as President of the Royal Astronomical Society, congratulated the Fellows that they had lived to see the day when the sounding line in the universe of stars had at last touched bottom.

It was only a few weeks later that the Scottish astronomer Thomas Henderson (1798–1844), who had been royal astronomer at the Cape of Good Hope, announced a parallax of just over one second of arc for the southern star, Alpha Centauri. This very bright star has a large proper motion, and is a binary star whose components have a wide angular separation. It therefore fulfilled all three of Struve's criteria for nearness. So far as is known, this star and its faint companion, Proxima Centauri, are the stars nearest to the solar system.

THE STRUCTURE AND HISTORY OF THE UNIVERSE OF STARS

While attempts were being made to measure the distances of our nearest neighbours among the stars, and to understand the changes taking place among them, speculators were trying to make sense of the cosmos as a whole. One of them provoked Isaac Newton to attempt an analysis of the structure of the stellar universe that was to have consequences for cosmology in our own day.

The Newtonian Universe and the Darkness of the Night Sky

If one reads Newton's *Principia* (1687) hoping to discover the author's conception of the universe of stars, one will be disappointed. He has in fact next to nothing to say about the stars, either as individuals or as a whole. To him they were of limited interest: despite nearly 2,000 years of observation there was not the least evidence to contradict the ancient Greek belief that the stars were 'fixed', motionless relative to one other.

One seldom sees what one expects not to see: Newton himself followed current practice in using for 'star' the Latin word *fixa* (that is, *fixa stella*), and this very term must have helped close his mind to the possibility that the stars might move. Although he was the first person in history to grasp the enormity of the distances that separate us from even the nearest stars, it never occurred to him that this might undermine the supposed fixity of the stars – that the stars might have seemed motionless, not because they were truly at rest, but because they were so very far away that their movements had so far escaped detection.

Nor had it then occurred to him that the fixity of the stars posed a threat to his law of gravity: for he claimed that gravity was a universal force, and forces generate motions – yet every one of the stars was apparently motionless. It was a young theologian, Richard Bentley (1662–1742), who in 1692 forced Newton to face up to this problem. Bentley had been commissioned to preach a series of sermons – in effect, lectures – on the compatibility of science and religion. He knew that the Lucasian Professor at Cambridge had written a work with major implications for cosmology, but the book was impenetrably mathematical, and so he plucked up his courage and wrote direct to the author.

Bentley rejected the Cartesian conception of a god who created the universe, set it in motion, and then left it to run its course without taking any further interest in its well-being. But he wanted to see what could be said in support of such a view, and so he asked Newton what would happen if matter were spread uniformly throughout infinite space, and thereafter allowed to move freely under the action of gravitational attraction. Newton, thinking that Bentley intended by 'uniformly' a more or less regular distribution of matter, replied that in any place where there was more matter than usual the force of gravity would be greater than elsewhere, and so the surrounding matter might be pulled in, thus adding to the existing concentration. This could lead to the formation of stars.

Bentley, however, had intended 'uniformly' in the absolute, mathematical sense; and Newton had to concede, when pressed, that symmetry would then ensure that the matter remained motionless, there being no reason why it should move one way rather than another. But such a universe, he admonished, would be artificial in the extreme – as implausible as having infinitely many needles all balanced on their points and standing on an infinite mirror. Bentley, thinking of the infinitely many stars all apparently at rest, very reasonably retorted: 'is it not as hard, that infinite such Masses in an infinite space should maintain an equilibrium...?'

The correspondence came to an end, but Newton could no longer close his mind to the challenge: if gravity is universal, how is it possible for the stars to be at rest? Fortunately for the historian, Newton seems to have hoarded every scrap of paper. He was currently at work on material for a second edition of the *Principia*, and in a succession of drafts of a new theorem we can watch him contriving an answer to the challenge.

His solution was that Providence in the beginning devised an infinite system of motionless stars, a system that was (almost) symmetric and hence (except in the very long term) stable; and when, after a lapse of time, the lack of perfect symmetry led to movements that became sizeable and threatened to bring about the destruction of the original order through what we would now term 'gravitational collapse', Providence intervened and pushed the stars back to their original positions. In this way, Newton maintained his belief in God as the great clockmaker, whose universe was a machinery that endured from age to age. Indeed, Newton was grateful to God for enabling him and other students of the Book of Nature to appreciate how – far from turning his back on his creation as the Cartesians supposed – he providentially intervened from time to time, maintaining the machinery of the stellar system, just as he maintained the machinery of the planets (see page 162). God, in Newton's view, had entered into a servicing contract with his creation.

Newton's great German contemporary, Gottfried Wilhelm Leibniz (1646–1716), agreed that God was a clockmaker. But a perfect clockmaker, he argued in a famous correspondence (1714–16) with Newton's spokesman Samuel Clarke, would make a perfect clock, one that had no need for repair and servicing. Newton's divine interventions were to Leibniz the miracles of a god driven to desperate remedies, and so he condemned Newton's conception of God as utterly inadequate. But to the Newtonians the interventions by Providence were not emergency, *ad hoc* miracles motivated by panic, but part of the divine plan intended from the beginning.

Isaac Newton at work

Newton's custom of drafting and redrafting, and then preserving every sheet of paper, allows us to look over his shoulder and watch him at work. These two passages are from his drafts for a theorem to show that the evidence from star catalogues was compatible with his model of the universe. In this model, the system of stars was unbounded and roughly symmetric, so that each individual star was pulled almost equally in all directions by the gravitational pulls of the other stars, and so could remain at rest ('fixed').

In the star catalogues were listed the observed position of each star (that is, its direction from Earth) and its apparent brightness. The model, however, involved not brightnesses but (relative) distances, and so the model could not be tested against observation without some assumption connecting brightness and distance. Newton therefore assumed that the stars were physically uniform, so that brightness and distance were directly related; specifically, he assumed that stars of 1st, 2nd, 3rd,... magnitudes lay at 1, 2, 3,... units of distance from the solar system.

The observer looked out from the solar system into the surrounding space, and so to make the test feasible, Newton had to modify his symmetric model of the universe (in which the solar system had no special significance). He substituted one in which the Sun was imagined as surrounded by spheres of radius 1, 2, 3,... units, on which were located stars of the 1st, 2nd, 3rd,... magnitudes, each star being at least 1 unit distance from its nearest neighbours.

The nearest (and brightest) stars were then 1 unit from the Sun, but they had also to be at least 1 unit from each other. Newton knew that on a sphere of radius 1, there could be 12 (or was it 13?) points separated in this way, and he therefore expected there to be 12 or 13 stars of the 1st magnitude. The next sphere, of radius 2 units, had 4 times the surface area and so room for 4 times 12 or 13 (say 50) points; and Newton therefore expected there to be some 50 stars of the second magnitude. Similarly, on the spheres with radii 3, 4, 5, and 6 units there was room for some 110, 200, 310, and 450 points respectively, and Newton expected there to be these numbers of stars of the 3rd, 4th, 5th, and 6th magnitudes.

The first passage shows Newton making two attempts to draft this. In the second attempt, he gives the numbers from the model and simply declares the number of stars to be about the same ('*Et tot fere sunt stellae primarum sex magnitud[in]um respective*'). In the first he gives the number of stars of the 1st magnitude ('*tredecim*', 13, a number that does indeed match his model) and leaves spaces for the numbers of stars of other magnitudes, meanwhile declaring without further enquiry that the numbers fit.

Newton, in other words, is so confident of his model that he sees no need actually to check the evidence.

In the second passage, Newton attempts another draft of the same argument, but this time he does check the evidence: there are, he says, 15 or 16 ('*quindecim vel*

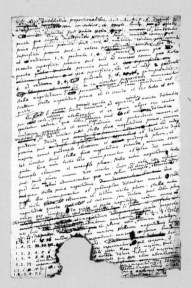

The exchanges between Clarke and Leibniz, like most such controversies, continued until the death of one of the parties. Eventually, the unchanging clockwork universe would fall out of favour, to be replaced by one in which developmental changes, brought about by gravitational and other forces, would be seen as natural and expected. But that was for the future, and mean-time Newton's symmetric universe was being challenged from a most unexpected quarter.

Above: Newton's first attempt to validate his model of the universe. He notes there are 13 (*tredecim*) stars of the 1st magnitude, a number that fits neatly with the model. He leaves spaces (highlighted here) for the other numbers and assumes they will fit his model.

Above: In the second draft, Newton fills in the numbers (highlighted in yellow) that his model would predict for stars of the 1st magnitude (12 or 13), 2nd (50), 3rd (110), 4th (200), 5th (310), and 6th (450). He claims that these are in fact the numbers of stars of the first six magnitudes: *Et tot fere sunt stellae primarum sex magnitud[in]um respective* (shown in green).

sexdecim') stars of the 1st magnitude, 63 of the 2nd, 220 of the 3rd, 500 of the 4th – at which point it dawns on him that these numbers are far too large.

His first reaction is to stop and cross out all mention of 5th and 6th magnitudes. His excuse would be that the star catalogues were incomplete when listing faint stars; yet the problem was that they were, one might say, too complete! It then occurs to him that his assumption that an *n*-th magnitude star lay at *n* units of distance was overhasty (and indeed the modern definition would place a 6th magnitude star at 10 units of distance). He therefore adopts a more flexible approach; and after a little creative manipulation, he persuades himself that he has rescued his theorem.

Below left: In a later version, Newton looks in the star catalogues to see how many stars there are of the first four magnitudes: *quindecim vel sexdecim* [15 or 16], 63, 220, and 500 (highlighted in yellow). Realizing that these numbers are too big, he stops at the fourth magnitude. While he ponders what to do, he goes back and crosses out all mention of the 5th and 6th magnitudes (highlighted in green) and changes a 6 (*sex*) into a 4 (*quatuor*).

In drafting his theorem in the 1690s, Newton had satisfied himself that there was an acceptable match between the numbers of nearby stars predicted by calculation from his geometrical model of a symmetric universe, and the numbers of bright stars actually listed in the star catalogues: in the neighbourhood of the Sun, there was reasonable accord between model and observation. He had, however, ignored the fundamental observational fact concerning the stellar universe in the large; namely, the concentration of stars in the Milky Way. (Newton was not alone in this: contemporary astronomers and speculative cosmologists alike displayed a strange lack of interest in the Milky Way.)

This observation-based objection to Newton's symmetric universe of stars was eventually put to him face-to-face by – of all people – the young physician and antiquarian, William Stukeley (1687–1765). Stukeley pictured the Sun and the stars we see as individuals as forming a spherical cluster. Surrounding this cluster, and separated from it by empty space, was a flattened ring in which the stars of the Milky Way were collected together. Stukeley's star system, then, seen from outside, would have looked not unlike Saturn (the spherical cluster) and Saturn's ring (the Milky Way). When Stukeley proposed this model of the universe to Newton in conversation in or about 1720, the great man responded by hinting cautiously at the merits of his own model, of a universe symmetrically populated with infinitely many stars.

Stukeley, however, was about to set cosmology on the path towards what has become known as 'Olbers's Paradox', by focusing not on the effect of *gravity* (something impossible to investigate in Stukeley's universe, for his God was forever extending the Milky Way by new creations of stars), but on the *light* the stars collectively emitted. He put it to Newton that if the system of the stars were symmetric and infinite, 'The whole hemisphere [of the sky] would have had the appearance of that luminous gloom of the milky way.' Having no answer to this difficulty, Newton, it would seem, did what all sensible people do in such a situation, and made no comment.

Early in 1721, Stukeley breakfasted with Newton in the company of Edmond Halley, the newly-appointed Astronomer Royal, and they discussed astronomical topics. What Stukeley contributed to the discussion we are not told, but it would have been natural for him to mention his theory of the universe. A few days later, Halley read to the Royal Society the first of two short papers on cosmology, remarking that 'Another Argument I have heard urged, that if the number of Fixt Stars were more than finite, the whole superficies of their apparent Sphere would be luminous' – a form of words so close to Stukeley's that the 'urging' must surely have come from him at their recent breakfast.

Halley then set out a (fallacious) argument purporting to show that, in a symmetric universe, the distant stars would – despite their numbers – send us only a negligible quantity of light. But more important than the details of his confused response is the fact that, with the publication of Halley's two papers in *Philosophical*

Transactions, the discussion of the Newtonian model of a symmetric stellar universe at last emerged – anonymously – into the public domain.

A succinct and accurate analysis of the behaviour of light in such a universe appeared in 1744, by the Swiss astronomer J.-P. L. de Chéseaux (1718–51). In a symmetrical universe of stars, Chéseaux pointed out, at twice the distance of the nearest stars from Earth there was room for four times as many stars (since the surface of a sphere is proportional to the square of the radius); but this increase in number was exactly offset by the fact that each star would appear only one-quarter as bright (because light diminishes with the square of the distance). That is, there was room for four times as many stars, but each was reduced to one-quarter of the previous apparent brightness. In aggregate, therefore, the nearest stars would contribute to the night sky exactly the same amount of light as the stars at twice the distance.

A generalization of this argument showed that the same was true of stars at each of the successively greater distances. The reader was, therefore, to imagine the night sky gradually filling up with light as the light of stars at greater and greater distances was taken into account, until at length the entire sky was ablaze with the equivalent of sunlight.

A glance at the night sky showed that this did not happen in the real universe. Why this should be so was obvious to Chéseaux: he had assumed that *all* the light setting out from a star reached its destination. If, however, the transparency of space was less than perfect – and surely this must be the case in the real universe – then a certain fraction of starlight would be lost as it travelled a given distance. A similar fraction of what remained would be lost over the next such distance, and a fraction of what then remained would be lost over the next, and so on. In consequence, even if the loss of light over one such distance was small, when this loss was repeated over and over it would be enough to cause nearly all the light from remote stars to be lost *en route*; and such stars therefore added little to the brightness of the night sky.

A similar view was taken by the retired German physician and amateur astronomer, H. W. M. Olbers (1758–1840), writing in 1823 and this time in a volume of the widely-read *Berliner astronomisches Jahrbuch*. Olbers showed that even if only 1 part in 800 of the light was lost in its journey from one star to the next, this loss would be sufficient to explain the appearance of the night sky.

By the mid-nineteenth century, enough was understood of the conservation of energy for physicists to realize that if light was absorbed by an interstellar medium, then this medium would itself heat up and begin to radiate light. In consequence, the Chéseaux–Olbers explanation was no longer adequate. But several other possible explanations were to hand – for example, the existence between one star system and the next of an etherless vacuum, across which no light could pass. And so it was that in the nineteenth century, as in the eighteenth, the darkness of the night sky was easily explained. Only in our own times has it come to play a significant role in cosmological thinking.

Speculative Cosmology in the Eighteenth Century

To Stukeley and other speculators in Royal Society circles early in the eighteenth century, it was not easy to make sense of the universe of stars. Except in the Milky Way, the stars seemed to be scattered across the sky without rhyme or reason. But perhaps the disorder was apparent rather than real. After all, the bizarre behaviour of the planets had been shown by Copernicus to be nothing more than orderly motions viewed from an unhelpful location, a moving Earth.

But what form might this order take? Stukeley, in private, made more than one attempt to identify it. Another attempt came from an equally unlikely individual: the self-taught itinerant lecturer, Thomas Wright (1711–86) of Durham in the

Thomas Wright of Durham

Thomas Wright, who styled himself 'of Durham' to distinguish himself from contemporaries of the same name, was born near that city on 22 September 1711. Apprenticed to a clockmaker at the age of thirteen, his involvement in a scandal forced him to flee his master. After some adventures he reached home, where he taught himself navigation, and in turn taught the subject to seamen in the port of Sunderland. This launched him on a career as an itinerant teacher, going from town to town offering public lectures in the physical sciences. At the same time he advised members of the aristocracy on the management of their estates.

Wright had an agile if eccentric mind, and he published a number of works, in astronomy, archaeology, and architecture. In 1762 he retired to his native village, where he died on 25 February 1786.

Left: Thomas Wright, from an engraving prefixed to some copies of his principal books. The serpent biting his own tail was a symbol of infinity. In the engraving Wright declares himself to be a professor of natural philosophy and mathematics. He was indeed invited in 1742 to be professor of navigation at St Petersburg, but negotiations over salary proved fruitless.

Opposite: four attempts by Wright to convey his earliest conception of the universe. All creation was centred on the Abode of God. The Sun, and the other stars, were distributed around this centre on every side, occupying a volume of space in the shape of a spherical shell. If any star had been motionless when first created, the gravitational pulls of the other stars would immediately have caused it to fall inwards into the Abode of God. To prevent this, at the Creation each star had been put into stable orbit, around the Abode of God. These figures are among the earliest to portray stars in motion.

north of England. In 1734 Wright prepared a public lecture on cosmology that was also something of a sermon. Wright was convinced that the Sun and the other stars occupied a space that had the form of a spherical shell, with the Abode of God located in the midst of the shell. Outside the shell was the Outer Darkness, where the damned got at best a distant glimpse of the stars clustered around the Abode of God.

Were the stars motionless, their system would collapse under their mutual gravitational pulls, and they would fall into the Abode of God. This being out of the question, the Sun and the other stars must avoid such a fate by forever travelling in orbit, circling this way and that around the Abode of God.

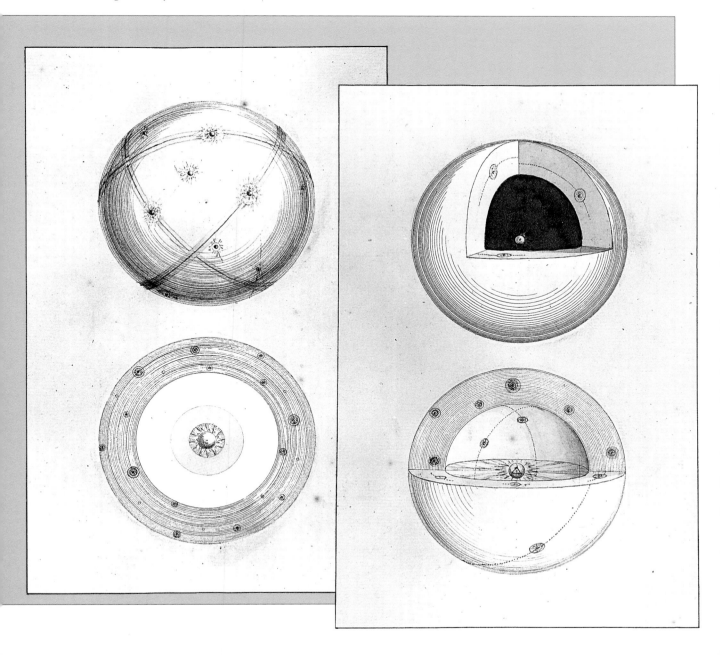

Opposite: A sketch used by Wright to help readers understand the explanation of his preferred model of the universe. In this sketch, the Sun and the other stars are sandwiched between two parallel planes (which his readers view edge-on, as vertical boundaries to either side of the stars). An Earth-based observer at *A* would see only a handful of nearby stars when looking in directions such as *B* or *C*. But he would see innumerable stars when looking in directions within the layer, for example towards *D*, *E*,..., and the light of these stars would merge to create the appearance of a milky way around the sky.

Below: In this alternative model of the universe, Wright portrays the stars as located in a flat ring that surrounds the Abode of God (shown as a sphere). The stars actually visible to us would form a disc-shaped system lying somewhere within the flat ring, and an observer would see a milky effect when looking in directions within the ring.

Wright attempted to explain the appearance of the Milky Way in the context of this model of the universe of stars, but he later realized a flaw in his explanation. Eventually he found a way round the difficulty, indeed two alternative ways; and in 1750 he expounded them in a handsome volume entitled *An Original Theory or New Hypothesis of the Universe*.

In his preferred solution, the Sun was again one of innumerable stars that together formed a spherical system surrounding the Divine Centre (or rather, since there were now many such systems and centres, our local Divine Centre). The spherical shell of space occupied by the stars of our system had an immense radius – so immense that the part of the shell immediately around us, whose stars astronomers could actually see, looked flat.

The shell was also thin, with the result that when, from their location within the shell, Earth-dwellers looked either inwards or outwards, their gaze quickly emerged from the layer of stars into empty space. The result was that in these directions they saw only a few stars, which were near and therefore bright. By contrast, when they looked around within the shell itself, their line of sight encountered first the nearby stars, and then more and yet more stars at ever-increasing distances; and the light of these innumerable stars merged to give the appearance of the Milky Way. The plane of the Milky Way, then, was the tangent plane to the spherical shell at the point where the solar system was located.

Wright recognized that there was an alternative model that could explain the Milky Way, one in which our star system formed a flattened ring that surrounded our Divine Centre. The stars actually visible to us on Earth would then occupy a disc-shaped space located to one side of the ring. But this model offered no explanation as to what had motivated God to select the particular plane in which the ring lay. As the spherical symmetry of its rival left no such loose end, it was the spherical model that Wright preferred.

Below: Wright's preferred model. Stars near enough to be seen by us form a small segment of a vast spherical system, whose radius is so great that the boundaries of the segment approximate to parallel planes. Again, an observer at *A* sees only a handful of nearby stars when looking towards *B* or *C*, but milkiness when looking towards *D*, *E*,...

Torn out of context, Wright's explanation of the Milky Way as the optical effect of our immersion in a layer of stars sounds strikingly modern. But this insight was embedded in a theological view of the cosmos that was little short of bizarre, and Wright's book would have had little influence had it not been summarized in a Hamburg periodical a few months after publication.

This summary chanced to come to the notice of the great German philosopher Immanuel Kant (1724–1804). Now, even with the illustrations to hand, historians have found it difficult to make sense of what Wright was attempting to convey; without the illustrations the task was well-nigh hopeless. Kant could not believe that Wright was seriously proposing a universe with innumerable Divine Centres, each surrounded by its star system; instead, he thought Wright must intend a single Divine Centre remote from our own star system – and therefore irrelevant to discussion of the structure of this system. Kant accordingly believed that, to explain the appearance of the Milky Way, Wright was offering two alternative explanations, both entirely in the natural order: one involving stars occupying a spherical shell of space, the other stars occupying a flat ring. To Wright, the centre of this ring was necessarily void of stars, because that was where our Divine Centre was located. Kant, misunderstanding Wright on this, saw no reason why the stars should not continue across the centre, so converting the ring into a continuous disc.

So was our star system spherical or disc-shaped? Kant believed there were other such systems in the universe, and that some of these had been observed (by the Frenchman P. L.M. de Maupertuis) to have an elliptical outline. Discs do have an elliptical outline when viewed slantwise on, but spheres always appear circular. The systems observed by Maupertuis, therefore, were disc-shaped rather than spherical, and the same must be true of our star system, or Galaxy. By such reasoning was the first essentially correct model of the Galaxy arrived at!

Meanwhile, another cosmological speculator was at work. Johann Heinrich Lambert (1728–77), an Alsatian who spent his life on the fringes of the scientific community, was a convinced believer in a stable, 'clockwork' universe. When he developed his conception of how our universe is structured, he was unaware of Halley's 1718 paper announcing the discovery of proper motions of certain stars. But, unlike Newton, he appreciated that the immensity of the distances separating the nearest stars from us had already undermined their supposed 'fixity'.

Lambert's universe had a hierarchical structure, with a large but finite number of steps in the hierarchy: moons, planets, stars, groups of stars, galaxies... At any given level, the unit consisted of a finite number of members from the level below, each in stable orbit about a central body. The Milky Way was composed of groups of stars, each group circulating about the central body of the Milky Way as do the planets about the Sun.

None of these speculations achieved wide circulation, and so their influence is not easy to assess. Yet the appearance in rapid succession of three works dedicated to speculative cosmology, shows that while professional astronomers were still preoccupied with the solar system, some outside their ranks felt a need to understand the universe 'in the large'. The man who attempted to answer this need was then in the north of England, scratching a living as a refugee musician.

William Herschel and the Construction of the Heavens

William Herschel (1738–1822) grew up in Hanover, and came to England as a refugee in 1757, in the aftermath of the Seven Years War. There he struggled at first to make ends meet, but in 1766 his fortune changed when he was appointed organist to a chapel in the spa resort of Bath.

His new security gave him the chance to broaden his interests. He explored the classic two-volume textbook of *Opticks* by Robert Smith of Cambridge, a work that instructed the reader in the theory of optics and in the practice of constructing telescopes and microscopes. Smith had concluded his work with a chapter on 'Telescopical Discoveries in the Fixt Stars', and these few paragraphs served to focus Herschel's developing ambitions in astronomy.

William Herschel

Friedrich Wilhelm Herschel, known to history in his anglicized name of William Herschel, was born at Hanover on 15 November 1738. At the age of fourteen he joined his father in the band of the Hanoverian Guards. Following the French victory in the Seven Years War, Herschel (who as a boy was not under oath, and so free to leave) fled to England, where he scratched a living, first by copying music in London, and then as organist and teacher of music in the north of England.

In 1766 however he was appointed organist to the fashionable Octagon Chapel in Bath, and despite his varied musical duties he was then secure enough to develop his other interests, especially in astronomy. His main leisure effort went into the construction of reflectors large enough to bring into view distant and therefore faint celestial objects, as part of his ambition to study the universe in the large. The turning point came in 1781 when he chanced upon the unknown planet Uranus (see page 187), and recognized it as no ordinary star. This discovery gave friends the opportunity to persuade the king to grant him a life pension, so that he could devote himself to astronomy. In 1782 he moved to Datchet near Windsor Castle, and then, four years later, to nearby Slough, where he lived for the rest of his life.

His loyal assistant in astronomy and, until his marriage in 1788, the mistress of his household, was his sister Caroline, who had joined him in England in 1772. In 1792 his only son, John, was born. William Herschel was knighted in 1816, and died at Slough on 25 August 1822.

In 1772, Herschel visited Hanover and persuaded his parents to allow his talented sister Caroline (1750–1848) to return with him to Bath. Caroline hoped for a career as a singer, but William was by now obsessed with astronomy, and her future role soon became evident: to be mistress of his household (until his eventual marriage in 1788), and his amanuensis and partner in the night watches at the telescope.

Next year Herschel bought a copy of *Astronomy Explained upon Sir Isaac Newton's Principles*, a best-seller by the leading popularizer of astronomy, James Ferguson (1710–76). It added little to his fragmentary knowledge of stellar astronomy, but it introduced him to some of the wider issues in cosmology. Ferguson, like many of his contemporaries, held that all planetary systems associated with stars are 'provided with accommodations for rational inhabitants'; and that even comets (once thought by William Whiston to be 'so many hells for tormenting the damned with perpetual vicissitudes of heat and cold') were probably peopled with beings capable of appreciating God's handiwork. Herschel was to extend this populating of the universe even to the Sun itself, which he came to believe was a cold body like a planet, but surrounded by clouds that protected the inhabitants from the exterior shell of fire; sunspots were glimpses of the clouds seen through gaps in the shell of fire.

Caroline Herschel

The 'comet sweeper' that William made for Caroline in 1783. 'Its distinctness is so perfect,' William later wrote, 'that it will shew letters at a moderate distance, with a magnifying power of 2000; and its movements are so convenient, that the eye remains at rest while the instrument makes a sweep from the horizon to the zenith'.

Caroline Lucretia Herschel was born at Hanover on 16 March 1750, and was brought to England by her brother William in 1772. She was his assistant in his astronomical work for half a century, the partnership ending only with William's death in 1822. In the early days she sometimes even put his food into his mouth, when he was grinding a mirror and dared not take his hands from it and allow it to cool. Later, on the nights when he was 'sweeping' for nebulae, she would sit at a table near the telescope, recording the descriptions as her brother shouted them out. So important was she to William's work that in 1787 the king awarded her a pension of £50.

She became a skilled comet-hunter in her own right, though her searches were restricted to what spare time she had. Despite this, she found her first comet in 1785, a second in 1788, and two more early in 1790.

The next year William made her a more powerful 'comet sweeper'. It was a sizeable

instrument, of 9-inches aperture and 5-foot focal length, but it had the low magnification and wide field of view appropriate to comet hunting. Like its predecessor, it had a mounting that was simple, but effective for its purpose. Caroline could sweep the whole of one vertical 'slice' of sky without looking away from the eye-piece, simply by pointing the instrument to the horizon, and then turning a handle that enabled her to lower the tube from the horizontal to the vertical. She would then turn the instrument to face in a new direction, and sweep through the new slice of sky. In this way she was capable of searching the entire visible sky in four nights. She found her fifth comet late in 1791, her sixth (which had already been seen by Charles Messier) in 1793, her seventh (a reappearance of what is now known as Encke's Comet) also in 1793, and her eighth in 1797.

Caroline worked equally hard at her desk in the daytime, often preparing fair copies

Ferguson was not a Newton, dogmatically confident that Providence would intervene whenever chaos threatened, and restore the structure of the solar system or that of the universe of stars: the universe, he wrote, 'will last as long as was intended by its Author, who ought no more to be found fault with for framing so perishable a work, than for making man mortal'. Herschel learned the lesson well.

To view the objects described by Smith and Ferguson, Herschel first experimented with refracting telescopes. But the aperture of refractors was severely limited by the great difficulty (and cost) of manufacturing lenses of appropriate quality, and a lens of the size to match Herschel's ambitions was a technological impossibility. For his cosmological artillery Herschel therefore turned to reflectors, in which the light falls on a mirror at the base of the tube and is reflected back to a focus. Mirrors were altogether more promising, and he had Smith's book to tell him how to do the grinding and polishing.

By November 1773, he had placed orders for a number of discs, one of which was for a reflector of 5^1/$_2$-foot focal length. And it was when using this instrument, on 1 March 1774, that he decided to open a 'journal', or observing book. The first page of this

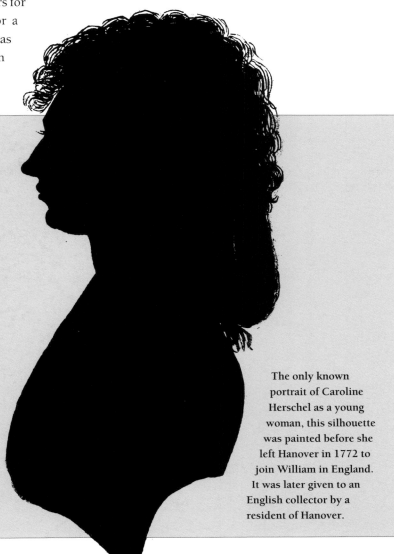

of the previous night's nebular observations. At William's request she improved the reliability of Flamsteed's 'British Catalogue' of stars, by compiling a list of stars observed by Flamsteed but omitted from his Catalogue (and a list of errata that included 'stars' never in fact observed). Her *Catalogue of Stars Taken from Mr. Flamsteed's Observations* was published by the Royal Society in 1798.

On William's death she returned to Hanover. Her last service to astronomy was to rearrange William's catalogues of nebulae into a form that her nephew John could use in his re-examination of them. For this she was awarded the gold medal of the [Royal] Astronomical Society. She died at Hanover on 9 January 1848, in her ninety-eighth year.

The only known portrait of Caroline Herschel as a young woman, this silhouette was painted before she left Hanover in 1772 to join William in England. It was later given to an English collector by a resident of Hanover.

An entry on the opening page of William Herschel's first 'journal' or observing book. On 4 March 1774 he observed the Orion Nebula, and compared what he saw with a sketch of its appearance many years earlier, as reproduced in Robert Smith's *Opticks*. The nebula, it seemed, had altered shape, and Herschel quickly realized that this fact had important implications for the understanding of its physical nature.

book (above) must be the most portentous beginning to any career in observational astronomy.

Six of the milky patches in the sky known as 'nebulae' had been listed by Edmond Halley in *Philosophical Transactions* in 1715, and these were discussed by Smith in his *Opticks*. The term 'nebula' referred simply to the object's milky appearance, and did not prejudge its physical nature.

On this there was a long-standing dispute. It was obvious that a distant cluster of stars would appear nebulous when seen in a telescope of insufficient power to 'resolve' the cluster into its component stars. The question was, were there also true nebulae, formed of some sort of diffuse luminous fluid; or were all nebulae merely apparent, nothing more than star systems whose true nature was disguised from the Earth-based observer by their great distances? Halley took the former view: nebulae 'in reality are nothing else but the Light coming from an extraordinary great Space in the Ether; through which a lucid *Medium* is diffused, that shines with its own proper Lustre'.

Observation could contribute to the debate in two ways. First, a more powerful telescope might succeed in resolving into its component stars a star system that had appeared nebulous when viewed with lesser instruments. Second, if a nebula was seen to alter shape from one decade to another, or even one century to another, then the nebula could not be a star system. After all, a star system so extensive as to appear to the observer as spread across the sky, and yet so distant that the component stars could not be detected, must be vast indeed – too vast to alter shape so rapidly.

Herschel was familiar with the crude sketch of the Orion Nebula that Smith had reproduced in his *Opticks*, from a drawing made by Christiaan Huygens in 1656. Looking at the nebula with his home-made reflector, Herschel decided it must have changed, and he at once saw the implications: '...from this we may infer that there are undoubtedly changes among the fixt Stars, and perhaps from a careful observation of this Spot something might be concluded concerning the Nature of it.'

But these were crowded years, and Herschel – organist, composer, conductor, and teacher of music – devoted most of the limited time he could spare for astronomy to improving his telescopes. But in 1779 he decided it was time he familiarized himself with the brighter stars, and so he systematically examined them one by one, using a portable reflector he had made himself of 7-foot focal length (see page 187). He then embarked on a second such 'review', much more thorough and this time with an additional goal: the identification of double stars that might be of use in the application of Galileo's method for the detection of annual parallax (see page 210).

Herschel harvested 269 double and multiple stars from this review, and a further 434 from a third review, thereby multiplying many times the number known to observers. He had introduced a new methodology into astronomy. Those who had received an orthodox education in the science knew that it was the job of an astronomer to study the familiar celestial bodies – Sun, Moon, planets and their satellites, comets, bright stars, each with its personal name and individual characteristics. Herschel, knowing no better, was beginning to play the natural historian, collecting specimens in great numbers, and counting and classifying

Attraction beyond the solar system

In his *Principia* (1687), Isaac Newton provided the strongest evidence that the force of gravitational attraction operated throughout the solar system, and the return of Halley's Comet in 1758 from beyond the outermost known planet was further confirmation of this. But Newton offered no evidence to show that attraction operated among the stars.

The first to do this was the Cambridge geologist and astronomer, John Michell (c. 1724–93). In a paper published in 1767, Michell used a mathematical argument to show that double stars were so numerous that they could not all result from mere accidents whereby the two stars lay in the same direction from Earth: most must in fact be physically connected pairs of companions ('binary stars'). The same applied to star clusters such as the Pleiades.

In 1802, William Herschel began re-examining double stars he had discovered two decades before. He found that in several of them, the two component stars had altered position relative to each other, in a way that showed they were indeed companions bound together by some attractive force.

But was the force that of Newtonian gravity? The necessary evidence was not available for another generation. At last, in 1827, the Paris astronomer Félix Savary was able to confirm that the two stars of Xi Ursae Majoris moved in elliptical orbits about their common centre of gravity, as required by Newtonian theory.

Black holes in the eighteenth century

The possibility of a celestial body that was invisible, because its mass was so great that its attractive pull prevented light from leaving it, was discussed several times in the late eighteenth century. The first to explore the question was John Michell. In a paper in *Philosophical Transactions* in 1784, Michell estimated that if a star of the same density as the Sun had a radius 500 times greater, 'all light emitted from such a body would be made to return towards it, by its own proper gravity'. However, such a body might betray its presence by the effects its pull was having on neighbouring bodies:

yet, if any other luminous bodies should happen to revolve about them we might still perhaps from the motions of these revolving bodies infer the existence of the central ones with some degree of probability, as this might afford a clue to some of the apparent irregularities of the revolving bodies, which would not be easily explicable on any other hypothesis.

Taking up this theme and applying it on the cosmological scale, William Herschel in 1791 wrote of 'the great counteraction of the united attractive force of whole sidereal systems, which must be continually exerting their power upon the particles [of light] while they are endeavouring to fly off'.

The idea received wide publicity in 1796, when Pierre Simon de Laplace (1749–1827) included an estimate similar to Michell's in his *Exposition du système du monde*. But Laplace dropped the subject from the 1808 edition of his book, possibly because it conflicted with the general view that the speed of light was a constant; and the concept of what is now termed a 'black hole' was relegated to the status of a far-fetched speculation, where it remained until recent times.

them. Soon he would be ordering nebulae according to the stage they had reached in their life-cycle.

Meanwhile, word of this extraordinary organist was spreading in astronomical circles. Leading astronomers called on him, recognized his great talent, and did what they could to smooth his path. A Bath neighbour communicated papers of his to the Royal Society in London. But some Fellows of the Royal Society – mingling incredulity at his claims with contempt for his ignorance of basic procedures and conventions – declared him fit for the mad-house.

However, his discovery in 1781 of the planet Uranus (see page 187), in the course of his second review, was a triumph that none could gainsay, and soon every astronomer in Europe had heard of Herschel. King George III, himself a Hanoverian, granted him a life pension that would allow him to give up music and devote himself to astronomy.

Finding a new planet had been far from Herschel's mind on that fateful evening, but its discovery was no accident: he was systematically searching the sky, and his dual skills as telescope-maker and observer enabled him to recognize at a glance that the object was no ordinary star. Telescope-making would always be the foundation for his success as an astronomer, and he now became a professional maker, supplementing his pension by manufacturing reflectors for sale.

A successful reflector embodied three key components: first, a well-shaped mirror – in fact, two mirrors, so that one might be used while the other was being repolished to remove the tarnish caused by exposure to the night air; second, a range of eyepieces for magnification; and third, a stable yet adaptable mounting. Even when at Bath, Herschel's ambitions to have large mirrors of great 'light-gathering power', to permit the study of objects that were very distant and therefore faint, had outrun the capacity of local foundries to cast the blanks. Nothing daunted, in August 1781 he converted the basement of his own home into a foundry, and twice attempted to cast a 3-foot disc, which would have made this organist the owner of the largest telescopic mirror in the world. On the first occasion the mirror cracked while cooling; on the second, molten metal poured out onto the flagstones which, expanding, began to fly about in all directions. At this even Herschel admitted temporary defeat.

In the making of eyepieces, success seems to have come easily to Herschel. Indeed his magnifications – of hundreds and even thousands – were cited by him without special comment, although to many these well-justified claims were simply incredible.

It was in the mounting of large reflectors that Herschel proved most innovative. A 20-foot reflector he had made himself in 1776 had been slung from a pole, like

Herschel's 'small' 20-foot reflector, completed in 1776. The primitive mounting is reminiscent of the long-focus refractors of the seventeenth century (see page 142). It was however possible for Herschel to adjust the vertical angle through the ratchet-wheel on the upper side of the tube. He could also move the telescope a little to left or right by turning the small wheel below the eyepiece, but for major changes in azimuth he had to descend the ladder, move the base, and climb once more to the eyepiece.

Opposite: William Herschel's 'large' 20-foot reflector, completed in 1783. Its mirrors were bigger than those of the 'small' 20-foot, but the major advances were in the stable and secure mounting, and in the controls available to the observer. This telescope was used by William throughout his twenty years of sweeping for nebulae. It was later refurbished by his son John under William's supervision, and after William's death John used the modified instrument, first in England and then in South Africa (see page 250), to examine the entire heavens.

the very long refractors of past generations. But once he was free to devote himself to astronomy, it took Herschel only a year to build himself a new 20-foot. This time the mirrors were 18 inches in diameter rather than 12 inches; but more importantly, the mounting was stable, and the observer stood in safety on a secure platform.

Herschel was now equipped to tackle the riddle of the nebulae. Despite his early look at the nebula in Orion, by December 1781 he had seen only three more. It was then that he was given a catalogue of some sixty-eight nebulae and star clusters. It had been assembled by the French comet-hunter Charles Messier (1730–1817), who had found these diffuse objects an unwelcome distraction in his searches for comets. (In fact Messier had already published an enlarged catalogue with just over 100 nebulae, and this catalogue is used today by astronomers when they refer to a prominent nebula by an 'M', followed by its Messier number.)

Herschel made the momentous decision to use the new 20-foot to sweep the entire sky visible from England, in order to collect as many specimens of nebulae as possible. On nights when the 'seeing' was good, the telescope was turned to the south, and the tube raised to some particular angle. Herschel then let the sky drift past, so laying an ambush for any nebula that came into the field of view. When he saw one crossing the meridian, he would call out its description, which Caroline

The Orion Nebula, drawn by Charles Messier early in 1773, at the Hôtel de Cluny in Paris. Like Herschel a year later (see page 234), he remarked on how much the nebula differed from its appearance in sketches by earlier astronomers, and he hoped that his drawing, made 'with the greatest attention', would allow future observers to decide whether or not it had in fact altered shape.

Herschel's map of the Galaxy

In a paper he published in 1785, Herschel pioneered the use of statistics in astronomy, by showing how the natural historian of the heavens can derive insights simply by counting stars. The problem he set himself was to determine the shape of our star system, the Galaxy.

Like Wright, Kant and Lambert, he realized that the Milky Way was the optical effect of our immersion in a layer of stars.

The Milky Way and its explanation. According to Herschel, the Sun and the other stars form a stratum; and the Milky Way is the optical effect of these stars, when seen by an observer on Earth who is looking in directions within the stratum. The bifurcation is to account for the division in the Milky Way in Scorpius and elsewhere.

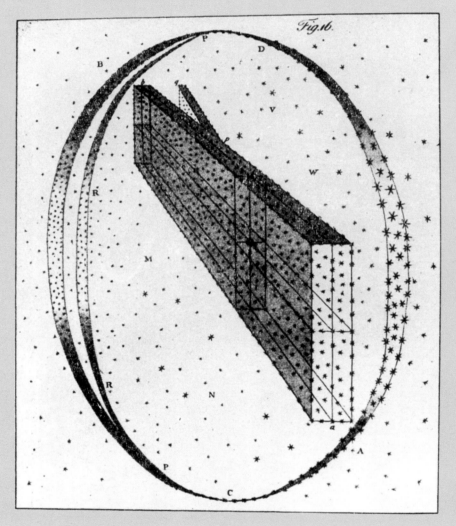

copied down, along with its angle from the celestial North Pole and the time. For two long decades the work continued. By 1802, the team of brother and sister had increased the number of known nebulae to 2,500.

But were all nebulae merely vast star clusters at great distances; or were some truly nebulous, and formed of a luminous fluid? Just before sweeping began,

But what was the precise shape of this layer? Clearly, Herschel could not attempt to answer this question unless he was allowed to assume that his telescope could reach to the limits of the Galaxy in every direction. But how then to proceed? Herschel decided that the way forward was to assume next that throughout the Galaxy, the stars were distributed uniformly: that the galactic space was uniformly stocked with stars. Obviously the assumption was not true in the literal sense; Herschel hoped it would be true enough for his purpose.

Granted this, the number of stars in Herschel's field of view in a given direction was proportional to the volume of galactic space within that field of view – that is, to a cone-shaped volume of space, whose vertex was at the telescope, and whose axis was the line of sight from the observer to the border of the Galaxy. Herschel would then count the stars to get a number proportional to the volume of the cone,

and a simple calculation would then give him the (relative) length of the axis. Time did not permit him to implement this programme in full, but to illustrate his method he made the counts for a circle around the sky, and sketched the resulting cross-section of the Galaxy.

In later life he was to abandon both of the assumptions on which this famous figure was based: he found that the new, monster 40-foot reflector that he completed in 1789 brought many more stars into view, and so his 20-foot had not after all penetrated to the borders in every direction; and increasing familiarity with star clusters brought home to him how very non-uniform is the distribution of the stars. But astronomers, like Nature, abhor a vacuum. Herschel's cross-section might have been disowned by its creator, but for long there was nothing to take its place, and so we find it being reproduced in books late into the nineteenth century.

The cross-section of the Galaxy resulting from Herschel's star counts. The Sun is the star near the centre, and the (relative) distances to the stars shown bordering the Galaxy are inferred from the star counts: the greater the number of stars in a given direction, the greater the distance. Note the bifurcation to the left. Herschel was involved in sweeping for nebulae, and could spare only enough time to give this limited example of his method.

Herschel had confirmed (as he thought) the variability of the Orion Nebula, which he found 'surprizingly changed'; this, then, must be a true nebula. But how in general was one to distinguish true nebulae from distant star clusters?

Herschel had noticed that some nebulous-seeming objects had a uniformly milky appearance, while others were mottled. The former, he decided, were true nebulae, while the mottled nebulosity of the latter indicated to him that they were 'resolvable' – in other words, they were clusters that with a sufficiently powerful telescope would be seen resolved into their component stars.

In June 1784, a paper outlining his current work on nebulae was read to the Royal Society. Within days, Herschel came across two nebulae that contradicted the very theory he had just published, for each seemed to contain both forms of nebulosity. Indeed, in one of them he believed he could also see stars mixed with the resolvable nebulosity. He interpreted the stars as being in the region of the nebula nearest to the observer, and the resolvable nebulosity as composed of stars that were further away, a little too distant to be individually visible. Must not then the milky nebulosity simply consist of stars at a still greater distance – rather than of the luminous fluid, or 'true nebulosity', that Herschel had hitherto postulated?

Ignoring the changes he believed he had earlier observed in the Orion Nebula, he now abandoned his belief in luminous fluid, and concluded instead that all nebulae were star clusters at great distances.

But clusters imply clustering, the assembling together of stars as a result of their gravitational pulls on each other. Accordingly, in an epoch-making paper 'On the Construction of the Heavens' published in *Philosophical Transactions* in 1785, Herschel examined the subject 'from a point of view at a considerable distance both of space and of time'.

He imagined a universe in which the stars were at first distributed with fair regularity; and he went on to outline how, in time, gravitational pulls would be likely to cause stars to assemble in places here and there, forming clusters of various types, examples of which he had already observed. He even envisaged the possibility that the continuing action of gravity might eventually lead to gravitational

Opposite: Herschel's 40-foot reflector, a cumbersome, scaled-up version of the 'large' 20-foot. It was completed in 1789, and remained unsurpassed in size until 1845 (see page 253). The second of its 4-foot-diameter mirrors weighed nearly one ton (1,000 kg), and to prevent it from bending under its own weight Herschel had increased the proportion of copper. As a result, the mirror tarnished easily.

An historic photograph of an historic instrument: the 40-foot in an advanced state of decay, photographed by John Herschel in September 1839. The tube had already been lowered for the last time, and the entire structure was to be dismantled during the winter. On New Year's Eve, 1839/40, John Herschel assembled his family inside the recumbent tube. He then read a poem written as an obituary for the instrument, before sealing the tube with his father's grinding and polishing tools inside. John Herschel was a pioneer of photography, and introduced the term into the English language.

The planetary nebula NGC 1514, which Herschel encountered in November 1790. The more distant planetary nebulae that he had earlier discovered and named were ambiguous in nature, but this object was clearly formed of a central star surrounded by a luminous atmosphere.

collapse, followed by some form of renewal. 'These clusters may be the *Laboratories* of the universe…wherein the most salutary remedies for the decay of the whole are prepared' – a far cry from the stable clockwork universe of the early years of the century.

In November 1790 Herschel was sweeping as usual for nebulae, when he came across 'a most singular phaenomenon! A star of about the 8th magnitude, with a faint luminous atmosphere…' It was in 1782 that he had encountered his first 'planetary nebula' – an object that was faint like a nebula but had the disc-shaped outline of a planet – and he had found several more since. What he had now encountered was in fact another of the class; but this one was unusually large in appearance, and he could see its structure and in particular its central star. The object must, he decided, be a 'nebulous star': a star surrounded by a cloud of (true) nebulosity, out of which the star was in the process of condensing. Faced with this new evidence, Herschel instantly reversed the position he had held since 1784, that all nebulae were nothing else but star clusters at great distances. Nebulosity existed after all, and it represented a pre-stellar stage in celestial development.

Herschel's final theory of the evolution of the universe therefore began with diffuse clouds of nebulosity, which gradually condensed here and there under gravity to form more concentrated nebulae, out of which in time individual stars began to form. These in turn would gather, at first into widely-scattered clusters, and then into more condensed ones. From the cataclysmic collapse of such clusters, and also from the light sent out into the universe from all luminous bodies, came the material to form new diffuse clouds of nebulosity, so that the cycle might repeat itself.

The status of our Galaxy was also changed by the 1790 observation. In the mid-1780s, Herschel had believed it to be a star system of known and therefore limited extent, and all nebulae to be similar star systems, though of varying shapes and sizes. The Orion Nebula, therefore, which he saw extended across the sky despite its being (supposedly) so distant that the individual stars escaped detection, had to be vast; indeed, it 'may well outvie our Milky Way in grandeur'. But in his post-1790 theorizing, the Orion Nebula reverted to being a nearby cloud of nebulosity located well inside our Galaxy, while the Galaxy became 'the most brilliant, and beyond all comparison the most extensive sidereal system'.

'A knowledge of the construction of the heavens', Herschel wrote in 1811, 'has always been the ultimate object of my observations…' Herschel had been able to make this into a true science because he combined to a unique degree the three talents necessary: those of instrument-builder, observer, and theorist. He built instruments that were ideal for his self-imposed task: they incorporated large mirrors, eye-pieces that magnified hundreds of times, stable mountings – all made with his own hands or under his direct supervision. With these instruments he played the natural historian of the heavens, collecting double stars by the hundred and nebulae by the thousand, in observational campaigns extending over many

The life story of stars and star systems

These sketches by Herschel show objects from his catalogues of nebulae and star clusters, arranged in sequence to illustrate (left) the stages by which nebulosity condensed into a star (or stars), and (right) the subsequent developments among the stars that led to the formation of condensed clusters.

In 1811, Herschel explained how astronomers – despite their own brief life-spans – might succeed in understanding the changes in nebulosity caused by gravitational attraction. They must select specimens of nebulae at different stages of development. They must then classify and order them by age (left): 'there is perhaps not so much difference between them [the nebulae in consecutive classes], if I may use the comparison, as there would be in an annual description of the human figure, were it given from the birth of a child till he comes to be a man in his prime.'

In 1814, Herschel published a second paper in which he continued the story, beyond the stage when scattered stars had been formed. In response to their mutual gravitational attraction, the scattered stars of a group would move ever closer together (right), until they formed a condensed cluster.

His son John would often be torn between scepticism, and the filial piety that made him hesitate to contradict the theories of his late father. In 1826, scepticism took control; how, he asked, could one argue from 'observed graduation' among nebulae and star clusters to 'concluding them to be in a course of progress from one state in the series to another'? 'So wide is the field for conjecture,' he concluded, '...that we shall do well for the present to dismiss hypothesis, and have recourse (perhaps for centuries to come) to observation.' Even the very existence of 'true nebulosity' continued to be a matter for debate. But John did come to adopt 'the theory of sidereal aggregation', according to which a star system passed through a period of collisions and disturbances, before eventually reaching a stable configuration.

years. And, unusually for one so dedicated to assembling facts, he saw it as his clear duty to speculate too much rather than too little.

Among his contemporaries, Herschel's impact was mixed. The nebulae in particular were his private domain: no one else had telescopes to equal his, so no

The national observatory of Spain

Madrid Observatory in 1848

Numerous observatories – most of them national, university, Jesuit, aristocratic, or the product of an amateur's enthusiasm – were founded in the eighteenth and nineteenth centuries. The observatory at Madrid (above), a national institution founded by the king of Spain in 1790, was equipped from the start with the usual transit instruments and clocks for positional measurements, but in 1796 two reflectors were commissioned from William Herschel. One was to be a small portable telescope of standard design (in the event, the obser-

vatory acquired two 7-foot instruments of slightly different mirror diameters), while the other was to be the largest instrument he ever made for sale, with 2-foot mirrors of 25-foot focal length. Right is the water-colour Herschel sent with the assembly instructions. This was perhaps his most successful telescope, more powerful than the 20-foot he customarily used but less cumbersome than his monster 40-foot. It was shipped to Spain in 1802 and erected in Madrid in 1804, only for the mounting to be destroyed by Napoleonic troops in 1808.

one else had access to the evidence. Few therefore knew what to make of him and his speculations. But his papers in *Philosophical Transactions* were readily available to the next generation of astronomers, who would be more receptive to his ideas. Prominent among them was his son, John.

John Herschel

John Frederick William Herschel was born on 7 March 1792 at Slough, near Windsor Castle, the only son of William. In 1809 he entered St John's College, Cambridge, where he was a leading figure in a campaign for the reform of British mathematics. He was elected to the Royal Society when he was only twenty-one, and in 1820 he took a leading part in the foundation of the future Royal Astronomical Society. He became perhaps the most prominent British scientist of his day, being knighted in 1831 and created baronet in 1838. So famous was he that a letter from abroad addressed to him in 'London' reached its destination.

After revising his father's observations of the skies visible from Slough, Herschel spent the years 1834–38 surveying the southern stars from the Cape of Good Hope. On his return he largely abandoned observational astronomy, though in 1864 he published a consolidated catalogue of over five thousand nebulae and clusters, the ancestor of the *New General Catalogue*, or 'NGC', that astronomers use today.

Herschel was of private means, though the source of this wealth is puzzling. Equally puzzling is his acceptance in 1850 of the post of Master of the Mint, which he filled for five unhappy years. He died at his home in Kent on 11 May 1871, and was buried in Westminster Abbey.

John Herschel and the Southern Skies

William Herschel had been a German-born provincial musician when he cut his astronomical teeth. His only son, John (1792–1871) – born when his father was already fifty-three – carried the most famous name in astronomy, and from his undergraduate days in Cambridge was a member of the scientific establishment. After a flirtation with law, John Herschel settled down to a teaching career in Cambridge. But in 1816 his father, whose strength was failing, prevailed upon him to return home, so that William could hand on his skills as a telescope-maker and observer before it was too late. 'My heart dies within me', wrote John as he left Cambridge. But, once he had made the sacrifice, the son saw himself as entrusted with a sacred mission, to complete his father's work and bring it to perfection.

Double stars were the most obvious place to begin: his father's telescopes had been designed as massive instruments of discovery, whereas the observation of double stars was best done with instruments of precision. Fortunately, among John Herschel's many scientific friends was James South (1785–1867). South's wealth through marriage had allowed him to give up surgery for astronomy, and he was a skilled observer and the owner of two exceptionally fine equatorials. From 1821 to 1823, though with interruptions, the two friends collaborated, often observing the same object with different instruments and then comparing notes. Their efforts resulted in a catalogue of 380 doubles, fully detailed, and conveniently ordered for the use of observers.

The principal legacy of Herschel's father, however, had been his studies of 'the construction of the heavens', and the catalogues of nebulae and clusters on which these studies were based. These catalogues listed the objects by class rather than

South's object glass: a price too high

In the early nineteenth century there was intense competition among observers to obtain large, high quality object glasses, for only rarely was one successfully made. The observer whose refractor was equipped with such a glass would have a clear advantage over his rivals.

Around 1830 the Paris maker Robert A. Cauchoix (1776–1845) produced three such glasses, all between 11¾ and 14 inches in diameter. One was bought by the Duke of Northumberland for the Cambridge Observatory – the Northumberland telescope was the instrument used by James Challis in his fruitless search for Neptune (see page 193). Another was acquired by Edward J. Cooper for the observatory on his estate at Markree Castle in northwest Ireland. The third was bought by James South for his private observatory at Camden Hill in London.

South was a leading observer of double stars, rivalled only by Wilhelm Struve, who had the Dorpat refractor (see page 217) at his disposal. In 1829 South heard that French astronomers were haggling with Cauchoix over the price of the lens. He rushed to Paris, paid the asking price, and returned in triumph with the precious glass. He then commissioned Troughton & Simms, the firm of the leading London maker, Edward Troughton (1753–1835), to make the mounting, without which the lens would be useless. South wished for a scaled-up version of the mounting of his 5-foot equatorial, but Troughton had other ideas.

Before long matters turned sour. South was notoriously impulsive and irascible, and he declared himself dissatisfied with the mounting supplied and refused to pay for it. Troughton & Simms took him to court, and after an expensive lawsuit lasting from 1834 to 1838 (during which time Troughton died) obtained judgment in their favour. Within months South, unhinged with rage, had smashed the polar axis to pieces and auctioned the fragments (engraved below). Three years later he auctioned the brasswork, advertising the sale with a poster of legendary vindictiveness. The lens, however, he preserved, and near the end of his life he presented it to Dublin University. But its moment had passed, and South's own career as an observer had been destroyed.

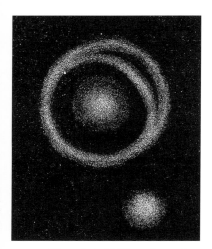

The nebula M 51, as drawn by John Herschel in 1828. An observer in the central cluster would see bright stars scattered all round the sky, and a fainter, bifurcated 'milky way' – much as we ourselves do.

position, and were therefore highly inconvenient for other observers to use. In addition, there had been ample scope for error when Caroline had copied down the observations shouted out by her brother, as the various nebulae came into view. Fortunately, John had himself refurbished his father's 20-foot reflector under the old man's supervision, so he possessed the ideal instrument with which to re-examine (and re-order) his father's nebulae. 'These curious objects...I shall now take into my especial charge', he told Caroline in 1825, 'nobody else can see them'. His immediate efforts resulted in a catalogue of 2,306 nebulae and clusters, published in 1833 in *Philosophical Transactions*; it became the standard reference work, and helped transform the study of nebulae from one of the maverick William's exotic pursuits, into a component of mainstream astronomy.

So far John Herschel's experience of nebulae had been little different from that of his father: the same skies, and much the same instrumentation. Not surprisingly, his theoretical stance was also the same. Only one significant new item of evidence had come his way: he had several times had a good view of the nebula M 51, which he saw as being composed of stars in the form of a central cluster surrounded by a divided ring. The sky as seen by an observer within the central cluster would, as he immediately realized, be strikingly similar to the sky we see from Earth: on all sides a number of nearby (and therefore bright) stars and, in and near the plane of the ring, a divided milky way of innumerable faint stars. 'Perhaps', he remarked, 'this is our Brother System'.

His father had seen only the skies visible from England, and to complete his father's work John Herschel would need to set up his telescopes south of the equator. And so, declining official offers of financial support, he set sail in November 1833 for the Cape of Good Hope. For deep-sky exploration he had the

The refurbished 20-foot reflector, at John Herschel's temporary observatory at Feldhausen, a few miles from Cape Town. In the sea air, a telescopic mirror tarnished in a matter of days and needed repolishing, during which time it was out of action. Anticipating this, John had brought with him three matching mirrors, one made by William, one by John, and one by the two in collaboration.

Early observations of the southern skies

The southern skies have much more to offer the observer than their northern counterparts, for they contain the brilliant Milky Way star clusters that lie in the direction of the galactic centre, as well as such wonders as the Magellanic Clouds. However, because astronomy developed in Europe, and travel south of the equator was difficult in past centuries, it is only in our own day that they have been given their due share of attention. No public observatory existed in the southern hemisphere until Fearon Fallows (1789–1831) arrived in 1821 at the Cape of Good Hope as His Majesty's Astronomer. Until then, knowledge of the southern skies had depended on sailors' lore, and on expeditions by two remarkable astronomers, Halley and Lacaille.

Edmond Halley made a reputation for himself in Royal Society circles when no more than a youth. Early in 1676, and still not yet twenty years old, he was corresponding about suitable sites for an expedition south of the equator; but some of these sites were in foreign hands, and his choice eventually fell on the south Atlantic island of St Helena, used as a waystation by the (British) East India Company. The King was persuaded to request the Company to give free passage to Halley and his colleague, and Halley's father agreed to contribute to the costs of the expedition.

Halley arrived at the island in February 1677, and stayed for nearly a year. He was equipped with a range of instruments, prominent among which was a sextant of 5½-foot radius with telescopic sights, for measuring the angular distance between pairs of stars. The climate was less favourable than Halley had hoped, but he managed to compile a catalogue of some 350 stars, listing their positions relative to two of Tycho Brahe's fundamental stars.

He also observed three 'nebulae', including the scattered cluster of stars known as M 7, and the fine spherical ('globular') cluster Omega Centauri. The Magellanic Clouds, he wrote, 'reproduce exactly the whiteness of the Galaxy, and,

examined through a telescope, they exhibit here and there small clouds and a few stars, from the concourse of which their white colour, like that of the Galaxy, is now believed to be produced'. On 28 October he saw Mercury move across ('transit') the face of the Sun.

It was to be the middle of the next century before the southern skies again came under scrutiny. In 1750, the abbé Nicolas-Louis de Lacaille (1713–62), who had built up a formidable reputation as an observer, secured government support for an expedition to the southern hemisphere, and he arrived at the Cape of Good Hope the following spring.

In a stay lasting less than two years, he measured the positions of nearly ten thousand stars. He achieved this extraordinary level of productivity by attaching a small telescope of wide aperture to a mural quadrant (a graduated quarter-circle mounted against a north–south wall), and placing a rhomboidal diaphragm in the field of view of the instrument. The telescope would be fixed at a particular elevation in the meridian. As a star drifted through the field of view, Lacaille noted the (sidereal) times at which it entered and left the rhombus; the average of the times gave one of the star's co-ordinates and the difference in the times gave the other. He also assembled information on no fewer than forty-two nebulous objects, and assigned names to southern constellations, most of which are in use today (see his 1752 chart, above).

In 1757 Lacaille published the positions of nearly 400 of the brightest of the southern stars, so establishing the framework for southern-hemisphere astronomy. His observations of 10,000 southern stars appeared posthumously, in 1763. However, he left the majority of his position measurements in their raw ('unreduced') state, and a definitive catalogue of his southern stars was not published until 1847.

The reflector with 3-foot mirrors, built in 1839 by the future Lord Rosse using the type of mounting developed by William Herschel. The first mirror was assembled in sixteen segments, but in 1840 Rosse, with the help of local labourers, successfully cast a solid mirror. Rosse persuaded himself that the telescope had 'resolved' a number of nebulae into their component stars, and that a larger instrument would resolve many more.

20-foot; for precision measurements, he had an equatorial he had bought from South; and for a preliminary reconnaissance, he had the largest of the 'comet sweepers' his father had made for Caroline (see page 232).

Over the next four years, and with no Caroline to help him, John explored the southern skies with a dedication surpassing even that of his father. The resulting volume, *Results of Astronomical Observations Made During the Years 1834, 5, 6, 7, 8 at the Cape of Good Hope*, appeared, after considerable delays, in 1847. Arguably the greatest single publication in the whole history of observational astronomy, it bore a proud subtitle: 'Being a completion of a telescopic survey of the whole surface of the visible heavens, commenced in 1825', for John Herschel was and would remain the only astronomer in history systematically to examine the entire sky with a major telescope. The book listed over 1,700 nebulae and clusters and over 2,100 double stars, as well as thousands of star counts, extensive sequences of the comparative brightness of stars, and much else besides.

In March 1838, his duty to his father's memory nobly discharged, Herschel took ship for England. His future work in astronomy would be done sitting at a desk. In any case, within a year or two the Herschel monopoly of great reflectors would come to an end, and with it the period in which, as Wilhelm Struve put it, the study of the nebulous heavens had seemed 'almost the exclusive domain of the Herschels'.

The Leviathan of Parsonstown

In 1839 William Parsons (1800–67), future third Earl of Rosse, built a large reflector in the grounds of his castle at Parsonstown (now Birr) in central Ireland. Though mounted in the manner of Herschel's 20-foot, its 3-foot mirrors had twice the diameter and four times the area. Scarcely was this instrument completed when Rosse began work on a monster with mirrors an incredible 6 feet in diameter. Casting of discs of this size had never before been attempted, and it proved necessary to use three huge crucibles simultaneously. The first disc was cast in April 1842, but thereafter all did not go smoothly, and five castings were needed to produce two serviceable mirrors, each weighing about four tons. With its south-facing tube crudely slung between two huge walls of masonry, the 'Leviathan of Parsonstown' was designed for a final assault on the classic problem of nebular astronomy: Are all nebulae merely distant star clusters?

In February 1845 the Leviathan first saw light, and within a few weeks Rosse was able to announce the discovery that some nebulae are spiral in shape. A comparison

The 'Leviathan of Parsonstown', with 6-foot mirrors each weighing four tons (4,000 kg). The instrument faced south, and was designed primarily to observe nebulae as they crossed the meridian, in the hope of 'resolving' them into stars. The first mirror was cast in 1842, work on the mounting began late the same year, and the telescope first saw light in February 1845. By April Rosse had discovered the spiral structure of M 51.

Above: Rosse's drawing of M 51, the first occasion on which a spiral form was recognized in a nebula. The drawing was made in April 1845, and circulated at the June meeting of the British Association for the Advancement of Science. John Herschel declared the building of the great reflector to be 'an achievement of such magnitude…that I want words to express my admiration of it'.

Right: Unlike Herschel's own primitive sketch of M 51 (see page 250), the Rosse drawing stands comparison with this modern photograph.

of the superb sketch Rosse made of M 51 with John Herschel's drawing (see page 250) and with a modern photograph offers proof of the Leviathan's power.

Yet in many ways it never fulfilled its potential. A terrible potato famine struck Ireland in the late 1840s when the instrument was in its prime; Rosse, a leading public figure and landowner, could scarcely devote himself to the stars while his tenants lay dying. In addition, for Rosse the construction of the great reflector – with all the technical challenges involved – was an end in itself. Rosse was no Herschel. He observed from time to time, his visitors were welcome to use the instrument, and he employed worthy if uninspired assistants; but this was no substitute for Herschelian dedication.

The decisive test came when the instrument was turned to the Orion Nebula. If every nebula was a star cluster, surely this nebula above all – so extensive as seen by us, and therefore relatively near – would reveal its starry nature when interrogated by the Leviathan.

In March 1846, Rosse announced that numerous stars were indeed to be seen in the nebula. In fact the stars that Rosse described are genuine enough; but they are embedded in what is generally a gaseous nebula. This was not appreciated at the time, with the result that the supposed resolution of this and other prominent nebulae persuaded all but a few sceptics that the hypothesis that true nebulosity existed and condensed into stars had been discredited.

But sceptics there were. Like William Herschel back in the 1770s, Mikhail V. Lyapunov of the university at Kazan, far to the east of Moscow, had convinced himself – and the influential Wilhelm Struve – that major changes were occurring in the Orion Nebula. This being so, the 'alleged miracles of resolution', as Struve termed them, 'are nothing but illusions'.

The reality of such changes was indeed debatable, but the complete disappearance of a nebula was not. In 1852 John R. Hind (1823–95), astronomer at George Bishop's private observatory in Regent's Park, London, reported the discovery of a small nebula in Taurus. The nebula was observed several times in the following years; but in October 1861 Heinrich Louis d'Arrest (1822–75), co-discoverer of Neptune (see page 192) and now of the University of Copenhagen, who was making a careful study of the positions and appearances of nebulae, could find no trace of it whatever. News spread of the vanishing of what would become known as 'Hind's wonderful nebula'. But by the end of the year, d'Arrest and Otto Struve (the son of Wilhelm) had both seen it again.

This was an example at last of change of unimpeachable authenticity. Other such claims began to be taken more seriously, and a new hesitation crept into accounts equating nebulae with star clusters. But the decisive proof, in 1864, that true nebulae exist – that the light of certain nebulae originates in gas rather than in stars – was to come, not from traditional methods of observation, but through the use of a technique of laboratory physics that was opening a new chapter in the history of astronomy.

CHAPTER 8

The Message of Starlight: The Rise of Astrophysics

From Antiquity to the Renaissance, the principal task of the astronomer had been to devise geometrical models that would reproduce the movements of the planets against the unchanging background of the 'fixed' stars. In the seventeenth century, Kepler taught astronomers to investigate the forces causing these movements; and in the aftermath of the publication of Newton's *Principia* in 1687, the agenda of 'celestial mechanics' was to demonstrate that these and all the other movements within the solar system had a single physical cause – the pulls resulting from the inverse-square law of gravitational attraction. By contrast, even as late as the mid-nineteenth century, the astronomy of the stars still consisted mainly in the patient compilation of ever more accurate and comprehensive catalogues of positions and magnitudes, a humdrum task that had been begun by Hipparchus and Ptolemy in Antiquity.

All this was to change in the second half of the century, when prisms were fitted to the eyepieces of telescopes and the resulting rainbow-like spectra analysed for the information they contained. The astronomer of the Middle Ages had been a geometer, the astronomer of the eighteenth century a celestial mechanician; now, in the late nineteenth century, the astronomer had to master the skills of the laboratory physicists specializing in the study of spectra. The theories and techniques used by these spectroscopists were adapted to the study of starlight, and this enabled the 'astrophysicist' to penetrate the secrets of the physical processes at work among the stars, and to determine their motions in three dimensional space. The stars, and the mysterious nebulae, now became the focus of interest, and even the fringe discipline of cosmology was assimilated into mainstream astronomy.

PUBLISHING ASTRONOMY, ORGANIZING ASTRONOMERS, 1665–1950

A comparable revolution was taking place in the organization of astronomy. The invention of printing in the mid-fifteenth century had made possible the rapid publication of broadsheets and pamphlets expounding the latest astronomical wonder; but major advances in knowledge continued to be published in books, and books usually took years to write and to see through the press.

The pace of scientific advance accelerated dramatically with the publication from 1665 of the *Philosophical Transactions* of the Royal Society of London. This grew out of the international correspondence conducted by the Secretary of the Society, Henry Oldenberg, but was soon taken over by the Society as its official monthly publication.

The discovery that lies at the heart of spectroscopy provides a good illustration of the speed with which scientific news could now be disseminated. Early in 1672

Newton was persuaded to share with the readers of *Philosophical Transactions* his discovery (see page 261) that sunlight is not simple but a combination of many colours. In no time at all, a surprised and outraged Newton found himself under widespread attack, from critics at home and overseas.

Publication of the first year-book dedicated to astronomy, *Connoissance des Temps*, began in Paris in 1679. It primary purpose was to publish tables of forth-coming solar, lunar, and planetary positions, and other tabular material of interest to observers and navigators, but it soon began to include topical articles on related subjects. Before long, volumes of more permanent interest started to appear else-where, as the public observatories established in the late seventeenth and eight-eenth centuries came to see it as their duty to disseminate, in published form, the observations that were their *raison d'être*. By the nineteenth century it was normal for an observatory to present copies of its publications to other observatories of the appropriate standing, and to receive their publications in exchange.

A monthly journal, *Monatliche Correspondenz*, designed for the rapid dissemi-nation of scientific news in general and astronomy in particular, was founded in 1800 by Franz Xaver von Zach (1754–1832), astronomer to the Duke of Saxe-Coburg. The first rapid-publication journal dedicated exclusively to astronomy was *Astronomische Nachrichten*, begun in 1823 by the German astronomer and surveyor, Heinrich Christian Schumacher (1780–1850). Schumacher has been called 'The Postmaster General of Astronomy', who 'wrote to everybody and sent copies of everybody's letters to everybody else'. To ease the burden of his corres-pondence, he had it set in type by a local printer and circulated it to an international mailing list every few weeks. The journal, with the same title, has continued publication to this day, though telegrams and, more recently, electronic mail have taken over its most urgent duties.

The Royal Society of London had begun in 1660 as an amateur society catering for scientific (and quasi-scientific) interests of all kinds, and remained predomi-nantly amateur until the mid-nineteenth century. The social status of a member, or 'Fellow', and his ability to pay the contributions, were as important as his knowledge of science. By the early years of the nineteenth century, the number of committed astronomers in England – academics such as the professors at Oxford and Cambridge, professional observers like the Astronomer Royal, dedicated amateurs like the Herschels – had grown to the point where there was a need for the type of forum that only a specifically astronomical society could provide. In the teeth of opposition from the then President of the Royal Society, who was fearful of the competition, the Astronomical Society of London was founded in 1820, receiving its royal charter and so becoming the Royal Astronomical Society eleven years later. The RAS held regular meetings at which the latest discoveries were presented and discussed, its premises housed a research library, and outstanding achievements were recognized by the award of the Society's medals. The Society began publishing its *Memoirs* within months of its foundation, to provide a vehicle

The title-page of the first volume of *Connoissance des Temps*, for 1679. Ephemerides (tables of positions of heavenly bodies for a given period) had long been an essential tool for astronomers and astrolo-gers alike. Staff at the Paris Observatory decided to under-take the publication of annual volumes of ephemerides and related materials – but with no astrology; this first volume had been 'purged of all the ridiculous things with which works of this sort have been filled till now'.

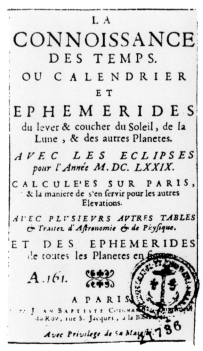

Right: The instrument installed at Sydney Observatory in 1890 for work on the *Carte du Ciel*. To ensure uniformity, the photographic refractors for the project followed closely the design developed by Paul and Prosper Henry for Paris Observatory.

Below: A small part of one of the many thousands of charts that constitute the *Carte du Ciel*, this one from the San Fernando Observatory. The accurate 'net' of 1-centimetre squares was imposed on each photograph to assist measurement.

CARTA FOTOGRAFICA DEL CIELO

Posicion del centro para 1900 $\begin{cases} \text{Æ} = 7^h 24^m \\ \text{D} = - 5° \end{cases}$

Zona - 5°N°56

for the dissemination of tables of observational results and material of similar calibre. There was also a need for a forum for the rapid publication of astronomical news, lesser research notes and the like, and for this the Society instituted its *Monthly Notices* in 1827.

Similar societies, both national and local, began to spring up throughout Europe and in the New World. In America the *Astronomical Journal* was founded in 1849 by Benjamin Apthorp Gould (1824–96), who had taken his doctorate in astronomy the previous year under C. F. Gauss at Göttingen and who had then returned home with the express intention of promoting astronomy in the USA. As the *Astronomical Journal* was not the organ of an established society, it was vulnerable to outside events, and publication was interrupted by the American Civil War; but Gould himself eventually revived it, after a gap of a quarter of a century.

On the continent of Europe, the Astronomische Gesellschaft was founded at Heidelberg in 1863. Although primarily German, its membership was explicitly open to other nationals, and the society took a lead in projects that were international in character. From 1881 it was involved in the publication of *Astronomische Nachrichten*, and in 1898 it helped in the foundation of *Astronomischer Jahrsbericht*, which aimed to abstract every publication in astronomy. This invaluable aid to research (known as *Astronomy and Astrophysics Abstracts* since 1969, when the German publishers bowed to the dominance of the English language among astronomers) now publishes over 6,000 pages each year.

The corps of twenty-four 'celestial police' organized at Lilienthal in 1800 to search for a planet in the Mars–Jupiter gap (see page 188) was perhaps the first true example of international co-operation in astronomy. The second half of the nineteenth century was to see several more schemes to achieve results that were beyond the resources of any one institution. Between 1859 and 1863, F. W. A. Argelander (1799–1875) of Bonn published a three-volume catalogue of nearly one-third of a million stars, known as the *Bonner Durchmusterung*, and a forty-plate atlas. It was the work of a tiny group of Bonn observers, and although the *Durchmusterung* ('Survey') was to prove invaluable to future observers, the listed positions were necessarily imprecise. In 1867, therefore, Argelander proposed to the Astronomische Gesellschaft that a project be organized to measure – this time with great accuracy – the positions of the *Durchmusterung* stars down to the ninth magnitude. The task, it was agreed, would be shared among seventeen observatories, all operating to a common plan.

Work, however, proceeded slowly. Not for the last time did some contributing institutions fail to honour their undertakings; and the project itself was to some extent overtaken by a new development, the application to astronomy of photography (see page 288). In 1885 this led the director of the Paris Observatory, Admiral E. B. Mouchez (1821–92), to suggest the possibility of a great photographic star chart. First responses from foreign colleagues were encouraging, and so the Paris Academy of Sciences issued invitations to a meeting in Paris in

George Ellery Hale

Hale was born in Chicago on 29 June 1868, the son of a prosperous businessman. He studied at the Massachusetts Institute of Technology, and before he finished his studies he had devised a 'spectroheliograph' for photographing the solar prominences in full daylight. In 1892 he was appointed associate professor of astrophysics at the University of Chicago. There he founded *Astrophysical Journal*, which today is by common consent the leading journal in the field, and persuaded the wealthy Charles T. Yerkes to endow the Yerkes Observatory. Next, in 1904 he secured an endowment from the Carnegie Institute of Washington for a solar observatory on Mount Wilson, near Pasadena in California. Four years later, the biggest reflector in the world was inaugurated on Mount Wilson using a 60-inch disc provided by Hale's father. (Andrew Carnegie – with cane – appears left at Mount Wilson with Hale.) In 1917 this was followed by the 100-inch, paid for by a California businessman, John D. Hooker.

Before long Hale was campaigning for a 200-inch, to be built on Palomar Mountain, funded in part by the Rockefeller Foundation, though Hale had died before it was completed in 1948. Hale was thus the presiding genius over the construction of three successive reflectors, each in turn the largest in the world.

Hale took a leading role in the foundation of what later became the American Astronomical Society, and of the International Union for Co-operation in Solar Research, which was later subsumed into the International Astronomical Union. Meanwhile he pursued his own fundamental researches into the physics of the Sun. Despite precarious health, Hale arguably did more for the promotion of astronomy than any other person in history. He died in Pasadena, California, on 21 February 1938.

April 1887. Fifty-six scientists from nineteen nations attended, and a permanent commission was set up to promote a *Carte du Ciel* – a photographic map of the sky (of stars to the fourteenth magnitude) – and, from the measurement of the photographs, a precision catalogue of stars to the eleventh magnitude. In order to secure the necessary uniformity, many decisions were required, some of which called for intensive research; and until the First World War the Permanent International Committee of the *Carte du Ciel* occupied a unique position in the international astronomical community. The work itself, however, ran into difficulties, as some observatories dropped out and others produced sub-standard results, and publication of the catalogue was not completed until 1964.

The International Astronomical Union

The IAU, founded in 1919, is organized into numerous 'commissions', each dedicated to a field of astronomy and comprising the IAU members working in that field. These commissions encourage collaboration in research, and adjudicate on standards and terminology. For example, the constellations, which formerly merged vaguely into each other, were given precisely defined boundaries in 1930. Every three years there is a General Assembly of the IAU at which astronomers of almost every nation meet to discuss the latest research in their field, and as a complement to these large gatherings, the IAU has developed the sponsorship of frequent specialist symposia and colloquia.

The IAU can speak for astronomy on the international scene, defending the interests of the science against (for example) excessive pollution of the night skies by city lights, or the cluttering of the outer atmosphere of the Earth by unnecessary debris that interferes with research. After a beginning marred by the exclusions of astronomers from the defeated nations in the First World War, and despite numerous problems in securing the necessary finance from the adhering countries, the IAU is the focus of the organization of astronomy worldwide.

In 1904 a major advance in international co-operation came about with the formation, at the instigation of the American, George Ellery Hale (1868–1938) and with the active support of the US Academy of Sciences, of the International Union for Co-operation in Solar Research. Experience of the benefits resulting from the work of this organization and of the *Carte du Ciel* Committee in their respective fields made astronomers receptive to the proposals for organized international co-operation, advanced by scientists of many disciplines in the aftermath of the First World War. In 1919 the International Astronomical Union was formed, and although at first the defeated nations were excluded – Germany did not become a member until 1952 – the IAU has come to play a central role in the organization of astronomy.

THE SUN AND THE ORIGINS OF ASTROPHYSICS, 1800–90

Until the publication in 1672 of Isaac Newton's 'New Theory about Light and Colors', it was thought that white light was simple and basic, and that colours were 'modifications' of white light – that something happened to white light to turn it into red light, or into blue light.

There were various theories as to how this happened. Curious to test them, Newton bought a prism, retired to his room in Trinity College, and set up an experiment. He closed the shutters, except for a circular hole; the sunlight that entered the hole passed through the prism, and the resulting spectrum fell on a screen that Newton had placed at a considerable distance.

There was nothing novel about the rainbow-like colours of a spectrum of sunlight; they were, as Newton put it, 'celebrated'. It was the shape of Newton's spectrum that drew his attention, for it was incompatible with the usual theory of

Top: Newton's diagram showing how he combined the colours to make white light once more. Light from the Sun (left) enters the hole *F* and is separated into colours by the prism. The lens brings the colours together again. They fall on the screen *HI*, and if this is at *Q* they produce 'whiteness'.

Bottom: Part of a diagram of the solar spectrum (Fig. 5), published by Fraunhofer in 1817. He carefully mapped the positions of over 300 lines, and put the rest in by eye. Above it appears a curve indicating the brightness of the spectrum as judged by the eye (Fig. 6), peaking in the yellow-green.

modifications. After a series of further experiments, Newton concluded that white light, far from being simple, was in fact compounded from the colours; and he went on to demonstrate that his spectrum could be reassembled into white light by the use of a lens.

In 1802 the English chemist William Hyde Wollaston (1766–1828) repeated some of Newton's work but in a more refined way: for Newton's round hole he substituted a narrow slit only one-twentieth of an inch wide. The spectrum of the light, now freed from the overlapping colours, showed him some seven lines that were more or less dark and which he regarded as natural boundaries between the colours.

That this simple explanation was inadequate became evident when the Bavarian instrument-maker, Joseph Fraunhofer (1787–1826), in trying to improve the design of lenses, developed the first simple 'spectroscope'. Sunlight was allowed to enter through an extremely narrow slit; it then passed through a prism, after which the spectrum was examined by Fraunhofer with the aid of a telescope. He was astonished to see, not Wollaston's seven lines, but many hundreds. He counted some 600 of them and drew a map, giving to the darkest and most prominent of them letters of the alphabet starting with A, B, C... from the red end – letters that are still convenient to use today. The spectrum of the Sun came to be known as 'the Fraunhofer spectrum'.

Fraunhofer was a practical optician, not a 'scientist'. He noted with interest that the D line in the yellow seemed to coincide exactly with a bright line that appeared in the light of many flames. That there was such a ubiquitous yellow light of very pure colour had been observed in the middle of the eighteenth century by a remarkable young Scotsman, Thomas Melvill, who died in 1753 at the early age of twenty-seven. Although Melvill's discovery was published, it appeared in an obscure collection of essays and went almost unnoticed. The discoveries of Fraunhofer, by contrast, aroused widespread interest.

In the investigations that resulted, the crucial experiments were carried out by the Heidelberg chemist Robert Bunsen (1811–99) and his colleague, the physicist Gustav Kirchhoff (1824–87). The first of their discoveries, announced in 1859, was that particular sets of lines are associated with individual chemical elements. William Miller (1817–80) of King's College London had already noted that the light emitted by electric arcs struck between two metal rods differed from metal to metal. By making many measurements in the laboratory, Bunsen and Kirchhoff identified lines with metals, and dramatically confirmed their results by identifying two new elements – caesium (a Latin word that means 'bluish grey') and rubidium (red) – with lines in those parts of the spectrum. A perplexing problem for the many investigators had been that even when quite pure specimens of metals were introduced into the flames, Fraunhofer's D line appeared in the spectrum. It only gradually became apparent – and the discovery was itself an important one – that the spectrum test was of great sensitivity, and that even the most minute traces of sodium, present for example in common salt, would produce that line. The great

An illustrative plate from Lockyer's *Contributions to Solar Physics* (1874), showing the essential similarity of laboratory and astronomical spectra. At the top (1) is the continuous spectrum of an incandescent solid, next (2) the spectrum of the sodium flame, and at the bottom (10) the spectrum of (1) viewed through the sodium flame. (3) to (6) are various emission line spectra of elements in the laboratory, (7) that of a nebula, and (8) that of the solar chromosphere. (9) is the Fraunhofer spectrum of the Sun.

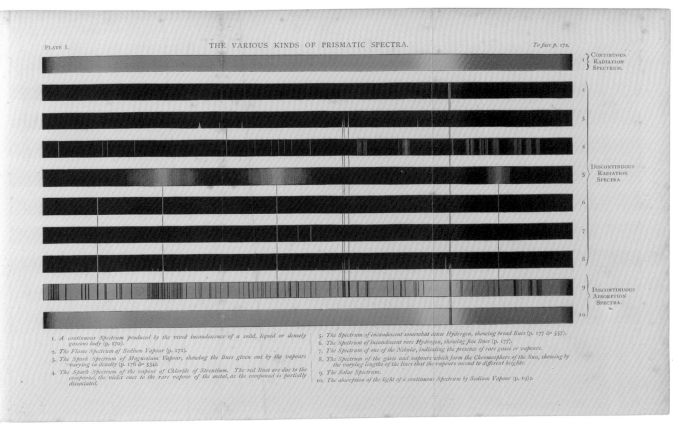

PLATE I.   THE VARIOUS KINDS OF PRISMATIC SPECTRA.   *To face p. 172.*

1 } CONTINUOUS RADIATION SPECTRUM.

2
3
4
5 } DISCONTINUOUS RADIATION SPECTRA.
6
7
8

9 } DISCONTINUOUS ABSORPTION SPECTRA.

10

1. *A continuous Spectrum produced by the vivid incandescence of a solid, liquid or densely gaseous body* (p. 170).
2. *The Flame Spectrum of Sodium Vapour* (p. 172).
3. *The Spark Spectrum of Magnesium Vapour, showing the lines given out by the vapours varying in density* (p. 176 & 534).
4. *The Spark Spectrum of the vapour of Chloride of Strontium. The red lines are due to the compound, the violet ones to the rare vapour of the metal, as the compound is partially dissociated.*
5. *The Spectrum of incandescent somewhat dense Hydrogen, showing broad lines* (p. 177 & 557).
6. *The Spectrum of incandescent rare Hydrogen, showing fine lines* (p. 177).
7. *The Spectrum of one of the Nebulae, indicating the presence of rare gases or vapours.*
8. *The Spectrum of the gases and vapours which form the Chromosphere of the Sun, showing by the varying lengths of the lines that the vapours ascend to different heights.*
9. *The Solar Spectrum.*
10. *The absorption of the light of a continuous Spectrum by Sodium Vapour* (p. 192).

purity of the metals and compounds coming from Bunsen's chemical laboratory was an essential feature of this work.

A further experiment at Heidelberg, carried out in 1859 by Kirchhoff in the physics laboratory, led to an understanding of how the lines were produced. He looked at a spectrum of the Sun through a yellow sodium flame, expecting the bright light of the flame to mask the dark line in the Sun: instead, it became even darker. A similar experiment, substituting an intense white incandescent lamp for the Sun as a light source, also showed a dark line. He inferred, and his interpretation was generally accepted, that Fraunhofer's dark D line arose because sodium vapour existed in a glowing atmosphere surrounding the Sun and absorbed light of that particular wavelength. Such lines became known as 'absorption lines'.

It followed that spectrum analysis could be applied to the Sun: the lines that were bright in the flames and arcs in the laboratory (that is, 'emission lines') were the same lines that appeared dark in the spectrum of the Sun – and Fraunhofer himself had been able to see a few of the most prominent lines, including the D line, in the spectrum of the very brightest stars. Earlier in the nineteenth century the French philosopher Auguste Comte had cited the chemical composition of the stars as an example of things that were inherently unknowable. Now it *was* knowable: several metals had been identified in the Sun, and within a few years the list was to be greatly extended. In particular, in 1862 A. J. Ångstrom (1814–74), professor of physics at Uppsala in Sweden, succeeded in showing that the naturally gaseous element hydrogen was also present in the solar atmosphere. By the end of the 1880s some fifty of the known elements had been identified in the solar spectrum. This assimilation, of spectroscopy in the laboratory and spectroscopy of the Sun, was one of the great leaps in the history of astrophysics.

Heinrich Schwabe and the Discovery of the Sunspot Cycle, 1843
We have seen in Chapter 5 that spots on the Sun (observed, rather dangerously, on the heavily reddened disc at sunset) were among the first objects studied with the newly invented telescope, and surprisingly early in the seventeenth century a number of basic facts about sunspots became established. They were not objects in interplanetary space but attached to the Sun in some way – being either dark clouds in the Sun's atmosphere, or rafts of cinders floating on its incandescent surface, or perhaps the dark surface of the Sun seen through gaps in a luminous atmosphere that was the source of its light. The Sun turned about an axis fixed in direction in space, with a rotation period variously estimated to be within a day or two of twenty-six days, and the spots rarely appeared more than 30° north or south of the Sun's equator.

But sunspots were sporadic phenomena from which little more could be learned; indeed already by 1618 they seemed to have become infrequent. Thereafter, except for occasional renewed interest during the eighteenth century when particularly large or numerous spots appeared, little systematic study of them was made.

A change was brought about by Heinrich Schwabe (1789–1875) of Dessau, a small town to the southwest of Berlin. An apothecary interested in astronomy, he bought a small telescope from Munich in 1826, and with advice from his friend K. L. Harding of the Göttingen Observatory began regular observation of the Sun whenever weather and time permitted. His interest was to look for a possible undiscovered planet between Mercury and the Sun. Eventually such a body must surely pass over the disc of the Sun as a small black dot, but only constant observation would trap it, and only regular recording would distinguish it from a small sunspot.

With commendable patience Schwabe observed and recorded. After seventeen years, he had still discovered no new planet, but he was able to announce that sunspots seemed to come and go in a fairly systematic way, the disc being more covered with spots at about ten-year intervals, with corresponding minima in between. His modest note in *Astronomische Nachrichten* in 1843 attracted relatively little attention, but he continued collecting observations, and in 1851 the famous German traveller and natural historian, Alexander von Humboldt (1769–1859), drew attention to their significance in his *Kosmos*. The amateur of Dessau was awarded the Gold Medal of the Royal Astronomical Society, then as now an accolade in the astronomical world. In the long term the average length of the sunspot cycle is in fact a little over eleven years, but the interval is not exact and was nearer to ten years when Schwabe made his early observations.

Sunspots and the Earth's Magnetic Field

To understand the importance Humboldt attached to Schwabe's discovery we must retrace our steps to the time when Schwabe was starting his observations, and to natural events that at first seemed to have nothing to do with astronomy at all. The north-seeking property of the needle of the magnetic compass had been used by seamen for centuries. But the needle points to a magnetic pole that is not the same as the north geographic pole, and furthermore the magnetic pole moves slowly over the Earth's surface. Nor are these the only complications: as early as 1722 it was known that there is a smaller but daily variation.

In 1828 Humboldt set in train a great international study of terrestrial magnetism, based at the Observatory in Göttingen University under the direction of C. F. Gauss, and within a few years a worldwide net of magnetic observatories was in operation. By 1851 John Lamont (1805–79), the Scots director of the Munich Observatory, discovered in the magnetic observations between 1835 and 1850 distinct evidence for a variation of the magnetic constants with a period of about $10^1/3$ years. Quite independently, and using different data, Sir Edward Sabine (1788–1883), a British army officer and future President of the Royal Society, found that the occasional sudden and more violent vibrations of the needle (which Humboldt had called magnetic storms) were more violent and more frequent at intervals of about ten years.

Schwabe's drawing of the solar disc on 3 December 1826, one of the many meticulous drawings, covering forty-two years of regular observing, that he later presented to the Royal Astronomical Society. Schwabe became interested in astronomy while studying pharmacy at Berlin University, and in 1829 he sold the family business in order to devote his life to scientific studies.

It was Sabine who announced in 1852 that between the sunspot cycle and these magnetic activities there was a remarkable coincidence, which he had found by patient deduction from many hundreds of thousands of observations collected for more than twenty years. His discovery elevated the study of the Sun from curiosity to practical utility, for if the Sun could influence a magnetic needle, what else might it not influence? The weather? The ability to predict famines from crop failure in India?

The existence of solar–terrestrial relationships was dramatically confirmed by two independent observations of the Sun made by the English amateur astronomers R. C. Carrington (1826–75) and Richard Hodgson (1804–72) one morning in September 1859. An outburst of two bright patches of white light near a large spot group, lasting for five minutes, was so remarkable that both noted its appearance carefully. Simultaneously there was a violent magnetic storm, telegraphic communication was interrupted, and that night one of the most magnificent auroral displays ever observed danced across the skies.

With the spectroscopic discoveries of the same year, solar physics had arrived. But the complex causes and effects – the emission of X-ray and extreme ultraviolet radiation from the flares that Carrington and Hodgson had been the first to notice, and the emission of streams of charged fundamental particles into space – were not to be unravelled for nearly a century.

Expeditions to Observe Eclipses of the Sun

The explanation that Kirchhoff had given of the dark absorption lines in the solar spectrum required the existence of some atmosphere, presumably faint and tenuous, of hot gases surrounding the Sun. The knowledge of physics at the time, however, made it very difficult to understand the nature of the Sun's surface. In the laboratory, the continuous spectrum of white light was produced only by white-hot

The English eclipse camp in Ceylon, 1871. Lockyer is at the central reflecting telescope, while on the left the upper telescope is fitted with a spectroscope. On the right, the time keeper calls the time at intervals during totality, to guide the observers through the carefully planned observing programmes, so that not a moment of the brief time available (in this eclipse, 120 seconds) should be wasted.

solid bodies or by liquids like molten metals, and the white light of the Sun presumably came from such a surface. Yet many of the things seen on the Sun's disc in white light – sunspots, the rather brighter areas surrounding them named faculae ('little torches'), and the fine details seen within the sunspots themselves – seemed to be of an atmospheric nature.

The resolution of these difficulties was to come from observations made during total eclipses of the Sun. By a remarkable coincidence in nature, the Moon and the Sun – the one very much smaller than the Earth, and the other very much larger – are at such distances from the Earth that their apparent sizes in the sky are very nearly the same. Because the orbits of the Moon (about the Earth) and of the Earth (about the Sun) are both elliptical, these apparent sizes change with time, and under favourable circumstances the Moon's disc is sufficiently larger than that of the Sun to eclipse the Sun for as much as seven minutes or so. As a total eclipse approaches the sky is darkened, and, at the onset of totality, the Moon slowly obscures the outer layers of the Sun seen at the edge of its disc, only to reveal them again progressively at the end of totality. This most beautiful and awe-inspiring event is rare at any one place on the Earth's surface, but every few years the narrow zone traversed by the Moon's shadow crosses some reasonably accessible part of the Earth.

By accidents of history, the latter half of the nineteenth century saw a number of favourable eclipses, and travel was then becoming easier and cheaper. Accordingly, it became usual for astronomers to set up temporary observatories, with increasingly elaborate instruments, along the line of the eclipse track, to study in scientific detail what had previously been regarded as little more than a spectacle.

The eclipse of 8 July 1842 brought astronomers from several European countries to sites in central and southern Europe. Most of them had never seen a total eclipse

Above left: The photograph of the solar corona during the eclipse of 7 August 1869, obtained by the Harvard College expedition to Shelbyville, Kentucky. This was one of the best images of the corona obtained to that time, but it over-exposed the lower region of the prominences.

Above right: The prominences in the same eclipse, shown hand-coloured in a depiction published by Angelo Secchi in 1875.

Norman Lockyer

The medal struck by the French mint to commemorate the independent discovery by Jules Janssen and Norman Lockyer of a method of generating artificial eclipses of the Sun. This allowed regular study of the chromosphere and prominences.

Joseph Norman Lockyer was born in Rugby, England, on 17 May 1836, the son of a surgeon-apothecary. In 1857 he became a clerk in the War Office, and began to devote his spare time to spectroscopy. Eventually his rise to scientific prominence despite his amateur status led to government appointment in 1885 as director of the Solar Physics Observatory established in South Kensington, a suburb of London. He was indefatigable in soliciting support for this highly innovative institution, but in 1911 and to his disappointment it was transferred to Cambridge. Lockyer then retired to Devon, where he built himself a private observatory.

Lockyer made fundamental contributions to many branches of astrophysics, and his other exploits included the founding in 1869 of the journal *Nature*, which he edited for fifty years. He was knighted in 1897, and died on 16 August 1920 in Devon .

before. They were all astonished at the extent and brightness of the white 'corona' surrounding the eclipsed Sun, and noted with even more surprise three small protuberances (later to be termed 'prominences') looking like mountains projecting from the edge of the Sun or the Moon, and of a striking colour generally described as red tinged with lilac.

Later favourable eclipses were to follow in Sweden and Norway in 1851, North America and Spain in 1860, and India and Malaya in 1868, to mention only the earlier and more significant ones.

Each eclipse brought important discoveries. It gradually emerged that all the significant phenomena arose in the various layers of the Sun's atmosphere (and none, as was once suspected, in a supposed atmosphere of the Moon). The red flames were seen to be sporadic extensions of a lower layer which surrounded the whole Sun, named the 'chromosphere' in 1868 by the English amateur (as he then was), Norman Lockyer (1836–1920); and the spectroscope showed that its lilac-red colour arose from several bright emission lines, especially those of hydrogen in the red (Fraunhofer's C line) and the blue. At the 1870 eclipse

Jules Janssen

Pierre Jules César Janssen was born in Paris on 22 February 1824. Lameness as the result of an accident in childhood prevented his receiving a conventional schooling, and it was 1852 before he completed studies in the University of Paris. Five years later he went to Peru on a scientific expedition, the first of many he was to undertake.

In 1859 Gustav Kirchhoff explained how the chemical composition of the Sun was revealed by its spectrum, and this persuaded Janssen to dedicate himself to astrophysics. In 1867 he announced the presence of water vapour on Mars. The following year he went to India to observe the solar eclipse, and while there he realized how a spectroscope could be devised to allow the solar prominences to be studied outside eclipses.

In 1876 the French government established an observatory for physical astronomy for Janssen at Meudon, his preferred site. The most famous of his projects there was an atlas of solar photographs that summarized the history of the Sun's surface between 1876 and 1903. To study solar rays at higher altitudes, he regularly made observations from Mont Blanc. He continued to direct the Meudon observatory until his death there on 23 December 1907.

the American C. A. Young (1834–1908) of Dartmouth College made the crucial spectroscopic discovery of an even lower and narrower 'reversing layer': in the last few seconds as the Sun's edge disappeared, the dark lines and the continuous spectrum also disappeared, and for the next two seconds hundreds of bright lines flashed out, apparently in the places in the spectrum where the dark Fraunhofer lines had been.

Such observations gradually built up a quite complicated picture of the structure of the Sun's outer layers, and in the laboratory the discovery that not only solids and liquids but also gases under high pressure could emit a continuous spectrum suggested that all the observed light from the Sun could come from gaseous outer layers. But puzzles still remained. A curious one was that with good telescopes under good observing conditions, the edge of the Sun at high magnification is always seen to be perfectly sharp – one might have expected it to fade out rather fuzzily into space. The true explanation would depend on a quite detailed understanding of the physical processes that occur in the outer layers of stars generally, and this was not arrived at until well into the twentieth century: briefly, the transition from the gaseous layer that is 'foggy' to the layer above that is clear is so abrupt that the edge of the Sun looks solid – rather as does the edge of a distant sunlit thunder cloud.

At the 1868 eclipse the French physics professor Jules Janssen (1824–1907) had been so impressed by the brilliance of the bright lines in the spectrum of the prominences that he was struck by the idea that, by use of a spectroscope of high dispersing power (which would spread out the continuous spectrum of the surface layer known as the 'photosphere' but not the monochromatic bright lines), it should be possible to see them without an eclipse at all. The same idea had also

Above: A 1907 photographic montage showing a spectro-heliogram of the solar disc combined with one of the prominences at the edge of the disc.

Right: Types of solar prominences, drawn using the artificial eclipse method of Lockyer and Janssen. This is the work of Etienne Trouvelot (1827–95), a French natural historian and astronomer who was at the Harvard College Observatory for some years. He was one of the most skilled scientific artists of the pre-photographic era.

occurred, independently, to Lockyer. Tests showed the idea to be well founded, and from then on it was possible to study the chromosphere and prominences from day to day. Soon such work was under way at a number of solar physics observatories.

The resulting ability to make accurate measurements at leisure, rather than in the hectic minutes of an eclipse, led to the discovery that the brilliant orange line which had been assumed to be the D line of sodium had in fact a rather shorter wavelength and could not be identified in any laboratory spectrum: it was attributed to an unknown element, which was given the name 'helium', from the Greek word for the Sun. The element was eventually isolated in the laboratory in 1895 by the Scots chemist William Ramsay (1852–1916), in radioactive minerals where it is formed as a decay product.

Similar unidentified bright lines in the spectrum of the corona were to prove less tractable: they were attributed to an unidentified 'coronium', but did not yield to explanation until 1941. It was then that the German astronomer Walter Grotrian (1890–1954) showed by theory that they are in fact due to iron atoms, but in extremes of high temperature and low density that are not reproducible in the laboratory. Indeed the nature of the corona remained an enigma for a similar length of time: its light is now known to be sunlight scattered partly by a true atmosphere of free electrons round the Sun, and partly by meteoritic dust particles in the space between the Sun and the Earth.

Photography enters astronomy

Most of the improved understanding about the nature of the Sun came about by the gradual addition of photography to the tools of research available to astronomers. Indeed the very brightness of the Sun made it an attractive subject for the very slow and inefficient early photographic processes. The first daguerreotype of the solar disc was made by the French physicists J.B.L. Foucault (1819–68) and A.-H.-L. Fizeau (1819–96) in Paris in 1845. The faster collodion process, invented by the English sculptor F.S. Archer (1813–57) in 1850, was used by his fellow-countryman, the photographer and amateur astronomer Warren De la Rue (1815–89), to photograph the sunspots on the solar disc with a specially constructed camera at the Kew Observatory, near London. This he did in 1858, only a few years after Humboldt's publicity of Schwabe's discovery of the sunspot cycle. The 'autographic' recording of the daily state of the solar disc started there almost immediately, and was continued until 1872, when it was further extended by large programmes of almost continuous recording on a worldwide basis, not only of the spots on the photosphere but also of the outer chromosphere over the disc and at the edge of it.

The greatly improved dry gelatine plate became available in the late 1870s. It was sensitive enough to record the flash spectrum of the Sun (which lasted only a few seconds during a total eclipse), so enabling the lines to be measured later in the laboratory. More importantly, these plates were at last sensitive enough for them to be applied to telescopes at night, to photograph the very much fainter stars and nebulae (see page 288).

FIRST PHOTOGRAPH OF THE NEBULA IN ORION.
TAKEN BY PROFESSOR HENRY DRAPER, M.D.
September 30th, 1880, Exposure 51 minutes.
The picture is an Artotype enlargement by Harroun & Bierstadt from the original negative. The large stars, being much brighter than the Nebula, are greatly overexposed. In the lower right hand corner is a photograph of the Trapezium alone,

Top: The first photograph of the Orion Nebula, taken by Henry Draper in 1880.

Bottom: The nebula as photographed by A. A. Common only three years later, showing the strides in technique.

The Source of the Energy of the Sun

Despite the improvement in knowledge of the workings of the outer layers of the Sun, a more fundamental question remained only partly answered: what was the source of the energy that the Sun poured out so prodigiously in its heat and light? By the middle of the nineteenth century, partly as a result of the studies of physicists and partly from the interest shown by engineers in the efficiency of steam engines, there was a fairly good understanding of the flow of heat and how to measure it. The German physician, Julius R. Mayer (1814–78) of Heilbronn, asked himself such questions about the Sun, and he calculated that even supposing the Sun to be made of coal and supplied with unlimited oxygen to burn it, the solar furnace could last for only a few thousands of years. He suggested another source of energy: the continuing infall onto the Sun of solid particles from interplanetary space, which would give up their energy on colliding with it.

The Irish-born physicist William Thomson (1824–1907), later Lord Kelvin, satisfied himself that this also was untenable, for even if sufficient material were available it would so alter the mass of the Sun that its greater gravitational pull would shorten the length of the year by weeks within a few thousand years, which clearly was not happening. An uneasy solution, of a more tenable nature, was worked out by the German physicist Hermann von Helmholtz (1821–94) in 1854 and was also studied by Thomson: the mass of the Sun was not changing, but the star was slowly contracting; the resulting change in the size of the Sun would be too slow to be measurable, and the output of energy could be maintained at least for several millions of years. But even this timescale was inadequate to meet the requirements of the geological record, and the problem remained unsolved until the discovery of sources of nuclear energy in the twentieth century.

Pulkovo Observatory at the end of the nineteenth century. The large central dome housed the 15-inch refractor by Merz & Mahler of Munich, while the meridian transit instruments by Ertel of Munich and Repsold of Hamburg are in the immediately adjacent wings. These were only some of the high-quality instruments installed at the foundation of the observatory in 1839 (see page 216). At the extreme right is the housing for the great 30-inch refractor.

DEVELOPMENTS IN TELESCOPE MAKING

The principal business of public – that is to say, state and university – observatories in the early nineteenth century continued to be the determination of the exact positions of the stars in the sky, to provide a framework against which the apparent motions of the Sun, Moon, and planets could be measured. The instruments used (see page 171) were 'transit circles', refracting telescopes of modest size mounted on massive masonry piers and fitted with accurately graduated circles to make the necessary angular measurements. But the interests of amateur astronomers like William and John Herschel, Lord Rosse, and William Lassell (1799–1880) called for larger telescopes than the professional telescope-makers could supply, and, as we have seen, they developed the reflecting telescope, using mirrors of speculum metal (an alloy of copper and tin) shaped in their own workshops. These were, however, temperamental and difficult instruments: unlike glass lenses, the mirrors tarnished easily, and to maintain them the astronomer needed to be skilled in the grinding and polishing of optical surfaces.

Fortunately a solution offered itself. Under pressure from the opticians and seeing some profitable new business, glass manufacturers (particularly in Switzerland and later in France) began to experiment with the fabrication of larger discs of good quality optical glass for telescopes, and the opticians in turn developed improved ways of figuring the perfectly spherical surfaces of the lenses. As we have seen (on page 216), before Fraunhofer's death in 1826 his workshops had completed for Dorpat Observatory a very fine refractor of 24 centimetres (9 1/2 inches) aperture. From this time onwards, many of the growing number of newly constructed observatories in Europe, and soon in America, began to include in their shopping list of desirable instruments as large a refracting telescope as their endowment or the generosity of their benefactors could afford. Sizes crept up a few inches at a time, from the 15-inch refractor by Merz & Mahler

(Fraunhofer's successors) for the Pulkovo Observatory near St Petersburg in Russia and a twin for the recently founded Harvard College Observatory in 1847, to a culmination in the 40-inch diameter lens for the telescope of the Yerkes Observatory of the University of Chicago in 1897. Many of these great telescopes still exist, now no longer suited to most of the needs of modern astronomy but preserved as superb examples of the instrument-maker's craft.

These great refractors were not to be without their problems when photography began to enter into stellar and planetary astronomy and into solar physics, with the rapid development of the dry plate process from the 1870s. The big doublet lenses (see page 153), although called achromatic ('without colour'), did not bring the light of all wavelengths to the same focus. They were designed for visual use and gave sharp images in the yellow-green part of the spectrum. But each star was seen surrounded by a coloured halo of out-of-focus light, the photographic image was less sharp and not at the same distance from the lens, and there was no position in which all the light from a star could be imaged to pass through the narrow slit of a spectrograph.

A discovery in chemistry presented a way out of the problems. In about 1853 the German chemist Justus von Liebig (1803–73) developed a process in which a very thin layer of metallic silver could be deposited on a clean glass surface from an aqueous solution containing dissolved silver nitrate. This was used by

Opposite: The 30-inch refractor installed at Pulkovo Observatory in 1885. The mounting was by Repsold & Sons of Hamburg, with optics by Alvan Clark & Sons of Cambridge, Massachusetts. The telescope was briefly the largest refractor in the world.

Below: The eye end of the Pulkovo 30-inch refractor. The complex array of controls for setting, focusing, adjusting, and recording are brought within reach of the observer. With such instruments, the management of the telescope reached a new level of sophistication.

The Melbourne Reflector, 'one of the greatest calamities in the history of instrumental astronomy'. In 1865, to enable the new Melbourne Observatory to study the southern nebulae, an order for a large reflector was placed with Grubb of Dublin. The telescope was installed in 1869, but the mounting proved unsatisfactory. Worse still, there was no optician in Australia with the skills to refigure the mirror when it tarnished. It was to be the last great mirror made of speculum metal.

K. A. Steinheil in Munich, and by Jean Foucault in Paris, to manufacture the first reflecting telescopes in which the mirrors were made, not of the customary speculum metal, but of glass with silver deposited on the front, optical surface.

The invention was at first viewed with suspicion: the silver film was fragile and also tarnished quickly, but its reflectivity was greater and the chemical process did not affect the optical figure of the mirror. The glass was easier to shape than speculum metal, it did not need to be of high quality since the light did not pass through it, and there was only one optical surface to be figured, as against at least four in an achromatic lens.

The last large telescope with a speculum metal mirror was a 48-inch made by the Dublin firm of Thomas Grubb in the 1860s for the Melbourne Observatory. It proved difficult to keep in good order, and thereafter the silver-on-glass reflector became the preferred instrument for astronomical photography and spectroscopy. The 36-inch Crossley reflector, donated by an English amateur to the Lick Observatory in California in 1895, first demonstrated the power of reflectors for the photography of nebulae, and the 60-inch reflector (commissioned in 1908) and the 100-inch (1917) of the Mount Wilson Observatory near Los Angeles were to establish the superiority of this form of telescope for the astrophysics of the twentieth century.

SOLAR SYSTEM STUDIES

As well as being suited for the exact measurement of double stars (see pages 216–17 and 249), the large refractors of the later nineteenth century were well adapted to the study of the physical nature of objects in the solar system. It chanced that this period witnessed the appearance of a larger than average number of spectacularly bright comets.

Comets: Heads and Tails

The nature of comets had always been enigmatical; now there were both comets and the telescopes to observe them. The 'great' comets, strikingly visible even to the casual observer and dominating the night sky for days or even weeks together, are nearly always those with orbital periods about the Sun of thousands of years; there

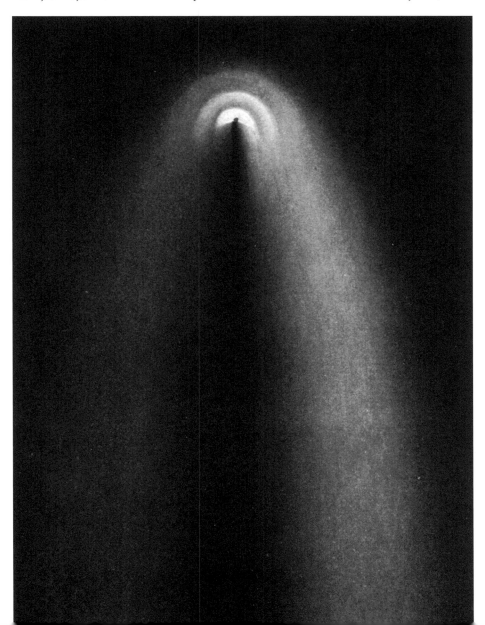

A drawing of the head of Donati's Comet of 1858, made by G. P. Bond of Harvard on 29 September. The comet was particularly well placed in the sky for the observatories of Europe and America. It still had to be observed visually, but telescopes of the calibre of the 15-inch refractors of Pulkovo and Harvard were available. Harvard produced in 1862 a volume of 362 pages devoted to the comet, with many accurate engravings. These skilful drawings are still of value in understanding the interaction between the material of the comet, the solar radiation, and the interplanetary material.

are therefore no records of their previous appearances, and their reappearances are unpredictable. There had been such comets in 1811 and 1843, but these were at least equalled by the comet first seen in the telescope by the Florence astronomer Giovan Battista Donati (1826–73) early in June 1858. As it approached the Sun in the following weeks it brightened to naked-eye visibility in September; by early October the head was brighter than the nearby star Arcturus, and the tail curved

A lithograph of Donati's Comet as seen over the bridges of Paris. Intended as a work of art rather than of science, this frontispiece to A. Guillemin's 1875 book, *Les Comètes*, remains one of the finest representations of a 'great' comet as seen by the naked eye. It is scientifically accurate: the bright star near the head of the comet is Arcturus, and the several components of the tail are correctly drawn.

A drawing of Encke's Comet of 1848, as seen through the 5.9-inch refractor at the private observatory of John Lee in Buckinghamshire, England. Most comets are unspectacular, and many are invisible without a telescope. The head of Encke's Comet is depicted with a nucleus of almost star-like appearance. This merges into a fainter coma, which in turn blends into the faint light of the night sky. There is no tail.

across one-third of the sky. It was well placed for the observatories of the northern hemisphere and was minutely observed.

As in other bright comets, the comet's tail and head both had complicated structures which changed from day to day and even from hour to hour. There was a straight tail, streaming away from the direction of the Sun, and a series of more or less curved tails. H. W. M. Olbers had suggested in 1812 that these shapes in the head and the tail were a consequence of different forms of material being ejected from the head of the comet as it approached the Sun and being swept away from the Sun along different paths that depended on the speed of ejection, the path of the comet in space, and the forces acting on the particles.

The heads of these bright comets were bright enough to enable the light to be examined with the spectroscope. This was first done by Donati on 5 August 1864, when he examined a comet that was then of the second magnitude. It had been thought that comets shone by reflected sunlight and would therefore have the same Fraunhofer spectrum as the Sun: Donati made the important discovery that there were three bright bands separated by wider dark spaces, suggesting that some of the light came from glowing gas of unknown nature in the comet itself. The bright comets of following years and the increasing use of the spectroscope both at the telescope and in the laboratory added considerably to knowledge about the nature of comets: by 1880 it was generally accepted that the light of the head of a comet came from reflected sunlight in part, but was dominated by broad emission bands (as distinct from narrow bright lines) which could be simulated in the laboratory by electrical discharges in tubes containing mixtures of carbon dioxide and simple hydrocarbons like methane and ethylene. The organic chemistry of the time was inadequate to understand exactly how the light arose, and only in the twentieth century was it understood that the bands arose from simple combinations of a few atoms, variously of carbon, nitrogen, and hydrogen, that are not stable under terrestrial conditions. It was sufficient to cause alarm, however, in the public press when it was learned that the Earth would pass through the tail of Halley's Comet in 1910, and that the tail presumably contained cyanogen, the poisonous compound of carbon and nitrogen.

Shooting Stars

In the late 1860s, the growing interest in comets was enhanced by the remarkable display of 'shooting stars' – meteors – visible from much of Europe on the night of 13/14 November 1866, when for some hours the number of meteors defied counting. The November meteors, known as the Leonids because their paths appear to radiate from the constellation Leo, had long been known, and similar spectacular displays had also occurred about the same date in the years 1799 and 1833. Although not characterized by outbursts at similar intervals, the regular displays of meteors in early August, appearing to radiate from Perseus, and in late April from Lyra, were also well known.

Several astronomers recognized that the position of the radiant point, combined with the date in the year of the maximum activity, provided precise information about the orbit in the solar system that the particles were following: John Couch Adams in Cambridge followed up the work of Hubert A. Newton of Yale and calculated a unique orbit for the November meteors, and others did the same for the August Perseids and the April Lyrids. These confirmed a prescient recognition by G. V. Schiaparelli (1835–1910) of Milan that the paths in space of meteoric particles were much like those of comets, and within a short time the identity was recognized of the orbits of the Perseids and the bright comet of 1862, of the Leonids with Tempel's Comet of 1866, and of the Lyrids with the comet of 1861.

The exact physical relationship between the numerous minuscule members of the solar system – the comets and asteroids, the

Opposite: Traces of the brighter meteors seen at Greenwich during the great meteor shower on the night of 13 November 1866. The tracks originate in the constellation Leo. Search of historical records showed that similar showers had occurred as far back as 13 October 902 (in the Julian calendar), revealing a shift of three weeks which J.C. Adams calculated was due to perturbations by planets.

G.V. Schiaparelli

Giovanni Virginio Schiaparelli was born in Savigliano, Piedmont, on 14 March 1835. After graduating at Turin, he studied astronomy in Berlin with J.F. Encke. A brief spell at Pulkova Observatory followed, after which he joined the staff of Brera Observatory in Milan in 1860, becoming director there two years later.

In his early researches, which were limited by the modest instruments available at Brera, Schiaparelli investigated comets and meteors; but after 1877, when he had a Merz refractor at his disposal, he worked mainly on planets, especially Mars. He also made major contributions to the history of ancient and Islamic astronomy, aided by the leisure resulting from his voluntary retirement in 1900. He died in Milan on 4 July 1910.

The rings of Saturn

Opposite: A drawing of Saturn, made in 1872 by Etienne Trouvelot (see page 270) at Harvard College Observatory. It is one of a series of the best such planetary drawings, published by Harvard in 1877. Even after the rapid improvements in photography in the 1870s, the human eye at the telescope remained the best way of recording fine planetary detail, and this continued to be the case until the development of interplanetary probes and space telescopes a century later.

The physical nature of the rings of Saturn – a system that remained until the space age unique in the solar system – had puzzled astronomers ever since their recognition in the mid-seventeenth century. As a visual spectacle arguably the most remarkable object in the sky, Saturn was a natural object of study for each of the new telescopes that came into operation, as observers used their increased optical power in the search for elusive detail. In 1850, William Cranch Bond (1789–1859) of Harvard, with the new 15-inch refractor, discovered the faint inner 'crepe' ring, seen independently a few days later by the Rev. W.R. Dawes in England. George Phillips Bond (1825–65), the son of William, suspected that the features of the system of rings changed slowly, and thought the rings must be fluid, not solid.

An important advance was made in 1857 by the young Scottish mathematician, James Clerk Maxwell (1831–79). The problem of Saturn's rings had been set as a prize essay at Cambridge, and Maxwell showed convincingly in his essay that neither solid nor fluid rings could exist, and that the rings were made up of millions of small particles, each pursuing its individual orbit in the plane of the planet's equator. His deductions from the mathematical theory accounted for most of the known observational features and was immediately accepted – though the crucial observation that the outer parts of the ring were moving more slowly than the inner, as the theory required, were made only in 1895, by James E. Keeler at the Allegheny Observatory, in Pittsburgh.

meteors, and the larger meteorites which crash through the atmosphere to land on the Earth's surface – long defied interpretation. It became recognized gradually that meteorite falls did not, as might have been expected, occur during great swarms of shooting stars and were perhaps more closely related to the asteroids. But the studies of the latter half of the nineteenth century were important foundations for later ideas about the history of the solar system itself.

Moons of Neptune, Uranus, and Mars

The increasingly powerful telescopes, sited, in the later years of the century, in better observing locations, added gradually to the number of known members of the solar system. The first satellite of the newly discovered Neptune was found by the Liverpool amateur William Lassell only seventeen days after the first observation of the planet; once the satellite's orbit about Neptune had been determined, this gave a value for the mass, and hence the density, of the new planet. Shortly afterwards, in 1851, Lassell added two satellites, Ariel and Umbriel, to the planet Uranus; these were interior to the brighter Oberon and Titania that had been discovered, as was the planet itself, by William Herschel.

In 1877 Mars was closer to the Earth than usual. Until then it had had no known satellites; but Asaph Hall (1829–1907) searched with the 26-inch refractor of the Naval Observatory in Washington, DC, and discovered two faint satellites (Deimos and Phobos) close to the planet, the inner indeed so close that its orbital period is

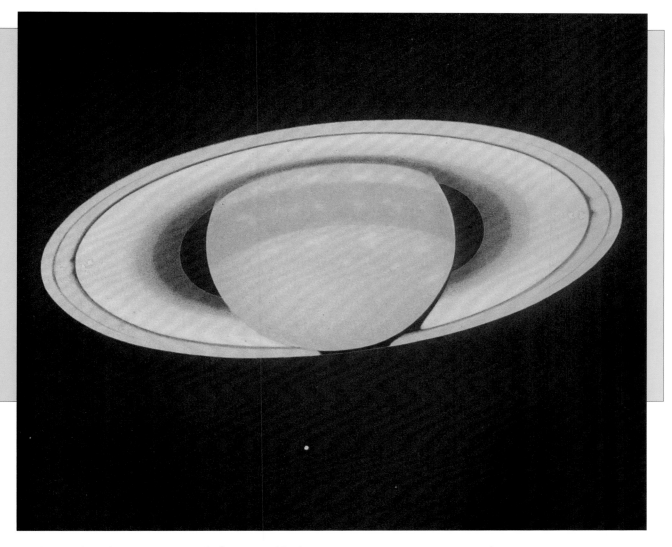

shorter than the rotation period of Mars itself. This was a unique circumstance, and the newly discovered satellites not only provided an accurate value for the mass of the planet, but were recognized as being more alike in size to the smaller of the asteroids than were the other, larger, planetary satellites.

Martian 'Canals'

Mars itself was also attracting interest. As early as 1783 William Herschel had commented that it was the most Earth-like of the planets, and by the mid-nineteenth century it was generally accepted that Mars – smaller than Earth but with a very similar rotation period and an axis of rotation that was similarly inclined to the plane of the solar system – had white polar caps that changed with the Martian seasons, and more or less permanent surface markings, some of which could be identified on the first telescopic drawings of the disc made in the seventeenth century.

Percival Lowell observing Venus in the daytime with the 24-inch Clark refractor erected at Lowell Observatory in 1896. He was born on 13 March 1855 into a prominent Boston family. After graduating from Harvard in 1876, he devoted himself to business and travel, but in the 1890s his interests turned increasingly to astronomy. In 1894 he selected a high-altitude observing site near Flagstaff, Arizona, and installed telescopes there. He soon became convinced that there was intelligent life on Mars, a view which he publicized extensively. Later he attempted to calculate the position of an undiscovered planet believed to lie beyond Neptune (the outermost planet then known), and the observatory staff devoted much effort to the search for 'Planet X'. But when Lowell died at Flagstaff on 12 November 1916, Planet X was still undiscovered.

Attention to these markings began to be paid by both professional and amateur astronomers from about 1870 onwards. Both good telescopes and a steady atmosphere were necessary; and also patience, for in some years Mars is not well placed for observation, and the closest approaches to Earth occur only at fifteen-year intervals. Even then the disc is less than one-seventieth of the apparent diameter of the full moon.

Schiaparelli embarked on a mapping of the surface at the close approach in 1877, and he noted that the dark markings (supposedly oceans) were the sources of hundreds of fine, barely perceptible, dark lines, running across the brighter, orange (continental?) areas. He named them *canali* ('channels'), claiming for them no more than a natural, topographical character.

He unwittingly launched planetary astronomy on a most curious controversy, that was to be resolved only a century later. From the start, some astronomers, with larger telescopes than Schiaparelli's, could see the channels, while others could not. The word *canali* was erroneously translated into English as 'canals', implying an artificial construction by intelligent beings, with the corollary that there was intelligent life on Mars. The controversy attracted the attention of Percival Lowell

The discovery of Pluto 1930

Percival Lowell accepted the 'planetesimal' hypothesis put forward in 1904 by the Americans T.C. Chamberlin (1843–1928) and F.R. Moulton (1872–1952), according to which the planets had formed from small chunks of matter or planetesimals, which had been pulled from the Sun by a passing star. Lowell thought he could calculate the whereabouts of the as-yet-undiscovered 'Planet X' immediately beyond Neptune, partly from speculations as to its period, partly from the supposed effects of its attraction on the known bodies of the solar system. Much effort was devoted at Lowell Observatory to the search for Planet X, and the search was continued after the death of Lowell in 1916. Eventually, in 1930, the planet now known as Pluto was discovered there by Clyde William Tombaugh (1906–), though it is now clear that Lowell's prediction had no scientific validity and that it was only by chance that it helped in the discovery.

Above: A region of Gemini, photographed by C. W. Tombaugh on 23 January 1930.

Right: The same region, six days later. When comparing the two plates on 18 February with a blink microscope (see page 327), Tombaugh noticed an object (arrowed) on the second plate that had moved position since the first plate was taken. This was Planet X.

(1855–1916) of Boston, a member of a powerful and wealthy New England family. In 1894, with a favourable opposition of Mars approaching, Lowell abandoned most of his business operations and embarked upon the construction of a well-equipped private observatory favourably sited near Flagstaff, Arizona. His later books, *Mars and its Canals* (1906) and *Mars as the Abode of Life* (1908), reveal by their titles Lowell's belief in the consequences of his observations.

Lowell was wrong: there are no artificial canals. But in a broader sense his work had great influence: his observatory made other discoveries that demonstrated the importance of good equipment placed at carefully selected observing sites; it kept

Above: Maps of two aspects of the surface of Mars, drawn by G.V. Schiaparelli in September 1877, when the planet was close to the Earth. From the dark 'oceans', natural *canali* ('channels') supposedly ran across orange areas.

Left: Mars as drawn by Percival Lowell in 1894/95, when Mars was in opposition to the Sun and favourably placed for observation. Lowell believed that intelligent beings on Mars, faced with a water shortage, would have irrigated the planet. He accordingly drew artificial canals, rather than the natural *canali* of Schiaparelli.

alive an interest in planetary studies (leading, for example, to the discovery there of the planet Pluto in 1930); and as we shall see later, it made important contributions to observational cosmology in the twentieth century.

THE STARS AND THE NEBULAE

The improved knowledge of the spectra of molecules, and refined optical methods of measuring the polarization of light reflected from planetary surfaces, led to further astrophysical studies of the planets in the early years of the twentieth century. These led to the discovery of methane in the atmospheres of the giant planets, and also of ammonia in those of Jupiter and Saturn. But increasingly the interest of the astrophysical observatories was turning to the more exciting problems of the stars and the nebulae. With a few exceptions, planetary observations were dropped. Partly because it was difficult to see what more could be done, and partly as a consequence of the Mars controversy, the study of the fascinating and ever-changing markings on the surface of planets like Mars and Jupiter was left, by tacit consent, to amateur astronomers.

Two drawings of Jupiter from the 1880s. Above left: Jupiter as drawn in 1885 by N.E. Green, an amateur astronomer observing from Madeira with an 18-inch reflector. Above right: Jupiter as drawn in 1889 by J.E. Keeler, a professional observing with the 36-inch refractor of Lick Observatory. The quality of Green's drawing is testimony to the high level of planetary studies by amateurs at this period.

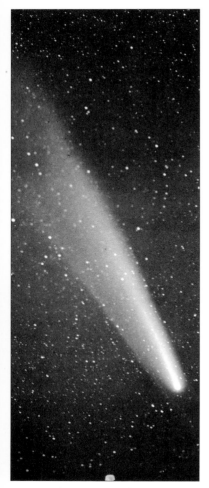

Above: A photograph by David Gill of the great comet of 1882. The stars visible (down to tenth magnitude) convinced Gill and others of the value of mapping stars by photography.

Right: Potsdam Astrophysical Observatory. The institute was inaugurated in 1874, and was one of the first observatories dedicated to the 'new astronomy', or astrophysics.

The Impact of Photography in the Late Nineteenth Century

We have already seen how rapidly the use of the spectroscope and of photography improved understanding of the Sun. It was to do the same for the stars and the nebulae, but later. When the limited light of even a bright star was passed through the narrow slit of a spectrograph and then spread out by the prism into the spectrum, it was barely visible to the eye, and much too faint to be recorded on the early photographic plates. In 1872 the New York astronomer Henry Draper (1837–82) managed to photograph a spectrum of the bright star Vega, and in 1879 the London amateur William Huggins (1824–1910) did the same for a few more stars, extending the recorded spectrum beyond the violet region into the ultra-violet, to which the photographic emulsion was sensitive and the human eye not. After about 1880 the newly invented dry plate was improved with every passing year, as the commercial manufacturers vied with each other to supply the rapidly widening trade in portrait and landscape photography with better products, and this increased the possibilities for the photography of stellar spectra.

Undoubtedly a key event was the photography of the great comet of 1882, which in September was well placed for observation in the southern hemisphere. The Scots astronomer David Gill (1843–1914), at the Cape Observatory, being anxious to secure a picture of the tail of the comet stretching across the sky, sought the help of a local portrait photographer. A studio camera was strapped to the counter-weight of one of the telescopes, and the telescope was then used to follow the comet during the time exposure. Gill was astonished by the result: the photograph not only had a superb image of the comet but also showed, on the single plate of a very large area of sky, the images of thousands of stars whose measurement and recording by the older methods of visual observing would have taken many weeks of night-time work.

Almost from this moment photography began to alter the very character of astronomy, and to change the nature of observatories too. On the one hand the large telescopes, both refractors and reflectors, were fitted with plateholders and spectrographs to photograph small areas of sky in great detail, and examine the spectra of individual objects. On the other, new types of instruments, developed from the studio camera, were introduced. Although by the standards of the big telescopes these had relatively small lenses – 5 or 10 inches in diameter – they had wide fields of view. Finally, by putting a prism of

Above: A measuring machine by Repsold & Sons for the study of photographic plates. The accurate measurement of such plates, carried out in a warm and comfortable office, transformed the positional astronomy of faint stars.

Left: The 24-inch reflector of Yerkes Observatory, with a skeletal tube of eight steel pipes connected by aluminium rings. The mounting proved rigid enough to permit high-quality photographs with long exposures, and was copied in other instruments.

small angle in front of the camera lens, each star image could be spread out into a little spectrum, so that a single photograph could give useful spectra of some hundreds of the brighter stars on the one plate (see page 298). Whatever the instrument, the astronomer was freed from the limitations of the human eye, and from trying to make difficult and delicate measurements in the cold and the dark of a telescope dome. When the photograph had been developed, the images could be measured more accurately, by day and in comfort.

It is fortuitous that these changes took place rapidly in the last quarter of the nineteenth century, when a number of observatories were built specifically for the study of astrophysics rather than the old astronomy of position, while some of the older observatories created astrophysical departments. By 1900 astronomers were provided with a new armoury of techniques that were to provide answers, over the next forty years or so, to problems that had tantalized them for centuries.

Angelo Secchi and the Classification of Stellar Spectra

The first photographic maps of the sky were at first only more complete and more accurate extensions of those that had been compiled laboriously by visual methods in earlier years. They were still no more than maps of the stars on the apparent surface of the sky, as only for a few stars had the distances – the third dimension – been determined by laborious and imprecise visual measurements.

But one difference between the visual and photographic maps soon became apparent. Although the positions of the stars agreed precisely, the apparent brightnesses of the stars differed: two stars that seemed equally bright to the eye in the telescope might differ quite strikingly on the photograph. It was known that the photographic plate was less sensitive to yellow and orange light than the human eye: evidently the stars differed in colour.

This was not a new discovery. As long ago as 1798 William Herschel, out of curiosity, had examined through a prism the light of six of the brightest stars. 'The light of Sirius', he noted, 'consists of red, orange, yellow, green, blue, purple, and violet... Arcturus contains more red and orange and less yellow in proportion than Sirius...' But these were facts for which he could offer no interpretation.

However, nineteenth-century physicists were well aware that as a piece of metal is heated in the laboratory it becomes first red hot and then white hot. This implied that, by examining the spectra of stars, one could learn something about their temperature as well as their chemical composition. Many astronomers contributed to this study, the most notable being William Huggins, Father Angelo Secchi (1818–78) of the Collegio Romano, H. C. Vogel (1841–1907) of the Potsdam Astrophysical Observatory, and E. C. Pickering (1846–1919) of the Harvard College Observatory.

In 1863–67, before the introduction of photography, Secchi had patiently examined the spectra of some 4000 stars, and at the meeting of the British Association for the Advancement of Science in 1868 he proposed that most stars could be grouped into four types of spectra. Vogel at Potsdam arrived independently at a very similar grouping. We no longer use Secchi's classification of Roman numerals, but his groupings are essentially correct and can be recognized in the later schemes. Secchi's types are:

I. The 'Sirian' (like Sirius), whitish or bluish, with a few dark bars due to hydrogen, and only faint metal lines.

II. The solar type – stars like Capella and Arcturus, sharing a predominance of light in the middle, yellow, part of the spectrum and with innumerable fine dark lines.

III. Red (and often variable) stars of long and irregular period like Betelgeuse (in Orion) and Mira (in Cetus), with broad bands of light each brighter at the red end and fading out at the blue end, and crossed by regularly spaced shadings giving a fluted appearance.

IV. Stars rather similar to III, but of an even redder colour and with the bands differently arranged. These stars are the smallest group, and none of them is bright in the sky; seen visually in the telescope they 'gleam like rubies among the other stars'.

These early studies of stellar temperatures and spectra led to the first attempts at a scheme of stellar evolution. In his book about measuring the brightness of stars published in 1865, the Leipzig astronomer Friedrich Zöllner (1834–82) suggested that stars were first formed hot and, as they naturally cooled, passed through the solar type to the red stars.

A coloured lithograph from a book published around 1870, showing Secchi's four classes of stellar spectra as they would be seen for the very brightest stars with a visual spectroscope on a large telescope. For most stars, the appearance would be much fainter and the observation more difficult. Under each spectrum the principal lines are identified by their Fraunhofer letters (see page 262).

William Huggins

William Huggins was born in London on 7 February 1824, and as a young man was forced by circumstances to devote himself to the family business. But in 1854 he succeeded in divesting himself of the business, so that he could indulge his passion for astronomy.

On learning of Gustav Kirchhoff's 1859 discovery that the chemical composition of the Sun was revealed by its spectrum, he instantly realized that the method could be applied to the stars and nebulae, and he formed a collaboration for this purpose with W. A. Miller, a professor of chemistry. In 1875 he married, and thereafter his young wife Margaret was his devoted partner in his researches. Huggins (seen here at the spectroscope attached to the 15-inch refractor lent him by the Royal Society) quickly became a world leader in the 'new astronomy', and was a pioneer in the field until failing health forced him to give up research in 1908. He died on 12 May 1910 at his home at Tulse Hill, south of London.

William Huggins and the Discovery of 'True Nebulosity', 1864

William Huggins, working with his friend the laboratory chemist William Miller, proceeded on rather different lines, examining and interpreting the spectrum of individual objects selected for their particular interest rather than surveying many objects as Secchi and Vogel were doing. He had already resolved, by one crucial observation, one of the great dilemmas of observational astronomy. We have seen that the great telescopes of William Herschel, Lord Rosse, and William Lassell had been built largely to determine the nature of the nebulae: were they distant, unresolved clouds of stars, or were they clouds of 'true nebulosity' – glowing gas from which stars might condense? On the night of 29 August 1864 Huggins turned his telescope to a planetary nebula in Draco which he thought bright enough to yield a spectrum. It did, but showed only a few bright lines in the green part, one certainly due to hydrogen, the others unidentified. By analogy with the spectra of gases in tubes in the laboratory the conclusion was immediate: there were two kinds of nebulae, gaseous (of which the planetary nebula was an example) and stellar, and the spectroscope could distinguish between them.

By 1868 Huggins had examined the spectra of some seventy nebulae of various shapes and sizes. About one-third of these were gaseous; two-thirds had faint continuous spectra which was probably the light of unresolved stars. Huggins's discovery was a diagnostic test of great value, and ranks as one of the great discoveries of the early period of astrophysics.

Of course there were complications. From the outset there were objects, some of them quite bright, that did not fit into the scheme. They were certainly stars as seen in the telescope, and with continuous spectra that sometimes had dark lines; but the spectra included bright, sometimes spectacularly bright, emission lines. Many of these objects were known variable stars; some were subject to spectacular out-bursts like that of Eta Argus (now Eta Carinae) observed by John Herschel at the Cape in the 1830s. They were to provide problems for the future. Were they perhaps stars that had most of their surface covered by the bright faculae seen on the Sun, or objects in transition from a nebular to a stellar state, or vice versa?

The First Radial Velocities of Stars

Spectroscopy also gave information of a rather unexpected nature: about the movement of stars in space. In 1842 Christian Doppler (1803–53), of the University of Prague, proposed that the colour of starlight would be affected by the velocity of a star; the wavelength of the light of a receding star would be lengthened and hence the light made redder. The Doppler Effect exists, but Doppler was wrong in detail; the typical velocity of a star relative to us (say 20 kilometres per second) is so much smaller than the velocity of light (300,000 kilometres per second) that the effect on the colour of the star is imperceptible; but the displacement of sharp defined features like the dark absorption lines can be measured, and this yields a velocity of the star towards or away from the Earth.

W.W. Campbell, the future
director of Lick Observatory,
in 1890 when he was a
volunteer assistant there to
J.E. Keeler. He is standing
beside the 36-inch refractor,
which had been installed two
years earlier. Fitted to the
eyepiece is a spectroscope by
John Brashear of Pittsburg.
This had a large effective
aperture, making it 'an
enormous thing to hang on
to the eye end of a telescope'.
It gave spectra with bright
images, which allowed Keeler
to study the line-of-sight
velocities of stars and nebulae.

This measurement was first attempted by Huggins but it proved too difficult for
his apparatus. The first measurements that agree with modern determinations were
made on a small number of bright stars by Vogel and Julius Scheiner at Potsdam,
and by James E. Keeler at Lick Observatory in California, in the years around 1890.
They were accurate to within a few kilometres per second. Such measurements
remained for many years of great difficulty, but their value was recognized by
astronomers like E.B. Frost of the Yerkes Observatory (who went to Potsdam to
study Vogel's methods) and by W.W. Campbell at Lick, who persevered with the
work. The difficulty lay in the smallness of the displacement of the lines. An error

of only a few thousandths of a millimetre in measuring resulted in an error of several kilometres per second. During the exposure of an hour or two the moving spectrograph sagged under its own weight, and contracted as the temperature fell in the dome. It was necessary to design the spectrograph's frame carefully and enclose it in a jacket with the temperature kept within a fraction of a degree of constancy. The patient work of a relatively small number of astronomers slowly accumulated lists of line-of-sight ('radial') velocities, and by 1950 these contained values for about 15,000 stars.

Spectroscopic Binary Stars

The Doppler Effect however explained a discovery made at Harvard in 1887 by E.C. Pickering, the director, and followed up by Antonia Maury (1866–1952). It concerned changes in the spectrum of the bright star Mizar in the tail of the Great Bear: sometimes the familiar sharp absorption lines were split into close twins. Mizar, it transpired, was not one star, but two. Studies of binary stars (pairs of stars in orbit about their common centres of gravity) had continued through the nineteenth century, but here was a new class: two stars so close together that visually they invariably appeared as a single star, but which, as one star approached the

The spectrum of Beta Aurigae, photographed at the Harvard College Observatory on successive nights in 1889. In the centre of the upper photograph is what seems a single, thick dark line (the K-line), but in the lower photograph this has split into two halves. One of these originates in a star now moving towards the observer, the other in a star now moving away. There are in fact two spectra, and these are 'Doppler shifted' alternately to the blue and the red as the two stars from which they originate orbit their common centre of gravity.

Earth and the other receded in the course of the orbit, the spectroscope revealed to be double. Other such close 'spectroscopic binaries', like Capella, were soon discovered and were later to prove of great importance. In special cases, where the Earth lies in the plane of the orbit of the close double star, one star can pass between the Earth and the other star. These are 'eclipsing binaries', like the star Algol (see page 203), and a detailed analysis of the variations in the brightness and spectrum of what always seems visually to be a single star can yield information about the sizes, temperatures, separations, and masses of the individual stars, that can be obtained in no other way.

The measurement of the distance of individual stars of particular interest was becoming increasingly desirable. After the successes in the 1830s (see pages 216–20) and 1840s for a very few of the nearest stars, the visual methods began to prove almost impossibly difficult, and for the same star different observers would get discouragingly different results. The combination of photography and the long focus refractor again came to the rescue, by transferring the process of measuring from the telescope to the laboratory. But until well into the twentieth century the determination of stellar distances remained on the verge of the possible, and even for stars only 75 light years away there were errors of 20 per cent or more. Knowledge of distances was limited to a tiny local pool of stars within the as yet unfathomed ocean of the Galaxy.

One remaining quality of the nature of the light of a star called for more exact measurement, namely the brightness of the star as seen in the sky. Since Classical Antiquity stars had been assigned a magnitude, the brightest in the sky being of the first magnitude and the faintest visible on a clear night of the sixth; the five intervals between were estimated. When the first telescopes revealed many still

Apparatus for creating an artificial star to permit the accurate measurement of starlight, the subject of an 1887 doctoral dissertation by Vitold Tserasky of Moscow Observatory. The star to be measured is observed through the eyepiece of the telescope (left), and compared with the light from the artificial star formed in the lamp (right), which is admitted to the eyepiece through a variable aperture (centre).

fainter stars, the scale was extended by what was little more than guesswork; a star classed by one astronomer as eighth magnitude might be described by another observer as eleventh.

This clearly would no longer do. Progress was made in two steps. Physics laboratories were experimenting in the measurement of the brightnesses of lights in the laboratory, and gradually such methods were adapted to the needs of the astronomer, for example by projecting into the field of view of the telescope an image of an artificial star, and then adjusting the brightness of that image until it matched that of the star being measured. The second step was taken by the English astronomer Norman Pogson (1829–91) in 1856. He recognized (as had Edmond Halley as long ago as 1721) that stars of the first magnitude were roughly one hundred times brighter than those of the sixth. By proposing that 5.0 magnitude steps should be taken as corresponding exactly to a ratio of 100 in brightness, he defined the scale precisely. Using measurements made with photometers on telescopes, and by the acceptance of certain well observed stars as standards, an accurate and internationally accepted system of stellar magnitudes had been built up by the turn of the century.

Annie Jump Cannon and the Harvard Catalogue of Stellar Spectra

These magnitudes were 'visual magnitudes', observed with the eye. We have already noted that the magnitudes measured on photographs were not the same, and a similar system of 'photographic magnitudes' was set up. Physics was far enough advanced for it to be understood that hotter stars emitted more of the blue light that was more easily recorded by the photographic plate. The difference between the two magnitudes, the 'colour index', measured the colour of the star's light, and this was directly related to temperature; a large colour index of say 1.5 magnitudes implied a red star of low surface temperature. Many refinements were introduced in later years, and in particular from the 1920s onwards sensitive photoelectric cells replaced estimates made by the human eye. But the principle was the same.

These developments made it possible to refine the rather simple classification of types of stellar spectra that had been introduced by Secchi and Vogel. The work was started by Henry Draper, who died prematurely in 1882. As a memorial to him, his family provided the Harvard College Observatory with funds for new instruments and additional staff, to continue Draper's work. For Secchi's four types Draper had substituted sixteen, lettered A, B, C... With a better understanding of what was going on, these letters, which denoted the occurrence of various lines in the spectra, were gradually rearranged into what was seen to be more nearly a sequence of decreasing surface temperatures of the stars, and simplified a little. The eventual sequence of classes was of types O B A F G K M R N S, and the description of the spectra was sufficiently minute to permit the division of some of the classes numerically, so that the Sun (for example) became a star of spectral type G2.

A team of women at the Harvard College Observatory engaged in the classification of stars. In the late 1880s, E. C. Pickering, the director (left), employed upwards of fifteen women at a time, under the leadership of Williamina Fleming (seen standing), remembered as a strict disciplinarian. The resulting first Draper catalogue, published in 1890, listed the spectral types and magnitudes of over 10,000 stars.

A typical photograph used in the preparation of the monumental Henry Draper Catalogue. The photographs were mostly taken with cameras of 8 inches aperture, through prisms of either 5° or 13° angle. This one was taken at the southern outstation in Arequipa, Peru, and is of a region of sky around the variable star Eta Carinae. Some of the spectral types have been labelled for this to appear as the frontispiece to the last volume of the main Catalogue, published in 1924.

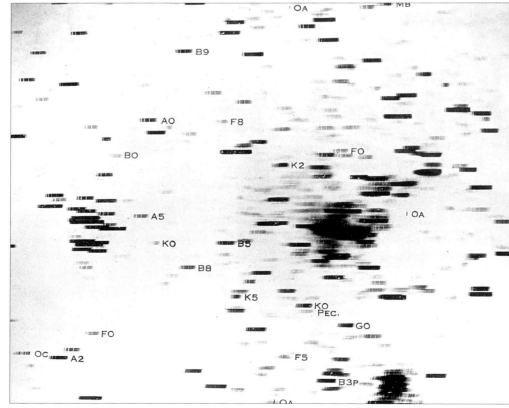

E.C. Pickering, the Harvard director, thereupon embarked upon a spectroscopic survey of the whole sky (Harvard set up outstations in the southern hemisphere). The work involved the use of wide-field cameras with an objective prism as already described. Pickering was fortunate that his skilled assistants included Annie Jump Cannon (1863–1941), for her prodigious industry was crucial in the publication, between 1918 and 1924 (with some later extensions), of the Henry Draper Catalogue. The catalogue, which contains the spectral type and magnitudes of some 225,000 stars, is still of value, 'the greatest single work in the field of stellar spectroscopy'. It is interesting to note that Zöllner's early speculation that the sequence followed also the life history of the stars was still thought useful, if not accepted as certain. The O B A... stars were referred to as 'early type' and G K M... as 'late type', and this usage curiously continued among astronomers long after matters were known to be much more complicated.

Plotting Magnitude Against Spectral Type: The H–R Diagram

By about 1910 two separate pieces of information were beginning to be available for a significant number of individual stars. The first was the spectral type, or the colour index, or the surface temperature of the star, all closely related to one another. The second, determined with greater difficulty, was the distance of a star, which enabled one to turn the (apparent) magnitude into a calculated 'absolute magnitude' or 'luminosity'.

The question to be asked was: can a star of a particular absolute magnitude be of any spectral type or surface temperature, or are only certain combinations of these quantities permitted in the universe? In other words, if one plots on a graph for each star the value of the absolute magnitude against (say) the spectral type, are the points scattered all over the diagram, or do they occur only in certain regions of it? Such a plot was made in 1913 by Henry Norris Russell (1877–1957) of Princeton University, using all the stars for which he felt there were reasonably reliable distances. He had been partly anticipated a year or two before by H. O. Rosenberg (1879–1940) and Ejnar Hertzsprung (1873–1967), who had plotted similar diagrams for the stars in each of the Pleiades and Hyades star clusters; the stars in any one cluster are so nearly at the same distance that apparent magnitudes are closely related to absolute magnitudes.

Such diagrams became of crucial importance in understanding the ways stars evolved: they became known as Hertzsprung–Russell diagrams, or 'H–R diagrams' in the jargon of working astronomers. Even on his first diagram, Russell felt that he could say that Nature does restrict the

H. N. Russell's 1913 diagram correlating the absolute magnitudes of stars with their spectral types. The Harvard spectral classifiers had noted subtle differences between otherwise identical spectra; in particular, in the later spectral types some stars had very narrow lines. In 1905 Ejnar Hertzsprung noted that these stars tended to have very small proper motions, and so were probably distant and highly luminous. This first indication of the existence of what later became known as 'giant' and 'dwarf' stars was strengthened by the distribution shown in Russell's diagram.

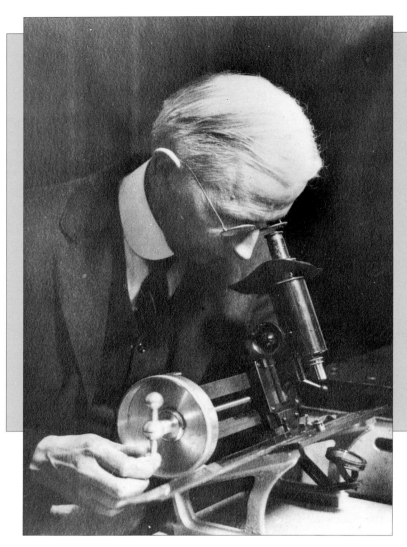

Henry Norris Russell

Henry Norris Russell was born in Oyster Bay, New York, on 25 October 1877. He studied astronomy at Princeton, and then was a research assistant at Cambridge. In 1905 he became instructor in astronomy at Princeton, being promoted professor in 1911; the following year he was appointed director of the observatory, a position he held until his retirement in 1947.

Despite indifferent health, Russell was an indefatigable researcher, and his interests ranged over several major fields of astrophysics. He was a brilliant communicator, who could be relied upon for a challenging, if not always correct, response to any new idea. Only G. E. Hale had comparable influence on the American astronomical community.

After retirement, Russell held research appointments at Lick and Harvard. He died in Princeton on 18 February 1957.

kinds of stars that can exist. There were two main bands populated by stars. One became known as the Main Sequence; it sloped down from highly luminous hot stars at the top left to intrinsically faint cool stars at the lower right. Across the top was the Giant Branch: the most luminous stars could be of any spectral type.

As the accuracy of measurements increased, these features of the H–R diagram were confirmed and more features became recognizable. An important discovery, made in 1914 at Mount Wilson Observatory by W.S. Adams and Arnold Kohlschütter, was that there were subtle differences between the spectra of the Main Sequence and Giant Branch stars of the same spectral type, revealed by the relative strengths of particular pairs of lines. Such differences had in fact already been noted by the Harvard classifiers, without being understood. Now, by a careful examination of the spectrum of a star, it was possible to identify to which part of the standard H–R diagram the star belonged, read off its absolute magnitude, and so, the apparent magnitude being known from observation, determine its distance. With careful refinement of the Harvard scheme of spectral classification in the 1930s by W.W. Morgan and his colleagues at the Yerkes Observatory, this inverse method of using the H–R diagram to determine distances from the spectral types and apparent magnitudes of stars ('spectroscopic parallaxes') became a powerful tool for determining the distances of stars far beyond the reach of the trigonometric method.

Atomic Spectra and the Chemical Composition of Stars

By about 1920 an enormous body of knowledge about spectra had accumulated. Meanwhile, in the physics of the Sun, new effects had been found. For example, the splitting of some spectral lines due to iron, when they originated in the laboratory in the presence of a strong magnetic field, was discovered in 1896 by the Dutch physicist Pieter Zeeman (1865–1943); this was quickly used to identify and measure the strength of magnetic fields in sunspots.

Now the spectra of hot blue stars showed only a small number of very black lines due to hydrogen and helium; in the cooler stars these were much weaker, but the lines due to metals like iron and chromium were innumerable. Did this mean there was no chromium in hot stars? And why, in the outer layers of the Sun, presumably made throughout of the same gases mixed together, did the cooler sunspots have a spectrum so different from that of the photosphere?

The explanation of this large body of empirical knowledge was at hand, but it came from parallel discoveries made in physics, about radiation and about the structure of the atom. In 1901 the Berlin physicist Max Planck (1858–1947) set out his quantum theory, by which light and other radiation like X-rays is emitted or absorbed in discrete packets of energy. In Manchester, in 1911, the New Zealander Ernest Rutherford (1871–1937) proposed a model of the atom, in which a positively charged nucleus was surrounded by a cloud of negative electrons. Two years later the Danish theoretical physicist Niels Bohr (1885–1962) combined features of both theories to develop a new model of the atom that was capable not only of explaining the sharp lines in spectra, but also of calculating their wavelengths and predicting the wavelength of new lines to be looked for. Although no more than a model, which served to calculate and predict, the 'Bohr atom' was enormously useful and continued to be used in the interpretations of astronomical spectra even when it had been realized, with the introduction of quantum mechanics and the wave theory of the electron, that the real atom was much more complicated.

In the Bohr model of the atom, the electrically charged nucleus was surrounded by a number of negatively charged electrons that exactly balanced the charge of the nucleus, so making the atom neutral. The electrons, arranged in shells, orbited the nucleus, and the number and arrangement of the electron orbits determined the pattern of lines in the spectrum of that element. When a gas was heated, however, an atom could lose one of its outer electrons to become 'ionized' and positively charged, and the changed electron orbits gave a completely different pattern of lines.

This kind of theory, combined with knowledge of temperatures and densities in the outer layers of the Sun and stars, was used by the Indian physicist Meghnad

Cecilia Payne-Gaposchkin
about 1946, seen comparing
two photographs with a blink
microscope (see page 327).
After graduating from
Cambridge University, she
went to Harvard College
Observatory for doctoral
research, and she was to
remain at Harvard for the rest
of her career. Her 1925 thesis,
on the abundances of the
elements in stars, was
described by Otto Struve as
'the most brilliant PhD thesis
ever written in astronomy'.

Saha (1894–1956) around 1920 to give a very satisfactory explanation both of the great differences between the spectra of different types of stars, and also of the subtler differences between the spectra of giant and dwarf stars of the same spectral type. In turn it underpinned the use of such methods as spectroscopic parallaxes, which previously had been based on empirical experience.

Cecilia Payne and the Abundance of Hydrogen in Stars

In a doctoral thesis completed at the Harvard College Observatory in 1925, Cecilia Payne (later Payne-Gaposchkin, 1900–79) established clearly the relationship between the temperature and the spectral class of a star, and went on to consider the relative abundance of the elements, suggesting that there was evidence that hydrogen was enormously more abundant in most stars than had been supposed. At first this idea was not accepted, but by 1929 H. N. Russell, for example, had been persuaded, and writing on the solar atmosphere concluded that it contained '60 parts of hydrogen (by volume), 2 of helium, 2 of oxygen, 1 of metallic vapors, and 0.8 of free electrons practically all of which come from the ionization of metals'. Later investigations would change these figures, but not the main conclusion, that much the most abundant element in the stars is hydrogen, followed by helium.

Further explanations followed from the use of physical theory. So, for example, in 1927 Ira S. Bowen of the California Institute of Technology showed that the lines supposedly due to the undiscovered element nebulium were in fact caused by oxygen in an unusual state of ionization, and (as already mentioned) in 1941 Walter Grotrian similarly explained the 'coronium' lines as caused by iron.

Eddington and the Mass–Luminosity Relation

There were other fruitful interactions between observation and theory at about the same time. The sizes of stars cannot be measured directly at the telescope by simply measuring their discs, for they are much too far away. They could be inferred in two ways, either by studying eclipsing binaries, or by calculating what the surface area (and hence diameter) of a star must be to emit the observed amount of energy (absolute magnitude) at the observed surface temperature (colour index). To determine the density of the star, and the attraction of gravity at its surface, required a knowledge of its mass – how much matter the star contained.

There was one, and only one, way in which the masses of stars could be determined, by the application of Newton's law of gravitational attraction to the measured orbits of binary stars. This was why the study of these stars was very important, the information from the close, spectroscopic binaries being particularly valuable. But such information was available for a relatively small number of stars. Nevertheless, it became apparent that stars could have a surprisingly large range of mass. In 1924 A. S. Eddington (1882–1944) of Cambridge collected the available information and found stars of as little as one-fifth the mass of the Sun, and as large as twenty-five times. More importantly, he showed convincingly that these values were quite closely related to the absolute magnitudes of the stars: there was a mass–luminosity relation in the sense that the most massive stars were the most luminous – stars of 25 solar masses were emitting about 4,000 times the energy of the Sun.

Fig. 1. – ORDINATES—ABSOLUTE MAGNITUDE. ABSCISSAE—LOG MASS
O First Class.   X Second Class.   ☐ Cepheids.   ⌁ Eclipsing Variables.

A. S. Eddington's 1924 mass–luminosity diagram for stars. The graph plots absolute magnitude (vertical scale) against the logarithm of the mass (horizontal scale), with each symbol representing an individual star. The Sun (not shown) would have absolute magnitude of about 5 and a mass of exactly 1 (logarithm = 0). The curve is drawn from theory.

The Source of Stellar Energy

Where did all this energy come from? At last the answer to this question, which had so perplexed the physicists of the nineteenth century, seemed to be in sight. It was becoming generally agreed that the clue lay in Albert Einstein's theory of General Relativity, which held that mass could be transformed into energy. In a semi-popular exposition, in his book *Stars and Atoms* (1927), Eddington speculated that four atoms of hydrogen (each of mass 1 unit) might be combined to form one atom of helium (with the slightly smaller mass of 3.97 units). 'To my mind the *existence* of helium is the best evidence we could desire of the possibility of the *formation* of helium... I am aware that many critics consider the conditions in the stars not sufficiently extreme to bring about the transmutation – the stars are not hot enough. The critics lay themselves open to an obvious retort; we tell them to go and find *a hotter place*.'

Eddington was right: the nuclear reaction that he suggested was taking place in the hot central cores, was eventually proved to be the source of energy of the Main Sequence stars. But at the time he could do no more than speculate, and he thought there were other possibilities. In the same passage he continued, 'The evidence, however, is not very coherent, and I do not think we are in a position to come to a definite decision'. It required more information from nuclear physics to understand exactly how the reactions could operate. It evidently had to be some step process, since the probability of four hydrogen atoms instantaneously colliding to form helium seemed very small. Several physicists and astronomers studied the problem through the 1930s.

R. d'E. Atkinson of England and F. Houtermans of the Netherlands, working together in Göttingen, had provided in 1929 an initial theory of nuclear interactions between the lighter elements in stellar interiors. At Rutgers University in New Jersey in 1931, Atkinson learned of the recently accepted great abundance of hydrogen in stars, and considered the mechanisms by which fusion of hydrogen into helium might proceed. Later, in Germany, C. F. von Weizsäcker worked out how further interactions could produce elements heavier than helium.

Finally, in 1939, the German-born H. A. Bethe (1906–), then at Cornell University, proposed a mechanism that was generally accepted as feasible and was consonant both with the known nuclear physics and with the accepted conditions of density and temperature at the centres of stars, stemming from Eddington's work in his classic book of 1926, *The Internal Constitution of the Stars*.

But that was 1939, and ironically many of those who had contributed to the solution of the problem were soon to find themselves engaged in using their knowledge to more practical, and grimmer, military ends.

Unusual Stars

In a volume of space around the Sun – say, a sphere of radius 50 light years – stars of much the most common kind are less luminous than the Sun itself. They are the dwarf stars at the lower end of the Main Sequence. On an H–R diagram, therefore, the giant stars, whether blue giants or red giants, were much over-represented, because although relatively rare they can be seen to very much greater distances. But whether rare or common, nearly all stars fell into the two main branches of the diagram.

A very few did not. Indeed, there was one already in Russell's first plot of absolute magnitude against spectral type: an intrinsically faint star of spectral type A in the lower left of the diagram (see page 299). It is the fainter companion of a nearby double star in the constellation Eridanus. A few similar stars were known. The bright star Sirius, about 9 light years away from the Sun, had a fairly rapid motion across the line of sight, and in 1844 F. W. Bessel announced that the motion was irregular and that the star moves in a wavy line. The simplest explanation was that Sirius was a double star with a very faint or even dark companion. Some twenty years later the American telescope-maker, Alvan Clark Jr, testing a new refracting telescope, first saw the faint companion, which was then favourably placed for detection in its fifty-year orbit about the brighter star.

These stars became known as 'white dwarfs'. For those that are members of double star systems, the mass of the white dwarf companion could be determined, as we have seen. They turned out to be paradoxical stars, with masses not too different from that of the Sun, but sizes (derived from their surface temperatures and measured brightness) not much greater than that of the Earth. The implied density was enormous: as Eddington said, a ton of the material could be put in a match box. The situation around 1914 was an embarrassment; astronomers felt that something must be wrong, but did not know what. The explanation had again to wait for a better understanding both of atomic physics and of the internal structure of stars; eventually, in 1926, R. H. Fowler of Cambridge used the recently developed theories of quantum physics to explain the existence of so-called degenerate matter of high density.

In 1931 the young Indian mathematician, Subrahmanyan Chandrasekhar, a student of Fowler's, first calculated an upper limit (later named the Chandrasekhar limit) for the masses of such stars (see page 361). White dwarf stars are examples of objects for which observation preceded understanding. The first were discovered in the nineteenth century; further observations early in the twentieth century revealed their incredibly high density, but it required new theories of the physics of the states of matter to explain them and to confirm that the observations had been right, and rightly interpreted.

But the alternative route to discovery can also occur. In 1932 James Chadwick identified in the laboratory a new fundamental particle of atomic structure, the neutron; and the Swiss-born astronomer Fritz Zwicky (1898–1974) and his German-born associate Walter Baade (1893–1960), both then resident in California, suggested that its known properties would permit the theoretical existence of stars, made essentially of packed neutrons, that would have a density many

In 1844, F. W. Bessel pointed out that Sirius was not travelling in a perfectly straight path as expected, but was following a wavy line (shown in the left-hand diagram with its position marked at five-year intervals). He inferred that Sirius was accompanied by a faint companion, and it was their common centre of gravity (about which Sirius itself orbited) that travelled in a straight line. 'Sirius B' was detected by Alvan Clark in 1862. Its path is also shown, with its positions relative to Sirius at five-year intervals. The smaller diagram shows the orbit of Sirius B relative to Sirius.

millions of times that of even the white dwarfs. This speculation, published in 1934, attracted little attention at the time, and it was not until 1967 (see pages 362–3) that such objects ('pulsars') were found to exist, discovered not as a result of deliberate search but serendipitously, in the course of unrelated researches in radio astronomy.

There were other classes of stars whose nature had also been enigmatic, that yielded to the improved understanding of the internal constitution of stars. In particular, the classes of variable stars with remarkably regular light curves and constant periods, of which the naked-eye star Delta Cephei (see page 314–15) is an example, had long presented a problem. The American Harlow Shapley (1885–1972, see page 313) had shown in 1914 that any explanation of them as spectroscopic binaries had to be abandoned, and suggested that these were single stars that were pulsating. Theoretical astronomers like Eddington developed this idea, and the theory of these stars was worked out in increasing detail during the 1920s and 1930s. Observations of individual stars agreed well enough with the theory: as the star oscillated in size in a regular cycle, the observed radial velocity of the star (the approach of its surface towards the observer as the star swelled), its surface temperature, spectral type, and observed magnitude, correspondingly changed through the cyclical period of the pulsation. This understanding gave astronomers increasing confidence that these stars, of high luminosity, were indeed of great utility in measuring distances in the universe at large.

Walter Baade

Wilhelm Heinrich Walter Baade was born in Schröttinghausen in northwest Germany on 24 March 1893. He studied briefly at the University of Münster, and then at Göttingen, where he received his PhD in astrophysics in 1919. He was a staff member at Hamburg Observatory throughout the 1920s, but a year spent in the USA confirmed his commitment to the 'extragalactic' astronomy being pioneered at Mount Wilson and elsewhere, and in 1931 he joined the Mount Wilson staff where Edwin Hubble was among his colleagues and Fritz Zwicky was at neighbouring Caltech. Baade worked at Mount Wilson, and later with the 200-inch reflector at Palomar Mountain, until his retirement in 1958. He then returned to Göttingen as professor, dying there on 25 June 1960.

A supernova explosion in a spiral nebula (galaxy). Left: NGC 4273 as it appeared in 1931. Right: the same object on 16 February 1936. Despite poor atmospheric conditions and the resulting indifferent quality of the photograph, a new star is clearly visible above the nucleus.

The more catastrophic variable stars like novae continued to puzzle. Routine photography of the sky greatly increased the rate of their discovery and of empirical knowledge of their behaviour; and from 1917 onwards (see page 327) faint stars with the same characteristics as novae in the Milky Way were discovered in the larger, and presumably nearer, of the spiral nebulae. It was only much later, in 1934, that Baade and Zwicky established that there were two very distinct classes of novae: the ordinary ones, which occur in galaxies like our own at a rate of ten or twenty times a year, and the much rarer and even more remarkable supernovae, whose light can approach that of all the other stars in the galaxy combined. The probable identification of the Crab Nebula, in the constellation Taurus, as a gaseous remnant of a supernova in our own Galaxy recorded in Chinese annals in AD 1054 had been suggested by Edwin Hubble in 1928 and was supported by further studies in 1942. But the cause of these stellar explosions remained a mystery, even when the general sources of stellar energy were known, and remained so until quite recently.

The Evolution of Stars

Much the greater majority of stars, however, were unchanging, and could be classified (as we have seen) into a limited number of spectral types. As soon as the process of classification began, in the 1860s, it was recognized that the classification must contain information about how the stars were formed and how they evolved from their initial state. With little physical understanding beyond the feeling that the source of stellar energy lay in the Kelvin–Helmholtz contraction (see page 272), and virtually no knowledge of the sizes or masses of the stars or of why stars had such widely differing spectra, early attempts to understand the life histories of stars could be little more than speculative.

There were two main schools of thought. One, following the suggestion of Zöllner in 1865 (see page 291), argued that the stars were formed as 'Sirian' – white, hot stars similar to Sirius. There was some support for this idea, for it was noticed

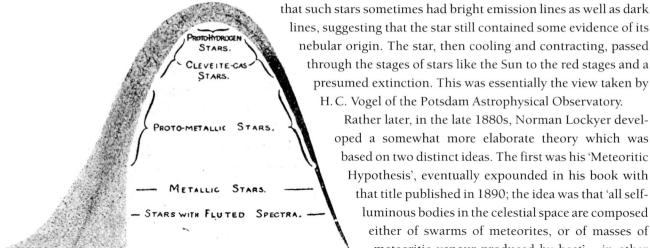

PROTO-HYDROGEN
STARS.

CLEVEITE-GAS
STARS.

PROTO-METALLIC   STARS.

—— METALLIC   STARS. ——

— STARS WITH FLUTED   SPECTRA. —

Norman Lockyer's theory of stellar evolution. This is only an illustrative diagram, but it can be regarded as a graph with temperature increasing upwards and the age of a star increasing from left to right. The star first forms from the initial meteoritic material at the bottom left, increases to a maximum temperature at the top, and declines to a final cool, condensed state. The spectral types corresponding to Secchi's, or Vogel's, classes are also indicated.

that such stars sometimes had bright emission lines as well as dark lines, suggesting that the star still contained some evidence of its nebular origin. The star, then cooling and contracting, passed through the stages of stars like the Sun to the red stages and a presumed extinction. This was essentially the view taken by H. C. Vogel of the Potsdam Astrophysical Observatory.

Rather later, in the late 1880s, Norman Lockyer developed a somewhat more elaborate theory which was based on two distinct ideas. The first was his 'Meteoritic Hypothesis', eventually expounded in his book with that title published in 1890; the idea was that 'all self-luminous bodies in the celestial space are composed either of swarms of meteorites, or of masses of meteoritic vapour produced by heat' – in other words, the primordial matter from which the stars formed was composed of chunks of rock, not gas. The second was an attempt to explain why the same chemical element, in laboratory spectra, gave different sets of lines in the electric arc (hot) and in the spark (still hotter). This postulated that, at high temperature, elements were dissociated into simpler building blocks he called 'protoelements'. Both ideas, though wrong in detail, foreshadowed later accepted ones: there are solid particles in interstellar space; and Lockyer's protoelements are akin to the ionized atoms of modern atomic theory. Lockyer's theory of stellar evolution started with large, cool stars condensed out of the interstellar material. These increased in temperature as the young stars contracted, until the rise in surface temperature was balanced by the loss resulting from radiation; thereafter a final stage of cooling set in.

Astronomers of the next generation, including both Hertzsprung and Russell, were quick to see that the distinction between giant and dwarf stars, apparent in the earliest H–R diagrams (see page 299), provided important new clues to stellar evolution. In a semi-popular article in 1914 Russell accepted, with everybody else, that the source of stellar energy was contraction, so that evolution proceeded in the direction of increasing density and decreasing size. Stars formed as red giants of spectral type M, moved along the Giant Branch in the direction of increasing temperature in the reverse order of the spectral types to classes A or B, and then moved down the Main Sequence, now in the order of the spectral types, to reach a cool, dwarf, state at spectral type M again. Although based upon a different and better established set of data, Russell's theory of stellar evolution had features not unlike that of Lockyer – for example, the affirmation that a star is hottest during the middle of its life history. It was simple, it accorded well with the observations, and it became generally accepted.

But the theory ran into trouble within ten years, when there were new ideas that we have already mentioned (see page 302) about stellar atmospheres, stellar structure, and the relation between mass and luminosity. In 1926 Russell produced a different theory, but this was more speculative and attracted little support. It did, however, contain a crucially important new idea: the life history of a star was primarily determined by its initial mass, which remained essentially unchanged, and stars were not constrained to move along the Giant Branch and Main Sequence but could move across them: these were rather resting places, where the star remained in a stable configuration for significant parts of its life. The absence of stars from some areas of the H–R diagram meant rather that stars moved through this combination of luminosity and spectral type quite rapidly, so that at any one time most stars were in the Giant Branch or Main Sequence areas of the diagram.

A fuller understanding of the life history of stars of different initial masses was to emerge only gradually from Russell's intuition of 1926, and Bethe's proposal of an acceptable source of nuclear energy in 1939. Even today it is not solved in detail, and there are still uncertainties in, for example, the ages of what are believed to be the oldest stars, in the spherical aggregations of stars known as globular clusters. But by about 1960 the main outlines of what turned out to be a very complicated situation had become understood, and have since been altered in detail rather than in fundamentals.

The life history of stars

The life history of the majority of stars may be summarized – in the framework of the H–R diagram – in four stages:

1 The star condenses from the interstellar medium and settles fairly quickly to a stable configuration on the Main Sequence – the more massive the star, the more luminous it is and the earlier (e.g. Type F rather than Type M) its spectral type.

2 The star spends much of its life in this stable configuration, deriving energy from the conversion of hydrogen to helium in the core; the more massive the star, the more rapidly it evolves.

3 When the hydrogen in the core is exhausted the star 'migrates' rather quickly to the Giant Branch of the H–R diagram, where energy is supplied by the further conversion of the helium core into heavier elements.

4 When finally these energy sources are exhausted the track of the star moves across the H–R diagram to the left, and the star collapses to the white dwarf state.

For the stars of the smallest masses, perhaps one-tenth of a solar mass, the evolutionary sequence proceeds very slowly, so that even the oldest stars are still in Stage 1; on the other hand, the most massive stars evolve with great rapidity and have total lifetimes of millions rather than billions of years, and the final stages of their collapse are marked by extreme phenomena like supernovae and the formation of exotic objects like neutron stars. The diagram, by Bart J. Bok, shows the evolution of a star with mass a little greater than that of the Sun, as it was understood in 1958. The theory on which the diagram is based has since been modified only in detail.

THE STRUCTURE OF THE GALAXY

In his attempt in 1785 to establish the structure of the Galaxy (see page 241), William Herschel had, like earlier speculators, taken the view that the Milky Way is the optical effect of our immersion in a layer of stars. But his bid to determine the outline of the layer had run into difficulties.

Nor had John Herschel fared much better. Greatly impressed by the Milky Way visible from the Cape of Good Hope, he had envisaged the Galaxy as formed of a thinly-populated central cluster surrounded by stars concentrated in meandering arms whose overall shape baffled him.

In the 1840s Lord Rosse's great reflector revealed the spiral shape of some of the nebulae (see page 254). Might the Galaxy itself be a spiral? Somewhat surprisingly, John Herschel never suggested this possibility, but in 1852 Stephen Alexander (1806–83), professor at what is now Princeton University, published a discussion entitled 'The Milky Way – a Spiral'. In the spiral M 99, the second to be recognized by Rosse, four branches curved out from the central cluster. Alexander argued that if the Sun and the brighter stars formed the central cluster in a similar spiral, and if the paths of the four branches were suitably chosen, then inhabitants of the solar system would see much the same night sky as we in fact observe from Earth.

There was too much contriving in Alexander's model for the taste of his contemporaries, and it was largely forgotten. But with spiral nebulae being discovered in ever-increasing numbers, the possibility that the Galaxy has a spiral structure was

Lord Rosse's sketch of M 99 in Coma Berenices, a nebula whose spiral shape he recognized in March 1848. Although three years had elapsed since Rosse's recognition of the first spiral nebula (see page 254), these years had witnessed a terrible potato famine in Ireland, and the great reflector had lain unused.

inevitably mentioned from time to time, notably by the Dutch journalist and amateur astronomer Cornelis Easton (1864–1929), who around the turn of the century published a series of influential drawings of the Galaxy imagined as a spiral structure seen face-on.

Another who sought to reproduce in detail the complexity of the observed Milky Way was the respected English amateur and popularizer, Richard Proctor (1837–88). His 1869 model, it was remarked, 'resembled a bent and broken ring, with long, riband-like ends, looped back on either side of an opening'. Most astronomers found this excessively complex and therefore unsatisfying.

Even less attractive – but for opposite reasons – was the model that had been proposed in 1847 by Wilhelm Struve, on the basis of statistical analyses of the star counts of William Herschel and of later star catalogues. He envisaged a universe in which there was a central plane (that of the Milky Way) of unlimited extent, in which the stars were everywhere distributed with a uniformly high degree of concentration. To either side of this central plane, the density of the stars fell off systematically with increasing distance from the plane. To explain the difference between his unbounded model and the limited number of stars in the observable universe, Struve invoked obscuring matter that rendered very distant stars invisible.

But whereas Struve was thought to have gone altogether too far in forcing the real star system onto the Procrustean bed of his model, many astronomers agreed that it was right to proceed statistically, working outwards from the solar system and examining first the nearer stars and then those further away. They hoped this would allow them to estimate the extent of the system of those stars that were near enough to be individually visible on telescopic plates, and they prayed that the riddle of the Milky Way would somehow solve itself. And so it was that by the turn

Above: Richard Proctor's sketch of the principal features of the Milky Way (marked 'Fig. 1'), and his final attempt (marked 'Fig. 2'), in the 1880s, to suggest a form of the Galaxy that could give rise to these features. For example, a bifurcation that is observed in the Milky Way (at 9 in Fig. 1) could be the result of the Galaxy's turning in on itself (near the Sun S, along the line S9, in Fig. 2).

Above left: Part of Cornelis Easton's 1900 drawing, showing the spiral shape he ascribed to our Galaxy. He used the lettering to illustrate his explanation of how various observed features of the Milky Way would arise, if the Galaxy had the shape as drawn.

J.C. Kapteyn

Jacobus Cornelius Kapteyn was born in Barneveld, The Netherlands, on 19 January 1851, one of fifteen children. He took a doctorate in physics at Utrecht, and then obtained a post at Leiden Observatory. At the age of twenty-seven he became professor of astronomy and theoretical mechanics at Groningen, a post he held until his retirement at the age of seventy.

As there were no major telescopes at Groningen, Kapteyn realized that the way forward was through collaboration with colleagues elsewhere. Assisted by convicts from a nearby prison, between 1886 and 1896 Kapteyn and his colleagues measured the positions of nearly half a million stars on plates of the southern skies taken by David Gill at the Cape of Good Hope. In 1906 he recruited the help of astronomers worldwide in assembling the data by means of which he planned to explore the Galaxy, through his *Plan of Selected Areas*; and although the First World War undermined the spirit of collaboration essential to the completion of the project, astronomers continued for a further fifty years to give preference to the sampling areas of the sky selected by Kapteyn.

Kapteyn retired in 1921, and died in Amsterdam on 18 June the following year.

of the century, when C. A. Young (1834–1908), then a professor at Princeton, came to write his textbook *Manual of Astronomy*, a consensus had emerged among the more conservative-minded. Most of the visible stars lay within a round, flat disc of space, whose diameter was about eight or ten times its thickness, and whose radius measured at least ten or twenty thousand light years. At right angles to the plane of the disc the stars were thinly scattered, eventually giving way to the region of the nebulae. 'As to the Milky Way itself, it is not certain whether the stars which compose it form a sort of thin, flat, continuous sheet, or whether they are ranged in a kind of *ring* or in *spires*, with a comparatively empty space in the middle where the Sun is placed.'

Kapteyn's 'Plan of Selected Areas', 1906

Continuous sheet? A ring? Spires? The reason for the uncertainty was obvious: our location inside the Galaxy puts us at a grave disadvantage when we try to picture the Galaxy as a whole. Some astronomers – notably Hugo von Seeliger (1849–1924), director of the Munich Observatory, and J. C. Kapteyn (1851–1922), professor at Groningen in The Netherlands – accepted that the secure way forward was through study of the individually visible stars. The trouble was that the necessary data concerning these stars – their positions in the heavens, apparent magnitudes, and proper motions – were accumulating much too slowly.

The basic problem, as we have seen (on page 296), was that only the very nearest stars were close enough for their distances to be found by straightforward triangulation, and the method of spectroscopic parallaxes lay in the future. However, it must in general be true that the nearer a star, the larger its proper motion; and when sufficient proper motion data were available, this truth could be exploited statistically to provide a measuring rod capable of reaching much further out into space.

How to acquire the necessary data? There was clearly insufficient manpower in the existing astronomical community to permit the investigation of the entire sky in the necessary detail, but the task could be reduced to manageable proportions by selecting representative samples of sky, and sharing these out among a large number of observatories. In 1906 Kapteyn published a *Plan of Selected Areas*, and he managed to persuade astronomers worldwide to make it the basis of an international campaign of data collection.

But there was a possible snag. The method involved the analysis of proper motions and apparent magnitudes, and it assumed that the apparent magnitudes could be relied upon – that they had not been distorted and the starlight dimmed by obscuring matter lurking in the interstellar spaces. The problem was recognized by those involved, but at the time the available evidence on obscuration seemed reassuring. However, it would eventually be found that dust was indeed present in the galactic plane, in such amounts that it not only affected the apparent magnitudes of the nearer stars, but concealed much of the Galaxy from inspection. Consequently, Kapteyn and his collaborators were unwittingly restricting themselves to a localized region within the Galaxy, while the great bulk of the galactic system lay hidden and beyond their ken. As a result, their investigations appeared to confirm the common view, that the Sun was located near the centre of the Galaxy, whose diameter was to be measured in terms of a few thousands of light years.

While this painstaking work was in progress, a daring young talent appeared on the American scene. In 1914 Harlow Shapley joined the staff of Mount Wilson Observatory, near Los Angeles. He had recently discussed his future plans with Solon I. Bailey of the

Harlow Shapley

Harlow Shapley was born in Nashville, Missouri, on 2 November 1885, and worked as a reporter before resuming his education. In 1911 he received a fellowship to work with H.N. Russell at Princeton on eclipsing binary stars, and in 1914 was recruited by G.E. Hale to the Mount Wilson staff. His career there, as observer and brilliant theoretician, was cut short by his appointment in 1921 to a position at Harvard, where the directorship was vacant. Russell had declined the post, Shapley greatly coveted it, and after a few months at Harvard was so appointed. He remained there until his retirement in 1952, creating a stimulating atmosphere and later playing a significant role in the international scientific community. He died in Boulder, Colorado, on 20 October 1972.

Harvard College Observatory. Also on the Harvard staff was Henrietta Leavitt (1868–1921), whose duties in recent years had included the painstaking examination of variable stars on the photographs of the Small Magellanic Cloud taken at Harvard's southern station in Peru. To identify a variable star by examination of the differences in a pair of photographs was one thing; to determine its period – the number of days between one maximum brightness and the next – was altogether more difficult. However, though the task was onerous, the effort was especially worthwhile because all objects in the Cloud lay at about the same distance from Earth. A difference in apparent magnitude therefore corresponded to a difference in absolute magnitude (luminosity).

In 1908 Miss Leavitt had published a memoir listing 1777 variables in the Cloud, and for sixteen of these she had been able to determine the periods. She remarked that the longer the periods, the brighter the stars. Four years later the sixteen had increased to twenty-five, and Miss Leavitt had found a mathematical relationship that linked the period with the apparent magnitude – and hence with the luminosity. This 'period–luminosity' relationship was to have momentous consequences for the history of astronomy.

In the variable stars in question (which Shapley had suggested were pulsating stars – see page 306), the brightness would rise rapidly to a maximum and then fade away slowly over a number of days, in the manner of the star Delta Cephei whose variability had been discovered by John Goodricke as long ago as 1784. Such 'Cepheid variables' are highly luminous and so very conspicuous, and their characteristic light curve makes them relatively easy to identify. They stand out from a cluster of stars much as a lighthouse beacon stands out from the ordinary house and street lights of a port. The 60-inch reflector at Mount Wilson was the most powerful telescope in the world for the study of distant and therefore faint objects, and Bailey – who had himself managed to identify a number of Cepheids in 'globular' star clusters – suggested that Shapley pursue this work with the 60-inch.

Globular clusters are striking objects, spherical assemblages of hundreds of thousands of stars. Edmond Halley had seen the brightest of them all, Omega Centauri, in 1677 when visiting St Helena (see page 251), and many more had been discovered by William and John Herschel in their sweeps for nebulae. Curiously, as John Herschel had noted, the globular clusters were by no means scattered uniformly across the sky. They were mostly in one half of the sky, and no fewer than one-third of them were concentrated in the Sagittarius region, which also contained some of the richest star clouds of the Milky Way. In 1909 the Swedish astronomer Karl Bohlin had suggested that the globular clusters formed a system that surrounds the centre of the Galaxy, which therefore lay far away in the direction of Sagittarius, at an enormous distance from the Sun. But cosmological speculations have a poor success rate and few took Bohlin seriously.

To measure the distances of the inconceivably remote globular clusters seemed a hopeless quest. But the resourceful Shapley saw a way of tackling this problem.

Above: The light curve of Delta Cephei, prototype of Cepheid variables. A Cepheid goes through a cycle in which it increases rapidly in brightness, and then slowly declines. Henrietta Leavitt found evidence that the number of days in the cycle of a Cepheid was correlated with the star's brightness.

Left: The 60-inch reflector at Mount Wilson, from 1908 to 1917 the largest telescope in the world. Solon Bailey recommended the young Harlow Shapley to seek a position on the staff there, so that he could use the reflector in the search for Cepheid and other variables.

Suppose Miss Leavitt's 'period–luminosity' relation for the Cepheids in the Cloud also held true for the Cepheids in the globular clusters. This being so, a Cepheid of a given period in one globular cluster was of the same luminosity as a similar Cepheid in a second cluster. Therefore, differences in the stars' apparent magnitudes must reflect differences in their distances – and therefore in the distances of the two clusters to which the stars respectively belonged.

But that would give only relative distances. To convert these to actual distances, it would be necessary to calibrate Miss Leavitt's relation – to establish the actual distance of one Cepheid somewhere. The only Cepheids near enough to offer any hope of this were among the isolated stars that are the Sun's neighbours in the Galaxy; and these galactic Cepheids would come into play only on the basis of a further assumption, that such Cepheids are likewise physically similar to those in the Cloud and in the globular clusters. Assumptions of uniformity of this type are inherent in any attempts in astronomy to measure great distances; the hope is that the assumption is true enough to give answers to otherwise unanswerable questions, and astronomers whose research depends upon distance measurements have their fingers permanently crossed. But such assumptions had never been made with the daring and virtuosity that Shapley was to display.

Unfortunately, even the nearest of the galactic Cepheids were too distant for the usual methods of measuring stellar distances. Shapley used ingenious – but questionable – statistical procedures to overcome the difficulty, and so achieved the necessary calibration. This done, he set about exploiting this veritable yard-stick for the universe. He was now able to obtain a distance for any globular cluster in which he could manage to determine the period and apparent brightness of a Cepheid variable.

This was easier said than done; even on a plate taken with the 60-inch, the individual stars in a globular cluster were very faint, if indeed they could be seen at all. To identify a star as a Cepheid variable, and to determine the number of days that it took to complete a cycle from one maximum brightness to the next, would require many such plates. With the dedication of a young man driven by ambition, Shapley worked and worked, until he had identified the periods of Cepheids in a dozen of the nearest globular clusters.

He then encountered the next problem: in the more distant clusters, Cepheids were too faint to be any longer visible. However, Shapley had noticed that the brightest stars in any one cluster were similar in luminosity to the brightest stars in other clusters. This he made the basis of another uniformity assumption, one that took him from clusters where he could see Cepheids in addition to the brightest stars, out to clusters where he could see only the brightest stars.

Finally, at distances where even the brightest stars could no longer be distinguished, he made yet another assumption, that the clusters themselves were physically uniform. Accordingly, comparison of the apparent diameter of a remote cluster with the apparent diameter of a cluster of known distance yielded the distance of the remote cluster.

By this succession of daring steps, Shapley in 1917 arrived at a distance for the most remote of the globular clusters, of some 200,000 light years. This was many times the accepted diameter of the entire Galaxy. To be so distant and still appear of significant size, the clusters themselves must be vast, perhaps (Shapley at first thought) comparable to the Galaxy itself.

But why were the clusters concentrated in the direction of Sagittarius, and why were they distributed symmetrically to either side of the galactic plane? Could it be that, as Bohlin had suspected a decade earlier, the Galaxy is enormously bigger than previously thought, and that its centre lies far away in the direction of Sagittarius, in the midst of the system of globular clusters? If so, the diameter of the Galaxy must be a staggering 300,000 light years, and its volume a thousand times greater than other astronomers believed.

Whereas practitioners of the 'selected areas' approach were studying in painstaking fashion the stars near enough to be individually visible, Shapley had leap-frogged further and further into space, from globular cluster to globular cluster. His resulting model for the Galaxy, announced in 1918, was too daring for his contemporaries. To most of them, the distribution of the globular clusters to one side of the sky – the fact that had led him to his Big Galaxy – was a single anomaly in an otherwise coherent picture.

Shapley was to see himself as a latter-day Copernicus, who had dethroned man and banished him to the margins of the Galaxy. But such banishments had been commonplace down the centuries, and had presented no problems. If anything, his estimate for the galactic diameter tended to make man special (and thereby added to the problems of acceptance); for by multiplying the volume of our Galaxy a thousandfold, Shapley made it the dominant structure in the known universe.

Orbital Motions about the Galactic Centre

In 1925 the Swedish astronomer Bertil Lindblad (1895–1965) considered the consequences for the observed motion of the stars that would follow from Shapley's Galaxy in which the Sun was far removed from the galactic centre. In the solar system, the inner planets not only have less far to travel in their orbits than do the outer planets, but they move through space with greater velocities. The very same laws of dynamics apply to the stars. Suppose then that Shapley was right, and the solar system was located not at the centre of the Galaxy but at a great distance from it. The stability of the galactic system would require the Sun and the nearby stars all to be moving in orbit about the centre of the Galaxy. This being so, then because of the laws of dynamics, the stars in orbits inside that of the Sun would be travelling faster and so forging ahead of us, while the stars moving in orbits outside the Sun's would be falling behind. As viewed from the solar system, therefore, the inner stars would seem to be moving in one direction (as they forged ahead of us) and the outer stars in the opposite direction (as they fell behind); and the greater a star's distance from us in the direction towards (or away from) the galactic centre, the greater would be its apparent speed.

This theoretical pattern can easily be converted into an equivalent pair of patterns, one of motions at right-angles to the line of sight (the 'proper motions' that accumulate century by century), and the other of motions in the line of sight (radial velocities, as revealed by spectroscopic analysis of starlight). The demonstration that such patterns actually exist in the observed stellar motions called for painstaking work, carried out principally by Jan Oort (1900–92) of Leiden Observatory.

Oort's conclusions, published in 1927, offered valuable support to Shapley. Both believed that the Sun was located far from the Galactic centre. But because Oort was studying stars that were nearer than Shapley's globular clusters, his data were less affected by obscuration, and so the size he derived for the Galaxy was markedly smaller than Shapley's.

A Problem Recognized: Dark Matter in Interstellar Space

The explanation for the discrepancy was not long in coming. The application of photography to the mapping of the sky had rapidly increased the number of irregular, gaseous nebulae like the Orion Nebula that were found in the Milky Way.

Left: The region of the Milky Way around Theta Ophiuci, photographed by E.E. Barnard in 1905.

Opposite: The stars of the Pleiades enveloped in nebulosity, on a photograph taken at Yerkes Observatory in 1901.

They were regarded as individual objects in space, and were given names and catalogue numbers, and plotted on maps of the sky. The spectrograph showed them to be made of glowing gas.

A new type of nebula was recognized in 1913. The star Merope in the Pleiades had been found by Wilhelm Tempel in 1859 to be surrounded by a faint nebulosity. Later photographs showed all the stars in the cluster to be involved in a rather unusual nebula of faint, wispy streaks. V.M. Slipher at the Lowell Observatory photographed the spectrum of this nebulosity in 1912, and was surprised to find it a continuous spectrum crossed by dark absorption lines, which mimicked exactly the spectrum of the stars. The nebula was evidently made of dust particles which reflected the starlight; and more of these 'reflection nebulae' in the vicinity of bright stars were soon discovered.

At about the same time, E.E. Barnard (1857–1923) of Lick Observatory was using wide-field portrait lenses to take a superb series of photographs of areas in the Milky Way. They showed a complicated structure of star clouds, and rifts and holes where there were few or no stars. At first Barnard believed that this was the real distribution of the stars, but as he continued his work he almost reluctantly accepted that these were true clouds, not of bright gas but of dark obscuring material, and in his two-volume *Atlas* of photographs, published in 1927, he included a catalogue of the more prominent of them.

There was a curious reluctance on the part of astronomers to accept that the bright and dark nebulae might be indicative of a more general substrate of gas and dust in interstellar space; the general view was that these were isolated objects, and that the vast interstellar spaces were largely empty and quite transparent. The consequence of any significant amount of, in particular, absorbing dust, which would dim the light of distant stars and further complicate the determination of their distances and intrinsic brightnesses, would be very serious.

NGC 7814 (far left) photographed by James Keeler at Lick Observatory in 1899, and NGC 4565 (left), photographed by C.D. Perrine in 1901, after Keeler's premature death the previous year. Nebular photography was resumed at Lick in 1910, when Heber Curtis was assigned to the task. On these and similar photographs, evidently of spiral nebulae seen edge-on, Curtis was struck by the bands of dark matter in the planes of the nebulae.

Indeed, it almost seems that this was too great a problem to be contemplated. Eddington once remarked that astronomers were like the guest who refused to sleep in a reputedly haunted room, and who explained, 'I do not believe in ghosts but I am afraid of them'.

The evidence for the presence of a layer of obscuring material through the central plane of the Galaxy now seems to us overwhelming. In 1869, by painstakingly plotting onto sky charts the 4,000 'irresolvable' nebulae that had been catalogued by John Herschel, Richard Proctor had shown that the region of the sky close to the Milky Way contained very few of them. Most of these irresolvable nebulae later proved to be spirals. Heber D. Curtis (1872–1942) of Lick Observatory argued in 1920, in the 'Great Debate' with Harlow Shapley (see page 329), that the spiral nebulae that were seen edge-on had at least a peripheral band of dark matter, and that such a band in our own Galaxy would explain why the spiral nebulae appeared to avoid the Milky Way (see page 327).

The observations that finally convinced astronomers were the result of patient work by another staff member of Lick Observatory, Robert J. Trumpler (1886–1956), published as recently as 1930. Trumpler had for many years concentrated on the study of the hundreds of open star clusters, which are rather narrowly confined to the plane of the Milky Way, and of which the Hyades and the Pleiades are the nearest examples. He first grouped them into types characterized by similar structures and the similar shapes of their H–R diagrams. Within one type, the relative distance of the various clusters could then be measured by two distinct methods.

Opposite: The nebulae (indicated by dots) and star clusters (crosses) in the northern hemisphere, based on the entries in John Herschel's 1864 catalogue, and plotted by Richard Proctor and Sidney Waters in 1873. The star clusters are mostly in or close to the Milky Way, but the nebulae are conspicuously absent from this region. The tendency of the nebulae to be present in some parts of the sky and absent from others had been remarked by William and John Herschel and others, but the Milky Way as a 'zone of avoidance' for nebulae was graphically demonstrated by the maps prepared, first by Proctor, and then by Proctor and Waters.

On the one hand the Main Sequence stars of, say, spectral type F had the same absolute magnitude in all the clusters. The greater the difference in the observed apparent magnitudes of such stars in any two clusters, therefore, the greater was the difference in the distances of the two clusters (the 'fainter means more distant' method). On the other hand, for clusters of the same intrinsic size, the angular diameter of a cluster on the sky was also a measure of its distance (the 'smaller means more distant' method).

The two methods did not give compatible results: the distances given by the apparent magnitude method were greater than those given by the angular diameters, because the starlight had been dimmed in its passage to the Earth. This showed the presence of a general absorption at all wavelengths.

Trumpler also investigated the phenomenon of selective absorption. If the dust particles are about the same size as the wavelength of light, the blue light is scattered out of the line of sight more than the red light, so the star appears to be redder than its spectral type would suggest. (The setting Sun appears red for just this reason.) This increases the measured 'colour index' of the star by an amount that is called the 'colour excess'. It was difficult to measure the quantities exactly on photographs, and only when photoelectric methods of measuring stellar magnitudes became widely used after about 1950 did measurements of the colour excess become a powerful method for correcting for the effects of the interstellar dust. Trumpler, summarizing his discovery in his 1930 paper, wrote:

> We are thus led to the conclusion that some general and selective absorption is taking place in our Milky Way system, but that this absorption is confined to a relatively thin layer extending more or less uniformly along the plane of symmetry of the system.

No later discoveries were to change this view. Trumpler derived a general absorption averaged over different directions in the Galaxy of about 1 magnitude of dimming for every 5,000 light years, not much smaller than the presently accepted value. The concentration of the material to the plane of the Galaxy is very strong, so that the actual nucleus of the Galaxy some 30,000 light years away is invisible to the eye and the telescope, which explains the early failures to discern the spiral pattern of the Milky Way.

The Galaxy Takes Shape

In the years following, the essentials of the modern picture of the structure of the Galaxy finally took shape. It was summarized in 1938 by J. S. Plaskett (1865–1941), who had recently retired as Director of the Dominion Astrophysical Observatory in British Columbia. There was, he said, 'a great central disc of stars, irregularly distributed in groups or clusters, probably with a general underlying field of stars and with possible spiral structure'. The diameter was some 100,000 light years, and the thickness increased to around 16,000 light years at the centre. 'A stratum of

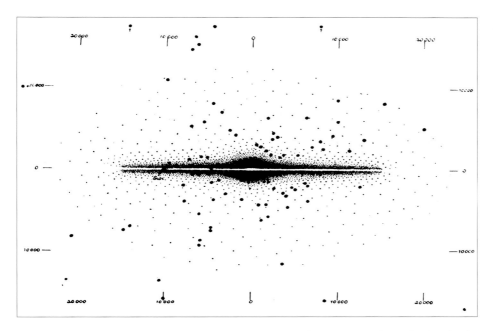

J.S. Plaskett's 1938 diagram to illustrate his picture of the structure of the Galaxy, here seen edge-on. The Galaxy has a diameter measuring some 100,000 light years, and the central bulge is 16,000 light years in thickness. In the galactic plane is a stratum of obscuring matter, while around the Galaxy is a halo of globular star clusters and special stars.

diffuse absorbing matter strongly concentrated to the galactic plane' had a thickness of some 1,000 light years, while around the centre was a spherical assemblage of globular clusters and stars of special types. After all the turmoil of recent decades, Plaskett could at last take comfort in the fact that 'the concept developed has a certain unity, completeness and probability'.

THE SPIRAL NEBULAE

Lord Rosse's reflector had revealed the spiral shape of M 51 within weeks of seeing 'first light' in 1845, and in the second half of the century, the number of known spirals grew by leaps and bounds. In 1898 James E. Keeler (1857–1900) became director of Lick Observatory, and began systematic photography of nebulae with the Crossley reflector. The distinction gradually became apparent between the irregularly shaped nebulae in the Milky Way and those with a more regular form. Moreover, in the parts of the sky where the spiral nebulae were common there were also many similarly sized nebulae which, although not showing a spiral structure, also had regular shapes, of circular or smoothly elliptical outline and, like the spirals, getting brighter towards their centres. It became common to call all of this class of regularly shaped nebulae 'spiral nebulae', even if a spiral structure was not recognizable. Keeler estimated that some 120,000 spiral nebulae were accessible to the Crossley, perhaps half of which would show an actual spiral form.

The 1900 Consensus

What were these mysterious objects? Many, including Keeler, thought they were planetary systems in the making. If, he wrote, 'the spiral is the form normally assumed by a contracting nebulous mass, the idea at once suggests itself that the

The observatory of Isaac Roberts. His original plan of preparing a photographic star chart was superseded by the *Carte du Ciel* project, and Roberts therefore concentrated on the photography of nebulae. The larger instrument, completed in 1885, was a 20-inch reflector for photographic work. The refractor, of 7 inches aperture, was for eye observations.

solar system has been evolved from a spiral nebula'. William Huggins, writing in 1889, had taken a similar view. Of a photograph of the Andromeda Nebula, he said that it 'shows a planetary system at a somewhat advanced stage of evolution; already several planets have been thrown off, and the central gaseous mass has condensed to a moderate size as compared with the dimensions it must have possessed before any planets had been formed'.

On the other hand, could the spiral nebulae be galaxies, vast star systems whose true nature was disguised from the Earth-based observer by their great distances? It was difficult to believe that such 'island universes' existed in their hundreds of thousands. And indeed there was compelling evidence to the contrary, some of it familiar in type, some new.

The long debate over the nature of the nebulae had witnessed many claims that change had occurred in a particular nebula, too rapidly for the nebula to be a distant and vast star system. One might have supposed that, once photographs superseded sketches made by observers of limited artistic ability working under difficult conditions, the reality or otherwise of such changes would be a question easily settled. Such was very far from being the case. As early as 1899, the leading English practitioner of nebular photography, the amateur Isaac Roberts (1829–1904), was claiming to have photographic evidence of the rotations of the Andromeda Nebula and of the Whirlpool Nebula, M 51 – rotations that we know to have been illusory.

There were also two important additional pieces of evidence against the identification of spirals with galaxies. One was the discovery by Richard Proctor (see page 321) that the region of the sky close to the Milky Way contained very few of them. If the spirals were galaxies comparable to our own, why did they distribute themselves in space so as to avoid our galactic plane?

Then, by a remarkable chance, in 1885 a star flared up in the Andromeda Nebula. It increased in brightness until it was equal to one-tenth of the entire nebula. If the nebula was indeed a galaxy containing millions of stars, then in a matter of days this one single star had increased its luminosity until it equalled the combined brightness of hundreds of thousands of ordinary stars. In fact this was exactly what had happened, for S Andromedae (as it is known) was a supernova; but no process known to nineteenth-century physics could have produced such dramatic celestial fireworks. It seemed far more likely that a single star had encountered a nebulous cloud of modest size, and flared up as it passed through the cloud.

Evidence of the striking advance brought about by the use of photography in the study of nebulae: the Andromeda Nebula, M 31, (below left) as drawn at the Harvard College Observatory in 1874, and (below) as photographed by Isaac Roberts in 1888. Whereas it was difficult to make much sense of the Harvard drawing, the Roberts photograph showed clear evidence of structure; in particular, the nebula lay in a plane tilted at an angle to the observer. A decade later, Roberts believed he had evidence that the nebula was rotating.

The majority opinion among astronomers in 1890 was summarized by the highly respected historian of astronomy, Agnes C. Clerke, in her book whose title, *The System of the Stars*, significantly used 'system' in the singular:

> The question whether nebulae are external galaxies hardly any longer needs discussion. It has been answered by the progress of discovery. No competent thinker, with the whole of the available evidence before him, can now, it is safe to say, maintain any single nebula to be a star system of co-ordinate rank with the Milky Way. A practical certainty has been attained that the entire contents, stellar and nebular, of the sphere belong to one mighty aggregation.

All scientists consider it their duty to work with theories as simple as the evidence will allow, and astronomers are no exception. The term 'nebula' had originated as a description of objects in the sky. Some (we know) are gaseous and so 'truly' nebulous, while others are stars systems disguised by distance; but over the centuries, astronomers had struggled to force them all into either one category or the other. Accordingly, when, in 1864, Huggins's spectroscope had demonstrated beyond doubt the existence of gaseous nebulae, this had sown doubt in the minds of those who had hitherto believed in a multiplicity of island universes. Now these doubts had ripened into a widespread conviction that astronomy knew but one island universe.

1900–20: The Consensus Called into Question

The pendulum had swung too far. Before long, new evidence suggested that the death of the island universe theory of spirals had been greatly exaggerated. Although the spirals were faint and what little light they sent was dispersed to the point of near-invisibility when passed through a spectrograph, by 1912 improvements in instrumentation and in photography allowed V. M. Slipher at the Lowell Observatory to secure spectrograms of a number of nebulae that were clear enough to reveal the more prominent spectral lines. The work was painstaking indeed, with twenty to forty hours of observing time required for each plate; but the results were sensational, and when he announced them to a meeting of the American Astronomical Society in 1914, he was rewarded with a standing ovation.

Not surprisingly, the Andromeda Nebula was the first that Slipher tackled. The continuous spectrum crossed by a few dark lines was characteristic of starlight, and by January 1913 he had secured four plates on which the shift in the spectral lines caused by the Doppler Effect (see page 293) could be measured. Stars were known to move with line-of-sight ('radial') velocities of perhaps 20 kilometres per second; Slipher found that the Andromeda Nebula was approaching at no less than 300 km per second, by far the greatest velocity known for any object in the universe. Word of his discovery was greeted by some astronomers with incredulity, but soon his results were confirmed by other observers. By the time of the 1914 meeting, Slipher had radial velocities for fifteen spirals, and by

1917 the total had grown to twenty-five. No fewer than four of these were in excess of 1,000 kilometres per second.

Slipher's twenty-five spirals were not uniformly distributed across the sky. Most were located to one side of the Galaxy and were receding; but some, including the Andromeda Nebula, were on the opposite side and most of these were moving towards us. To Slipher the explanation was clear: the Galaxy was itself 'a great spiral nebula which we see from within', and it was drifting among the other spirals with a velocity of some 700 kilometres per second.

Meanwhile, at Lick Observatory, Heber D. Curtis was continuing the programme of nebular photography begun by Keeler, and (as we have seen, page 321) he was encountering nebulae that were clearly edge-on to us. These had at least a periph-eral band of obscuring matter, and Curtis realized that similar obscuring matter in our own Galaxy would prevent our seeing spiral nebulae close to the galactic plane. He concluded that spirals did not in fact avoid the Milky Way as Proctor and others had thought; spirals were present in those directions as elsewhere, but concealed from our view. This confirmed him in his belief that the spirals were 'inconceivably distant, galaxies of stars or separate stellar universes so remote that an entire galaxy becomes but an unresolved haze of light'.

His conviction received a further boost in 1917, when he discovered three novae on his photographs of spirals. At about the same time, G. W. Ritchey (1864–1945) of Mount Wilson found a nova on a photograph of a spiral taken only a few days earlier, and this nova was still visible and so available for study by astronomers worldwide. Encouraged by these successes, others joined in the hunt, and yet more novae were discovered. Without exception they were altogether fainter than S Andromedae, which suggested to Curtis that novae might fall into two distinct classes (as later proved to be the case, see page 307). Suddenly the inference drawn from the exceptional brilliance of S Andromedae began to seem less secure.

But not all the new evidence favoured the island universe theory of spirals. At Mount Wilson Observatory a Dutch astronomer, Adriaan van Maanen (1884–1946), had won himself a reputation as a meticulous measurer of photographic plates. In 1916, to study possible changes in the spiral nebula M 101, van Maanen used a machine called a blink microscope that enabled him to switch between pairs of photographs of the nebula taken at different times. This made it relatively easy to detect alterations that had taken place in the interval between the two photographs. He concluded that the nebula was rotating. This being so, it could hardly be a distant galaxy of vast diameter; for as the spiral rotated, the outlying parts of the supposed galaxy would have had to travel through space with incred-ibly large velocities.

It is difficult to imagine any investigation in astronomy that could stay closer to the plain facts. The photographs represented objective evidence; and van Maanen had in effect superimposed them, noted the changes, and described the rotation that had given rise to these changes. By 1921 he had similar results for three more

Evidence that the spiral nebula M 101 was rotating. In 1916, Adriaan van Maanen used a 'blink microscope' to compare pairs of photographs of the nebula taken at different times. The machine allowed him to switch instantaneously from one photograph to the other, so that differences were immediately apparent. Van Maanen concluded that the nebula was rotating, in directions shown by the arrows. To rotate perceptibly in so short a time, the nebula had to be small and near.

spirals. 'Congratulations on the nebulous results!', Shapley wrote. 'Between us we have put a crimp in the island universes, it seems, – you by bringing the spirals in and I by pushing the Galaxy out. We are indeed clever, we are.'

Another development concerned the radial velocities of spirals. As the number of known velocities grew, it became clear that Slipher had been premature in interpreting the evidence as revealing a drift of the Galaxy among the spirals. Spirals that were approaching the Galaxy proved to be exceptional; for the most part, the spirals were receding, in all directions. Shapley even suggested that they might be little more than wisps of matter driven off from the Galaxy by radiation pressure from the

The 'Great Debate'

In 1920 a meeting in Washington was arranged, at which Harlow Shapley and Heber D. Curtis were to argue for their contrasting positions on the structure of the universe. But their attentions were in fact focused in different directions. Shapley was committed to his Big Galaxy; the spirals were to him of minor importance, though they could scarcely be comparable galaxies with diameters of 300,000 light years. Curtis was studying spiral nebulae, which he believed to be island universes; the structure of the Galaxy was of secondary interest, though he required it to be of the manageable size accepted by nearly everyone except Shapley.

Though the meeting has since been built up into legend as 'The Great Debate',

this is the result of confusion between what actually happened in Washington and what was published many months later. At the meeting itself, Shapley was uncomfortably aware of the presence of representatives from Harvard Observatory, whose vacant directorship he coveted; and so he made sure he did not suffer defeat, by reading a text so elementary as to verge on the trivial. Curtis, who spoke second, thought about abandoning his technical material, but decided it was then too late to change. In the published version of the 'debate', however, each man gave a powerful summary of the arguments in favour of his position. Astronomers found it hard to decide whose case was the stronger.

combined light of the galactic stars. And so the movement in favour of the 'island universe' theory of spirals lost a little of the momentum it had recently acquired.

Since 1917 the 100-inch Hooker reflector had been in service at Mount Wilson. It had the largest mirror of any telescope in the world and this was of superb optical quality, and so the 100-inch was uniquely powerful for the photography of faint objects such as nebulae. In 1920 J. C. Duncan (1882–1967), director of the small observatory at Wellesley College in Massachusetts, had noticed three faint variable stars in the spiral nebula M 33, on photographs that included one taken with the 100-inch. They were encouraging indications of what might be achieved through a determined campaign with the 100-inch.

Hubble's Discovery of Cepheids in the Andromeda Nebula, 1923–24

Release from army service in 1919 had enabled Edwin Hubble (1889–1953), a lawyer turned astronomer, to take up an earlier invitation to join the Mount Wilson staff. In October 1923, Hubble embarked on a systematic search for novae in the Andromeda Nebula, biggest of all the spirals and presumably one of the most accessible. He found one on the very first of his plates.

This at least was his initial verdict. But as he worked his way through the Mount Wilson collection of photographs of the nebula, mostly taken with the 60-inch and going back to 1909, he realized that this star was no newcomer but a variable. When the search was complete he had found over sixty plates on which the star was present, sometimes fainter than nineteenth magnitude, at other times as bright as

Opposite: The Mount Wilson 100-inch reflector under construction. In the present configuration the telescope would be pointing towards the zenith (and the workman would be standing on the mirror). The polar axis is in the form of a rectangular steel cradle. This type of equatorial mounting prevents the telescope examining the sky in the region of the North Pole, but this was considered an acceptable price to pay for stability.

Edwin P. Hubble

Edwin Powell Hubble was born in Marshfield, Missouri, on 20 November 1889. He studied law at Oxford but soon turned to astronomy, becoming a graduate student at the Yerkes Observatory of the University of Chicago. The war prevented his immediate acceptance of the offer of a post at Mount Wilson, but he joined the staff there in 1919. His discoveries with the 60-inch and 100-inch reflectors on Mount Wilson, and later with the 200-inch on Palomar Mountain, can be said to have changed the face of the universe; but his vanity and poor judgement led to his being passed over for the post of observatory director. He died of a heart attack in San Marino, California, on 28 September 1953.

Edwin Hubble at the controls of the 48-inch Schmidt telescope on Palomar Mountain. A short-focus reflector of wide field suffers from one form of distortion if the mirror is spherical and another if it is paraboloidal, but about 1930 the Estonian Bernhard Schmidt devised a correcting plate to overcome the former problem. `Schmidt telescopes' with this device can photograph a wide area of sky on a single plate. They have been used extensively in photographic surveys, in particular for the Palomar Sky Survey of the 1950s and the more recent Southern Sky Survey, which together cover the entire sky down to stars of about the 21st magnitude.

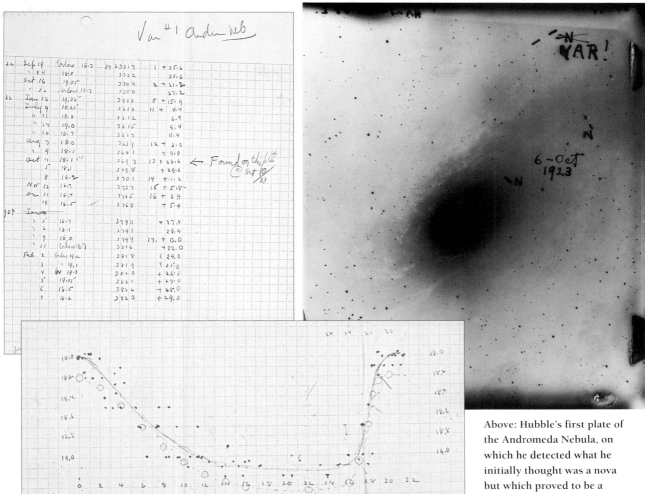

Above: Hubble's first plate of the Andromeda Nebula, on which he detected what he initially thought was a nova but which proved to be a variable star.

Above left: Part of Hubble's search through the Mount Wilson plate collection for data on the fluctuating brightness of the variable.

Left: The results of the search: the fluctuations were cyclic and the light curve was of characteristic Cepheid shape, showing rapid increase and slow decline (see page 315), with period 31.415 days.

eighteenth. The plates were numerous enough to allow him to plot the pattern of the rise and fall of the star's light.

It began to look as if the variable was the greatest prize possible: a Cepheid, one of the 'lighthouse' stars that Shapley had used in his distance determinations of the globular clusters. If so, its light should leap rapidly from minimum to maximum. To test the prediction, Hubble took a series of plates during the first week of February 1924. The brightness of the star rose rapidly as expected: it was indeed a Cepheid.

Hubble must have had difficulty in containing his excitement as he confirmed the characteristic shape of the light curve and its lengthy period of over thirty-one days. The longer the period of a Cepheid, the greater the luminosity of the star; yet this star appeared as only eighteenth magnitude at best, and was so faint that it had

An illustration of the power of the Mount Wilson 100-inch: a 1925 photograph of the Andromeda Nebula, as used by Hubble as frontispiece to his 1936 classic, *The Realm of the Nebulae*. 'The observer looks out through the swarm of stars which surrounds him, past the borders and across empty space, to find another stellar system... The brightest objects in the nebula can be seen individually, and among them the observer recognizes various types that are well known in his own stellar system. The apparent faintness of these familiar objects, as they lie in the nebula, indicates the distance of the nebula – a distance so great that light requires seven hundred thousand years to make the journey.'

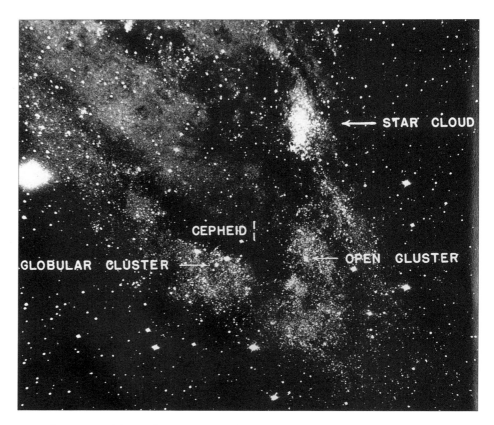

passed unnoticed by earlier Mount Wilson investigators. For it to be so luminous and yet appear so faint, its distance – and the distance of the nebula of which it was part – must be enormous, indeed approaching one million light years. Even on Shapley's estimates, the nebula lay far outside the Galaxy.

Furthermore, the Cepheid proved that the nebula contained not merely star-like objects of doubtful interpretation, but a true star that varied in familiar fashion. And not just one such star: by the time Hubble was confident enough to break silence on 19 February with a letter to Shapley, he had found a second variable, as well as nine novae.

It was a momentous step in the history of astronomy. Hubble had used the same yardstick as Shapley, but in far more convincing and direct fashion. Shapley, to whom the Big Galaxy was paramount and the status of the spirals a secondary consideration, immediately accepted the implications of Hubble's discovery.

Not so van Maanen, who was Hubble's colleague but no friend – the pair of them communicated only via intermediaries. Hubble hesitated to go into print, troubled by the incompatibility of his results with Van Maanen's alleged rotations in spiral nebulae. Eventually, and only at the insistence of Shapley and others, Hubble allowed a paper summarizing his discoveries to be read in his absence at a meeting of the American Astronomical Society, on New Year's Day, 1925. Those present knew the long debate had ended: there are indeed many 'island universes'.

Hubble's Demolition of van Maanen

Two serious anomalies remained: the van Maanen rotations, and Shapley's estimate for the diameter of our Galaxy, which implied that if the Andromeda Nebula was an island, the Galaxy was a continent by comparison. The van Maanen problem became acute as the Dutchman pointedly persevered with his comparisons of pairs of photographs of spirals, concluding in every case that the nebula was rotating at a speed that made it physically impossible for it to be an island universe.

Hubble, his irritation ripening into anger, finally decided on an ingenious strategy. During 1932 and 1933 he retraced Van Maanen's steps, comparing the very same pairs of photographic plates (though with significant differences in technique). In addition, he had new plates specially taken for the purpose. He also enlisted two experienced colleagues, who were to make their own independent measurements.

All three agreed that in the nebulae in question, 'internal motions of the order predicted by van Maanen do not exist'. It could hardly be supposed that Hubble and his colleagues had all made errors in their measurements, errors that by the most amazing coincidences invariably cancelled out the 'rotations' and left them with null results. Clearly, the rotations were illusory.

Hubble drafted papers for publication embodying his demolition of van Maanen, but the director of Mount Wilson would not permit a public display of the feuding within his staff. A compromise was imposed, and in 1935 readers of the *Astrophysical Journal* were no doubt intrigued to find there a two-page paper by Hubble delicately outlining his conclusions, immediately followed by a two-page paper by van Maanen, conceding that 'it is desirable to view the motions [the rotations in the spirals] with reserve'.

The Galaxy and the Andromeda Nebula: Comparability at Last

The second anomaly created by Hubble's discovery stemmed from Shapley's estimate of the size of the Galaxy. According to Shapley, the Galaxy was some 300,000 light years in diameter; Hubble's distance for the Andromeda Nebula implied that in diameter it was only one-tenth as big as the Galaxy, and so a mere one-thousandth in volume. The discrepancy was diminished in 1930 by Trumpler's discovery of the obscuration that dims the light from distant objects in the galactic plane, which had misled Shapley into thinking galactic Cepheids more distant than they really were. This reduced the Galaxy's diameter to 100,000 light years. And meanwhile, increasingly sensitive photographic plates showed that the diameter of the Andromeda Nebula was more extensive than had been thought. Yet even so, the Galaxy remained a super-system and the Andromeda Nebula small by comparison.

The instincts of many astronomers made them uncomfortable over any theory that privileged our location in the universe. The discrepancy was the more puzzling in that the Andromeda Nebula resembled the Galaxy in so many ways. Each contained numerous bright Cepheids, a system of globular clusters, and (in all probability) spiral arms outlined by layers of dust and highly luminous blue stars. Yet the novae in the Andromeda Nebula were supposedly fainter than their counterparts in the Galaxy, as were the most luminous stars, and the globular clusters.

The globular clusters were particularly puzzling, since here uncertainties as to the effects of loss of light through the presence of obscuring matter in interstellar space could play no part: obscuration would affect equally a globular cluster and the distance-measuring stars in it, and their relative brightnesses would be unaffected. Yet the globular clusters in the Andromeda Nebula were on current figures some four times fainter than the globulars in the Galaxy.

One of the few astronomers to point out repeatedly that these anomalies would disappear if the Andromeda Nebula were assigned a distance twice as large as that currently accepted was Knut Lundmark (1889–1958) of Lund in Sweden. Also, in 1922 the Estonian astronomer Ernst Öpik (1893–1985) used an ingenious new method to determine the distance of the Andromeda Nebula. A few years earlier F. G. Pease of Mount Wilson had used the Doppler Effect (see page 293) to measure the rotation velocities of different parts of the nebula. Öpik showed that if, as seemed reasonable, the mix of stars was the same as in our own Galaxy and their average ratio of mass to luminosity was the same, the Andromeda Nebula had to be about 1.5 million light years away if Pease's observations were to be explained. This was also much greater than Hubble's 1925 value of 900,000 light years, and would imply that the Andromeda Nebula was bigger than Hubble thought. But so much had been invested in the accuracy of the Cepheid method of establishing distances, that it had become almost an article of faith among astronomers, and Hubble's values prevailed.

One obstacle to progress was the lack of information about the bright central region of the Andromeda Nebula; its outer parts had been 'resolved' into stars on photographic plates, but not so the central condensation. With existing instrumentation the task seemed hopeless. The Mount Wilson 100-inch was the biggest reflector in the world, but even so, theoretical considerations put the resolution of the central bulge at the very limit of its powers. In any case, the city lights of nearby Los Angeles made the quest hopeless.

But as luck would have it, these lights were dimmed as a security measure when the United States entered the Second World War; and although most of the astronomers were away on war service, one who remained was the German Walter Baade (see page 306) who had a physical handicap (and who in any case had overlooked the formalities of taking out American citizenship, and so was exempt from military duty).

Baade was exceptionally skilled in long-exposure photography with the 100-inch, and his first efforts in the autumn of 1942 brought him to the verge of success. Unfortunately, although the city was relatively dark, his fast blue-sensitive plates registered the faint background glow of the night sky, and this limited the useful

exposure time to about ninety minutes. Baade therefore turned to the newly introduced red-sensitive plates. These would be less sensitive to the background glow,
which he further reduced with coloured filters. They would also be more sensitive
to red stars; and the fact that Hubble had resolved the spiral arms of the nebula,
where there were very bright, blue stars, but had failed to resolve the central bulge
suggested that red stars might predominate there.

In the autumn of 1943, Baade photographed the central region of the Andromeda Nebula with exposures of about four hours each. His control of the telescope
had to be meticulous in the extreme, not least because changes in air temperature
as the night progressed constantly threatened to throw the instrument out of
focus. He was rewarded with photographs that showed stars in their 'thousands
and tens of thousands', as was also the case with the two smaller elliptical companion nebulae.

Large numbers of the stars were indeed red, but they were much brighter than
the red stars Trumpler had found in the open clusters within our Galaxy, and so they
occupied different positions on the H–R diagram. It eventually occurred to Baade
that the red stars in the Andromeda Nebula had, instead, positions on the diagram
similar to the red stars in the globular clusters that surround our Galaxy.

Could the comparison be extended? Did the central bulge of the Andromeda
Nebula also have short-period variables of the type known as RR Lyrae stars, similar
to those so common in our globular clusters? Unfortunately, there was no way of
answering the question, for such stars would be altogether too faint for detection
with existing instrumentation.

But Hubble and Baade found an alternative way forward. Some years before,
Shapley had announced the discovery of a new type of stellar system, examples of
which were in the constellations Sculptor and Fornax. They were located outside
our Galaxy, but were much smaller than the elliptical galaxies. Yet while they were
much bigger than our globular clusters, their populations of stars resembled them:
lots of red giants and RR Lyrae variables. Astronomers tended to think of them as
overgrown globular clusters, but after Baade found numerous red giants in the
centre of the Andromeda Nebula, the penny finally dropped: the Sculptor and
Fornax systems were intermediate between the globular clusters and the elliptical
nebulae, and similar populations of stars were to be found in all three.

In 1944, Baade announced his conclusion that stars belonged to one or other of
two types. Type I stars populated the central plane of the Galaxy. They included
stars like the Sun and most of its near neighbours, and the stars that are found in
open clusters like the Hyades and the Pleiades. They had been formed from the
interstellar material, gas and dust, that also occupied the central plane of the
Galaxy, and their average chemical composition was similar to that of the Sun. The
most luminous blue giants at the top end of the Main Sequence had formed only
recently from the interstellar material, and all – stars, gas, and dust – followed
nearly circular orbits about the centre of the Galaxy.

Type II stars, on the other hand, were older stars, found in the gas- and dust-free elliptical galaxies, in the equally dust-free globular clusters associated with spiral galaxies, and in the central bulges of the spirals. In our Galaxy, the globular clusters and the isolated Type II stars move in elliptical orbits inclined at all angles to the galactic plane. The stars of Type II on average do not share the rapid circular motion of the Sun and the other Type I stars in the central plane of the Galaxy. When the orbit of a Type II star carries it into the neighbourhood of the Sun, its measured velocity consequently appears to be high.

The true complexity of nature had once again frustrated astronomers' attempts to impose a simple pattern. And in time, even Baade's two stellar populations would themselves prove to be an oversimplification.

Another anomaly that concerned Baade involved RR Lyrae variables. These variables were especially numerous in globular clusters, and for this reason were often spoken of as 'cluster variables'. They had much in common with the Cepheids of shorter periods, with which they had earlier been conflated. But whereas the luminosity of a Cepheid depended on the number of days in its period, the luminosities of RR Lyrae stars in our Galaxy were all much the same, irrespective of the length of their periods.

In his estimates of the distances of globular clusters (and hence of the diameter of the Galaxy), Shapley had sometimes been forced by the lack of Cepheids into using RR Lyrae stars instead, even though RR Lyrae variables with periods of half a day were about two magnitudes fainter than Cepheids with periods of 10 days. In photographs of the Andromeda Nebula taken with the 100-inch, these Cepheids appeared to have magnitude 20, in which case the RR Lyrae variables would presumably have magnitude 22 – beyond the reach of the 100-inch, but within reach of the 200-inch being built at Palomar Mountain, far enough south of Los Angeles to avoid the worst of the glare of the city lights.

Baade started to use the 200-inch as soon as it came into service in 1948. He expected the telescope would reveal RR Lyrae variables in the Andromeda Nebula, but, as he told the 1952 meeting in Rome of the International Astronomical Union, the first plates taken of the nebula with the 200-inch 'showed at once that something was wrong'. The reason was to be found in Baade's two populations of stars. The nearby Cepheids that Shapley had used to calibrate the period–luminosity relationship were in the spiral arms of the Galaxy and so of Type I, and these now proved to be more luminous – and therefore more distant – than had been thought. So too were the remote Cepheids that Hubble had detected in the spiral arms of the Andromeda Nebula, which likewise had to be removed to a greater distance. But the Cepheids that Shapley had used to determine the distances of the globular clusters, and hence the diameter of the Galaxy, were of Type II, and their brightnesses had been correctly assessed. It followed that the diameter of the Galaxy remained as before, while the distance, and therefore the diameter, of the Andromeda Nebula was to be doubled. As Baade put it:

Opposite: The 200-inch pyrex mirror in the optical shop at the Mount Wilson Observatory headquarters at Pasadena. Construction of such a large mirror presented enormous problems, whose solution was costly in time and money. After several attempts, a perfect mirror was cast in December 1934, but cooling and annealing took a further year, and polishing was not completed until after the Second World War. But eventual success opened the way to the construction of large reflectors worldwide. Note the ribbed structure; this kept the weight below 20 tonnes, and limited the problems caused by fluctuations in temperature while the instrument was in use.

Moreover, the error must be such that our previous estimates of extragalactic distances – not distances within our own Galaxy – were too small by as much as a factor 2. Many notable implications followed immediately from the corrected distances: the globular clusters in M 31 [the Andromeda Nebula] and in our own Galaxy now come out to have closely similar luminosities; and our own Galaxy may now come out to be somewhat smaller than M 31.

Our Galaxy thus lost the pre-eminence with which it had been endowed by Shapley, and was given the status it has today, that of a somewhat inferior sister to the Andromeda Nebula.

EXPANSION OF THE UNIVERSE AND THEORIES OF RELATIVITY

In his address to the Rome meeting in 1952, Baade added a further comment: 'Above all, Hubble's characteristic time scale for the Universe must now be increased from about 1.8×10^9 years to about 3.6×10^9 years.' How had Hubble succeeded in assigning an age to the Universe?

To answer this question we must go back into the nineteenth century. In astronomy the application of Newton's inverse-square law of gravity, applied to increasingly difficult problems in celestial mechanics with ever increasing mathematical power, had been triumphantly successful. It worked in the context of a simple and familiar universe: a framework of three dimensions of space in which the geometry of Euclid applied, and in which a uniform flow of time – past, present, and future – could be measured everywhere by clocks which, in theory if not in practice, were uniformly perfect. But one irritating observation remained: the anomalous advance of the perihelion of Mercury, discovered by Le Verrier in 1859 (see page 195), obstinately refused to yield to Newtonian theory.

In the physics laboratory, the middle decades of the nineteenth century had produced a vastly complex body of knowledge about the relations between electricity, magnetism and matter, which were largely explained by Clerk Maxwell in the 1860s, in a unifying theory that ranks in importance with Newton's theory of gravity. There were features of Maxwell's equations that made other theoretical physicists, like the Dutchman H. A. Lorentz (1853–1928), think deeply about the concepts of space and time and their relation; and geometers such as Gauss and his pupil G. F. B. Riemann (1826–66) had already explored geometries in curved spacetimes that were logically consistent even though they disregarded one of Euclid's postulates.

In 1880, the Polish-born American physicist Albert A. Michelson (1852–1931) was working at the Astrophysical Observatory in Potsdam, which had a vibration-free cellar. This fortunate circumstance allowed him to carry out a delicate optical measurement suggested by Maxwell's theory. The puzzling result was, in brief, that the electromagnetic theory of light required an ether, but the Earth could not be detected moving through it.

These and other investigations show that cracks were appearing in what had seemed a firmly-established structure of physics and mechanics, and that concepts that had once seemed beyond challenge were now being questioned. The greatest contributions to the creation of a new set of concepts of space and time were to be made by the German-born Albert Einstein (1879–1955) in the first two decades of the twentieth century. As Newton had said of himself, Einstein stood on the shoulders of giants, for Einstein developed his own ideas within the context of this new spirit of enquiry, by building on the work of Lorentz and numerous others.

Einstein's first important contribution came in 1905 with the recognition that the measurement of quantities like length and time were made, not in an absolute Newtonian framework, but relative to the observers. This Special Theory of Relativity explained, for example, the puzzling result of Michelson's experiment.

The General Theory, of 1915, went far beyond. It was essentially a theory of gravitation – not a modification of Newtonian theory, but a quite new way of thinking about the relation between matter, space, and time. The equations permitted the calculation of planetary motions (as Newton's had done) and almost immediately gave a new theoretical value for the motion of the perihelion of Mercury which now agreed with the observed one. They also predicted that a ray of light would be deflected when it passed near a massive body like the Sun by an amount that was significantly greater than the Newtonian value. The test was made by two expeditions from the Greenwich Observatory, under the direction of Eddington, at a particularly favourable eclipse of the Sun in 1919, when the new value was verified.

The Redshift in the Spectra of the Nebulae

What concerns us here is that the equations also permitted various solutions (depending on the assumptions that were made, for example, about the amount of matter in the universe and the geometry of space-time) as to how the universe would behave 'in the large'. Some of these solutions predicted that the universe would not be static, but would expand.

This brings us back to the observations of the large radial velocities of the spiral nebulae by Slipher from 1912 onwards (see page 326). It is clear that at first Einstein was not aware of Slipher's work. Equally, the difficult concepts, and advanced mathematics, of Einstein's General Theory were beyond the reach of most practising astronomers. But from the middle 1920s, particularly following the theoretical work of Willem de Sitter (1872–1934) of Leiden University and A. A. Friedmann (1888–1925) of St Petersburg, it became apparent that cosmology was no longer a speculative subject but a new branch of science with both theoretical and observational aspects.

By 1925, a total of forty-five radial velocities of nebulae were available, most of them determined by Slipher. A few, however, had been checked at other observatories, and the reliability of Slipher's measurements was generally accepted. The

Milton Humason

Milton Lasalle Humason was born in Dodge Centre, Minnesota, on 19 August 1891. As a young man he worked for a time as a mule driver on Mount Wilson, and then in 1917 he became a night assistant for the telescopes there. A student taught him the art of astronomical photography, and Humason showed such talent for the work that in 1922 he was appointed assistant astronomer. Before long he was unrivalled in the photography of the spectra of faint nebulae, and the former mule driver proved an invaluable assistant and partner to Edwin Hubble. Humason retired in 1957, and died on 18 June 1972.

largest were over 1,000 kilometres per second, suggesting that the nebulae were independent bodies outside the gravitational control of the Galaxy, consistent with Hubble's newly demonstrated theory of island universes.

The great range in the observed velocities made them difficult to interpret. In particular, it was difficult to disentangle the component of the observed motion that was due to the combined motions of the Sun and of our own Galaxy and not to that of the nebula under investigation. In 1918 Carl Wirtz (1876–1939) of Kiel, Germany, had attempted a method that had proved useful in the study of the motions of stars, and had introduced into the equations an additional term, known as the K term. It was in effect a quantity to be subtracted from all the measured velocities before one searched for the motion of the solar system. The K term for the nebulae also turned out to be enormously higher than that for stars, being measured in hundreds of kilometres per second rather than only a handful. It did not escape Wirtz's attention that the K term implied a remarkable expansion of the system of the nebulae, with nebulae rushing away in all directions.

In 1924, now using forty-two nebulae, Wirtz tried to see if the K term depended on the distance of the nebulae. These distances were of course unknown; but if the nebulae were supposed to belong to a single class of similar objects, then the smaller the apparent diameter, the greater the distance. A rough relation between distance and velocity away from the observer did emerge, in the sense that the more distant the nebula, the faster it was receding. The cosmological implications were now clearly understood: the title of Wirtz's paper (translated from the German of the *Astronomische Nachrichten* of 1924) is 'De Sitter's cosmology and the motion of the spiral nebulae'.

Hubble and the Law of the Redshifts

Hubble was well aware of Wirtz's work and acknowledges it freely in his writings. From 1925 he devoted a generous ration of his observing time on the 100-inch telescope (and, in the very last years of his life, on the 200-inch) to this problem of observational cosmology – the relation between the velocity of recession and the distance of the spiral nebulae. Over the years the development of thinking about these objects is reflected in the term used for them: 'spiral nebulae' become 'extragalactic nebulae' – that is, nebulae outside our own Galaxy – and finally the modern 'galaxies'.

Hubble shared the work with his colleague Milton Humason: Humason concentrated his observing skills on the determination of the radial velocities of ever more distant (and fainter) galaxies, and Hubble devised and put into practice methods of determining the actual distances from the brightness of the images on the photographic plates. But even with the 100-inch telescope, Cepheid variables were detectable only in the nearest galaxies. Hubble developed a 'stepping stone' method which later astronomers followed. In the galaxies where Cepheids were just visible, he identified the most luminous individual stars, which were some 50 or 100 times brighter than the Cepheids, and then used the same types of stars as standard objects in more distant galaxies. It was tediously slow work, in which small errors could easily multiply, as Hubble recognized.

By 1929 he had radial velocities and independent distance determinations for twenty-four galaxies, and he published a graph of the velocities (up to 1,100 kilometres per second) as a function of distance (up to 2×10^6 parsecs, or a little over 6×10^6 light years). Though the scatter was large, there was clearly a relation between velocity and distance; the best straight line showed that for every million parsecs (that is, 1 Mpc) the velocity of recession increased by kilometres per second. The quantity for which Hubble gave this first estimate, 500 km/sec/Mpc,

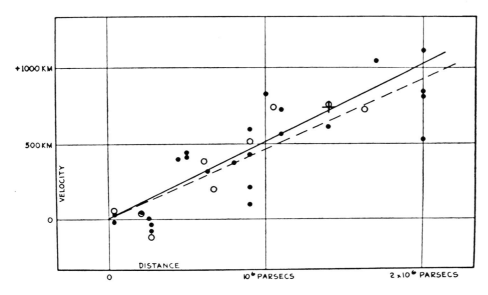

Hubble's velocity–distance relation, for galaxies at distances up to 2×10^6 parsecs. The vertical axis represents radial velocities in kilometres per second (corrected for the motion of the solar system itself), the horizontal axis the distances in parsecs. The black discs and circles indicate the observational data, corrected by alternative methods of calculating the solar motion. The (alternative) lines indicate the simple relation inferred from the data.

became known as Hubble's constant, and is one of the fundamental constants of cosmology. That the velocity of recession of galaxies is proportional to distance is known as Hubble's Law, or – since the velocities are determined from the Doppler shift to the red – the Law of the Redshifts.

How far out into the universe did the law hold? This had to be determined by a quite different method, for redshifts could be measured for galaxies in which even the most luminous stars could not be discerned. Hubble developed an ingenious method. Just as there are clusters of stars in the Galaxy, so also there are readily

Right: The cluster of galaxies in the constellation Coma Berenices, photographed with the 100-inch in 1933. Some four hundred galaxies are visible in an area of sky equal to that occupied by the Moon.

Below: the velocity–distance relation extended out to remote regions occupied by clusters of galaxies. The distance of each cluster has been inferred from the apparent brightness of the fifth brightest nebula in the cluster.

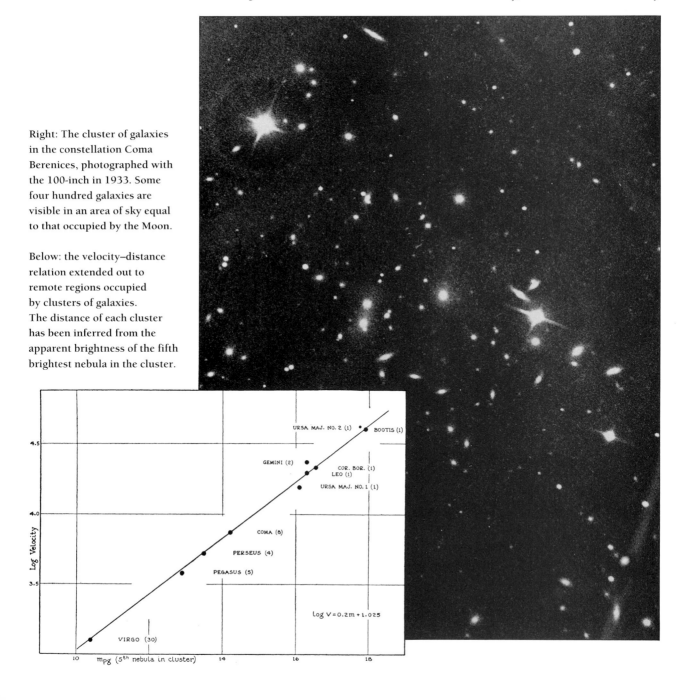

recognizable clusters of galaxies in the universe. Humason took examples of these clusters, ones whose size and apparent brightness showed they were at very different distances, and measured the redshift of the brightest galaxies in each of the clusters. Hubble recognized that a cluster of galaxies might contain one or two individual members that were freakishly bright, but he considered that the risk of such anomalies could be avoided if one measured in each cluster the apparent magnitude, not of the brightest galaxy, but of the fifth brightest. Working along these lines, Hubble and Humason showed that the law of the redshifts held to great distances. This implied that if Hubble's constant was known by other means, the redshift of a faint galaxy could be used to calculate its distance.

In 1935 Hubble was invited to give a course of lectures at Yale University, and he published them under the title of *The Realm of the Nebulae*. It clearly summarized the knowledge of observational cosmology at the time, and influenced a whole new generation of astronomers.

The Interpretation of the Redshift

It is interesting to note that the redshift was, from the first, interpreted as a Doppler shift – that is, due to the movement of the galaxy away from the observer – and measured in kilometres per second. The implication was that the universe was expanding. There is no doubt that the nearly simultaneous detection of the redshifts and the derivation of solutions of Einstein's equations that suggested that the universe would be expected to expand greatly encouraged this interpretation.

It followed from the expansion interpretation that as one went back in time, the galaxies were closer together: at some early time (which we may simplistically term 'the beginning of the universe') the universe would have been extremely dense. The interval from then to the present could be termed 'the age of the universe', which in the simplest interpretation was inversely proportional to the value of Hubble's constant – the larger Hubble's constant, the younger the universe. This already presented something of a problem in the 1940s: the Hubble age of about 1.8×10^9 years was less than the accepted age of the Earth required by geologists.

We have already seen (on page 338) that Baade's version of the Cepheid calibration reduced the value of Hubble's constant by a factor of two, and a further downward revision to about 100 km/sec/Mpc followed in 1958 when Hubble's former pupil, Allan Sandage (1926–), showed that what Hubble had thought to be the individual brightest stars in galaxies were in fact aggregations of highly luminous stars embedded in gaseous nebulae. Both these reductions eased the conflict with the geological time scale.

It was already recognized, however, that this definition of the age of the universe was an oversimplification, for the value of Hubble's 'constant' would change with time if the expansion slowed down. In the 1950s and 1960s, Hubble and his successors expected these problems to yield to an early resolution. They did not, and uncertainties and disagreements remain to the present day.

CHAPTER 9   *Astronomy's Widening Horizons*

Where does history end and astronomy begin? Historians of every scientific discipline feel less and less secure as they approach the present day, and the historian of astronomy is no exception. The majority of astronomers who have ever lived are alive today, and the avalanche of publications is so great that to read even the title of every article would be well-nigh impossible. To make matters worse for the historian, the written communications between astronomers in times past – documents that were often preserved for posterity – are being replaced by telephone calls, conversations in the conference coffee queue, electronic mail messages, and so forth, few of which leave any trace. But in seeking to look at the present with an historian's eyes our task is not completely hopeless, for it is already clear that certain contemporary developments will be seen by future historians as of fundamental and lasting significance. Those that involve theory often require a good knowledge of physics for their understanding and so for the most part are beyond our present scope. But data must be obtained before they can be interpreted; and in the collection of data it may even be that readers – as taxpayers and voters – have themselves a role to play.

EXTENDING THE HUMAN SENSES: 'MORE LIGHT!'

Unlike many natural scientists, the astronomer does not have Nature on the rack – he cannot 'put Nature to the question', in the seventeenth-century phrase of Francis Bacon. The physicist or chemist can set up a controlled experiment, and like a torturer compel Nature to answer the questions put to her. Only occasionally is the astronomer able take the initiative. There have been visits to the Moon by so-called 'astronauts', and voyages of exploration by spacecraft to other planets in the solar system, while in 1986 no fewer than five spacecraft were sent to intercept Halley's Comet – the European Space Agency's *Giotto* came within 600 kilometres (375 miles) of the nucleus and spent ten hours gathering data and taking pictures. But for the most part the task of the astronomer is as it always has been, to collect and make sense of whatever information Nature is prepared to supply to Earth-based observers.

The invention of the telescope provided Galileo and his contemporaries with an extension of the human eye, and by the end of the eighteenth century, observers had become acutely aware that the bigger the objective lens or primary mirror of a telescope, the fainter the objects it could bring into view. The clamour for ever-larger telescopes has grown in volume ever since.

By the middle of the present century, American philanthropy had provided the more fortunate of American astronomers with privileged access to the secrets of the universe. The astronomer with a 'state of the art' instrument at his disposal has an

The nucleus (frozen core) of Halley's Comet, as seen in March 1986 by the European spacecraft *Giotto*. The photograph shows a dark, potato-shaped nucleus measuring some 16×8 kilometres (10×5 miles), and activity around it.

Lick Observatory, on Mount Hamilton, California, about 1910. Lick was the first major observatory sited so as to take advantage of the better observing conditions possible at high altitude. The principal dome houses the 36-inch refractor, and on the right are the main observatory building and (in the foreground) the astronomers' residence.

Opposite: Telescope domes on the summit of the Mauna Kea volcano in Hawaii, at a height of nearly 14,000 feet, and so above almost one-third of the Earth's atmosphere. Among the instruments in the picture are (anticlockwise, from the far left) the 3.8-metre (150-inch) UK infrared telescope, the 2.2-metre (87-inch) University of Arizona telescope, the 3.6-metre (141-inch) Canada–France–Hawaii telescope, the 3-metre (118-inch) US infrared telescope, the 10.4-metre (33-foot) dish of the Caltech Submillimeter Observatory, the first of the huge twin 9.8-metre (387-inch) Keck telescopes (here seen while under construction), and the 15-metre (49-foot) dish of the UK James Clerk Maxwell Telescope.

advantage over his rivals, not only in the disinterested pursuit of new knowledge, but in the competition for status and salary within the astronomical community; and in the last chapter we saw something of the achievements of observers blessed with time on the 100-inch on Mount Wilson or, later, the 200-inch on Palomar Mountain. Unsurprisingly, their colleagues in other countries demanded comparable facilities, but the cost of constructing and running a major telescope had reached a level that often called for governmental, and sometimes even international, funding.

There was a further problem, created by an unhelpful Nature. Until the 1970s most of the major telescopes of the world were located in the northern hemisphere, where most of the manufacturers and most of the astronomers were based. There are rich skies, however, south of the celestial equator: they include the dense star clouds in the direction of the centre of the Galaxy, and the two Magellanic Clouds (which are irregular galaxies, and our nearest extragalactic neighbours). It is as though the fates had conspired to make astronomical research as difficult and expensive as possible. However, as we shall see, in the last quarter of a century the increasing speed and economy of modern air travel and ease of communication have permitted the development of southern sites with facilities at least equal to those in the north. But the costs have continued to escalate.

In the US one response was the foundation in 1957 of the federally funded Association of Universities for Research in Astronomy, to fund three major observatories: for the northern skies, at Kitt Peak in Arizona, where a 3.8-metre (150-inch) telescope began work in 1973; for the southern skies, at Cerro Tololo in Chile, where a matching instrument came into service in 1976; and a solar observatory, at Sacramento Peak in New Mexico.

Other countries in many cases relied on international collaboration, often with time-sharing on the resulting facilities. Thus a consortium of nations of continental Europe sponsored the European Southern Observatory at La Silla, north of Santiago in Chile. The 3.6-metre (147-inch) telescope there came into service in 1971, and there are now more than a dozen telescopes on the site. Such international collaboration involves bureaucracy, not to say diplomacy; but in times of economic stringency, member states find it harder to renege on their promises of funding for international projects than they would with purely domestic schemes.

The 1970s saw two other major southern telescopes come into operation: the 3.9-metre (153-inch) Anglo-Australian telescope at Siding Spring Mountain in New South Wales, built by Australia and the United Kingdom, and the 3.6-metre (141-inch) telescope on Mauna Kea in Hawaii, created by a partnership between France, Canada, and Hawaii. Nor was the northern hemisphere neglected. The United Kingdom, Spain, Denmark, and Sweden combined to build an observatory on La Palma, in the Canary Islands, where conditions were more favourable than anywhere on the mainland of Europe, and they were soon joined by the Netherlands and Ireland.

All these instruments were located on sites chosen so as to reduce as far as possible the limitations inflicted by the presence of an atmosphere around the Earth: sites with less frequent cloud cover, little air movement, a reduced thickness of the atmosphere, and also greater remoteness from city lights – in brief, carefully selected mountain-tops. It is more than a century since the establishment of the first mountain-top observatory of note, Lick Observatory on Mount Hamilton near Santa Cruz in California, where observations began in 1881 at an altitude of over 4,000 feet. Today the highest major observatory is on Mauna Kea, at an altitude of nearly 14,000 feet, a height that poses a challenge to the physiology of the observers and staff. Needless to say, the remoteness of such sites has added still further to the costs.

Meanwhile, the collection and analysis of information has been transformed for the better by the technological revolution of recent decades. Photographic plates, which near the end of the nineteenth century brought about their own revolution in the way astronomers gathered data, are in fact not very efficient as light collectors. The great majority of photons striking a photographic plate are reflected away, and at best only about 2 per cent serve to impress an image on the sensitive surface. The rapid development in the 1980s of photosensitive charge-coupled devices, or 'CCDs' (which have wide-scale commercial importance for video cameras), enabled astronomers to assimilate photons with efficiencies exceeding 70 per cent. This meant that a 30-inch telescope in 1990 could record more photons than the 200-inch could in 1960. Furthermore, by acquiring the images electronically, observers could inspect progress as an observation was being made, thereby eliminating the wasteful problems of under- or over-exposure. And the CCDs provided yet another advantage – they could be flown in planetary probes or orbiting observatories such as the Hubble Space Telescope, and the images could be transferred to Earth by radio links.

The high speed of modern computers serves not only to process electronic images, but also to drive telescopes and to permit 'active optics'. One long-standing problem in the construction of large reflectors has been the distortion of the primary mirror as the instrument is tilted in different directions and the mirror bends under its own weight; today this distortion can be controlled mechanically, the pressure at various positions on the back of the mirror being continu-ously altered to compensate for changes in internal stress. In a similar way the constant changes in the atmosphere above the telescope can be compensated for by introducing into the light path a thin, deformable mirror that responds to these changes many times per second. Tracking an object as the sky rotates is another job a computer can undertake, and therefore equatorial mountings (see page 218) may now be dispensed with in favour of the simpler and cheaper up/down, left/right of the altazimuth.

As a result, the days when amateur astronomers – men like William Herschel or William Huggins – could be at the forefront of the science are long gone; amateurs still have a role to play, but it is in areas neglected by the professionals, such as the search for arriving comets, the observations and counts of meteors during meteor showers, or the monitoring of variable stars. Gone too are the days from the recent past, when an astronomer on his observing nights could have undisputed command of a great instrument. The modern telescope depends for its continued operation on teams of experts. Observing has been turned into a collaborative exercise, in which the astronomer finds himself in the hands of specialist support staff whose expertise is as likely to be in engineering or computing as in physics or astronomy. National observatories that once undertook their own research now find themselves occupied in maintaining facilities for a succession of visiting astronomers from academia.

THE NEXT GENERATION TELESCOPES

Most of the telescopes recently constructed have had mirrors with diameters a little less than the 200 inches of Palomar Mountain. In 1976 a 6-metre (336-inch) telescope was completed at Zelenchukskaya in what was then the USSR; but it proved a disappointment, and some thought the limit of useful mirror size (imposed by the movements in the atmosphere above the telescope) had already been reached. But experience of the new telescopes on high mountains showed that much of the disturbance came from air currents near and within the dome, and even within the instrument itself. Something could be done to alleviate these problems, and so the construction of telescopes with larger mirrors became attractive once more. Even so, for such a massive instrument to work effectively, it would need to be supported by the very latest technology.

The Multiple Mirror Telescope at the Whipple Observatory on Mount Hopkins, near Tucson, Arizona, which became fully operational in 1980. It comprises six identical mirrors, each 1.8 metres (6 feet) in diameter. All six images are brought to a common focus, so giving the equivalent of a single mirror 4.5 metres (15 feet) in diameter.

The 91-centimetre (3-foot) telescope of the Kuiper Airborne Observatory, first operational in 1975, was carried in a Lockheed C-141 transport aircraft. This normally flew at 36,000 feet, above most of the water vapour of the atmosphere, and the telescope (which looks out through the square port) was used for infrared and sub-millimetre observations. It was retired in 1995, to release funds for a more ambitious project.

The bulk of the daunting cost of one of these 'Next Generation Telescopes' (or NGTs) is incurred in the fabrication and shaping of the primary mirror, in the construction of machinery for controlling the telescope, and in the erection of the building with its rotating dome. The latter two are governed by the mirror's weight and by its focal length. Deeply hollowed mirrors of short focal length (and therefore requiring a relatively short tube and a dome of modest dimensions) are difficult to shape by conventional means; but a mirror can actually be cast with approximately the required shape, if the material of which it is made is rotated while still molten, and slowly allowed to cool. This is because a liquid under rotation in a circular container piles up towards the outside of the container, and its surface thereby acquires (approximately) the parabolic shape that is ideal for a telescopic mirror.

The largest NGT, the Keck I telescope on Mauna Kea, came into partial operation in 1990. Its mirror is 9.8 metres (387 inches) in diameter, and is formed of thirty-six hexagonal segments, nine of which were in place when the telescope first 'saw light'. A matching telescope, Keck II, is located nearby, and when used in conjunction the resolving power of the two instruments becomes equivalent to that of a telescope as large as the distance between the two domes.

THE HUBBLE SPACE TELESCOPE

Today's astronomer, however, is not wholly dependent on terrestrial observatories, for rocketry, orbiting satellites, and spacecraft – developments funded for military use or for reasons of national prestige – have opened the way to observation with instruments located above the atmosphere. By far the most substantial of these efforts is the Hubble Space Telescope (or HST); launched in April 1990 from the Space Shuttle *Discovery*, it orbits 600 kilometres (370 miles) above the Earth. Its primary mirror is of diameter 2.4 metres (94 inches), medium-size by the standards of terrestrial instruments, but capable in outer space of a resolution far surpassing its Earth-based competitors. The smoothness of the HST's mirror far exceeded the standards of any previous large astronomical instrument, but unfortunately, the surface was made highly accurately but to an erroneous shape, and the telescope fell far short of meeting expectations. A servicing mission from a Space Shuttle in 1993 successfully placed correcting elements in the optical path. Subsequently, the exquisite smoothness of the mirror's surface has paid off in a spectacular series of high-resolution images, most notably allowing the discovery of hundreds of Cepheid variables in the distant Virgo cluster of galaxies.

The start of work on the mirror of the Hubble Space Telescope. By the time the work was finished, the cutting wheel (on the right) had removed over 130 kilograms (285 lb) of glass from the disk, which is of 2.4-metre (94-inch) diameter. The mirror was then polished and coated. It was expected to have a resolution of one-tenth of an arc second or better, but the mirror had been ground to an erroneous shape; after correcting elements were placed in the optical path by a Space Shuttle three years after launch, the results exceeded expectations.

The antenna constructed by Karl Jansky (seen in the foreground). It had wheels that permitted it to be rotated in any horizontal direction. Jansky found that in addition to irregular sounds, such as the crackle of lightning, he was receiving a hissing that varied in intensity with a period four minutes short of 24 hours – that is, the period of the rotation of the heavenly sphere. The hissing proved to be coming from the Milky Way.

THE INVISIBLE UNIVERSE OF RADIO ASTRONOMY

Visible light, the electromagnetic radiation to which the human eye responds, is only one of the bands in which rays succeed in penetrating the atmosphere. Another, much larger band consists of radio waves with wavelengths between about 1 millimetre and 30 metres. The first clue that waves in this range were arriving from celestial bodies, as 'starry messengers' bearing new information, came in 1932. In that year, an engineer at the Bell Telephone Laboratory named Karl G. Janksy (1905–50) was using a steerable antenna to study interference on the recently inaugurated trans-Atlantic radio-telephone service. In addition to thunderstorms, Jansky met with a background hiss that varied with the time of day. The signals turned out to come from the Milky Way, and were strongest in the direction of the centre of the Galaxy.

Jansky's results attracted little attention, but in 1937 another American radio engineer, Grote Reber (1911–), built himself an antenna with a paraboloidal shape and began to devote his spare time to following up Jansky's work. For nearly a decade he produced contour maps of the intensity of radio waves across the sky. This span of activity included the war years, and wartime radar operators on the lookout for hostile activity also found themselves encountering 'interference' coming from outer space. In 1942, James S. Hey (1909–) and his colleagues at the British Army Operational Research Group were investigating what was thought to be German jamming of British radar, when they found that the intense radio emission was coming in fact from the Sun.

When the war ended, there were ex-service physicists with the expertise to develop this new science of 'radio astronomy', and discarded military equipment for them to use. Only a few months after the end of hostilities, Hey found the first localized ('discrete') source of radio emission, and two years later Martin Ryle

Radio astronomy and the spiral structure of the Galaxy

During the Second World War, Dutch astronomers were deprived of most of their facilities, and so found themselves forced to concentrate on theoretical questions. As we have seen (on pages 312–26), speculations that the Galaxy might have a spiral structure had proved difficult to substantiate, not only because the human inhabitants of the Galaxy would inevitably find it hard to discern the shape of the star system to which they themselves belonged, but also because dust clouds concealed most of the galactic stars from sight. Jan Oort of Leiden Observatory realized that, because of their longer wavelengths, radio waves might be able to penetrate the dust clouds, and he invited his student, Hendrik van de Hulst (1918–), to investigate what radio waves might be expected.

In a paper published in 1945, van de Hulst showed that when the spin of the electron in a 'neutral' hydrogen atom reversed itself, there would be an emission at a wavelength of about 21 centimetres. In any individual hydrogen atom, such a spontaneous reversal was to be expected only once in millions of years; but hydrogen is the most abundant element in the universe, and so there might be enough reversals occurring in the Galaxy to be detectable on Earth.

It was 1951 before the prediction was confirmed, first at Harvard, and later in the Netherlands and Australia. The latter two groups then embarked on a collaborative program to map out the intensity and velocities of the 21-centimetre line in different directions in the Galaxy. It was laborious work – the Leiden group used an old German radar antenna which for nearly two years they had constantly to reposition by hand-cranking every $2^1/2$ minutes – but in the resulting map the spiral arms of the Galaxy could be seen emerging at last.

Poignantly, William W. Morgan of Yerkes Observatory had recently completed an optical investigation of the structure of the Galaxy using the highly luminous stars of types O and B to trace out the spiral arms, and his short presentation of his results to the Christmas 1951 meeting of the American Astronomical Society convinced the audience that the spiral structure of the Galaxy had at last been demonstrated. But Morgan fell ill the following spring, and when he was able to resume work his draft paper had been superseded by the achievements of the radio astronomers.

The composite diagram prepared in 1958 to show the spiral structure of the Galaxy, as determined by radio astronomers in the Netherlands and Australia from the distribution of neutral hydrogen. In the middle of the diagram is the galactic centre, and the position of the Sun is marked above it by *S*, at an assumed distance of 8,200 parsecs. The left of the diagram is mainly the work of the Sydney observers, while the more detailed picture to the left is based on the Dutch results. Where the two halves meet, there is reasonable agreement.

(1918–84) and F. Graham Smith (1923–) at Cambridge located the most powerful such source in the northern hemisphere.

Could such 'radio stars' be identified with visual objects? Such identifications – or their lack – would clearly be fundamental to understanding the nature of radio stars. But there was a more specific reason: until identifications were made, radio astronomers would be ignorant of the third dimension, that of distance. The distance of a radio source would be unknown until it had been identified with a visual object, and the redshift (see pages 341–3) of the object's visual spectrum translated into distance, on the familiar theory that the greater the redshift, the greater the speed of recession, and the more distant the object.

Unfortunately the first radio telescopes were not sufficiently precise as to the position on the sky of a radio source to make such an identification possible. In 1949 came a significant advance, when three Australian radio astronomers, John G. Bolton, G. J. Stanley, and O. B. Slee, succeeded in identifying three discrete radio sources with objects familiar to the optical astronomers; these included the Crab Nebula, which Edwin Hubble had earlier recognized as the remnant of the supernova explosion of 1054 (see page 61).

The 250-foot-diameter radio telescope at Jodrell Bank, near Manchester, on 20 June 1957, when the still unfinished bowl was tilted for the first time. The dish weighed 1,500 tonnes, and was carried on turrets salvaged from battle-ships. The instrument was commissioned in October of the same year, just in time to enable the free world to monitor the movement of the first *sputnik*'s carrier rocket – an unforeseen use that earned the telescope worldwide fame. Sir Bernard Lovell's feat in overcoming the technical obstacles in the construction of this huge but fully steerable dish led to the adoption of steerable dishes as the standard form for large radio telescopes everywhere.

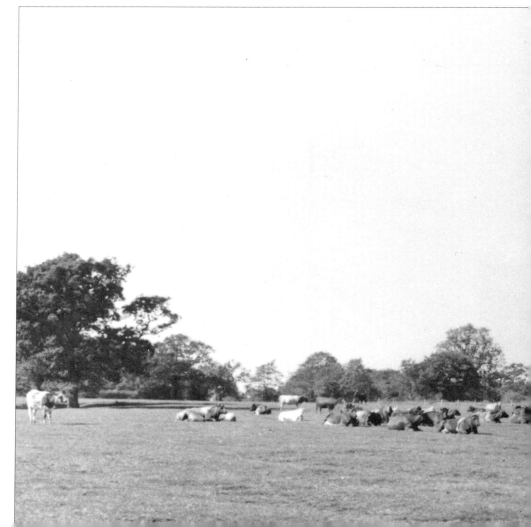

Two years later, Graham Smith was able to obtain an accurate position for the radio source known as Cygnus A; this allowed Walter Baade and Rudolph Minkowski at Palomar Mountain to identify it with a curious-looking object, which Baade interpreted as a distant pair of colliding galaxies. This interpretation had to be abandoned when the peculiar double appearance was shown to result from a band of dust running across the galaxy, but Cygnus A gave a foretaste of the cosmological potential of radio astronomy.

As the number of such identifications grew, it became clear that objects that lay at enormous distances, though very feeble emitters of light – so feeble that many may never be detected by optical telescopes – could nevertheless be strong sources of radio waves. Radio astronomers were 'seeing' these sources as they were a very long time ago, when their light set out on its journey to us. This meant that if the universe was indeed expanding from its origin in a superdense beginning or 'Big Bang', as most cosmologists believed, the radio astronomer could observe regions as they were when significantly closer in time to the Big Bang itself. This would permit evidence from radio astronomy to be used to test theories of cosmology.

RADIO ASTRONOMY AND COSMOLOGICAL CONTROVERSY

In 1948 three Cambridge astronomers, Fred Hoyle (1915–), Hermann Bondi (1919–), and Thomas Gold (1920–), none of them enamoured of the idea that the universe had an origin, proposed instead a theory whereby the universe appeared broadly the same, at all times and from all places. In this 'Steady State' cosmology there would be no beginning, and no end. That the galaxies were receding from each other was not the consequence of a Big Bang (Hoyle's term); rather, the galaxies were simply moving apart in all directions, much as the currants of a cake move apart as the cake expands during cooking. To fill the spaces left by the receding galaxies and so preserve the overall appearance of the universe, the three friends postulated the creation of matter in intergalactic space, matter that was required to appear on a scale sufficient to form new galaxies.

If the Steady State cosmologists were right, then in the remote past the distribution of the major components of the universe was broadly the same as it is today; but if the Big Bang theorists were right, then these components were more concentrated when the universe was younger and smaller. This was a question to which radio astronomy might supply the answer.

A radio telescope that seemed suited to the task was also in Cambridge, where Ryle had pioneered a technique whereby several matching radio telescopes, one or more of them mobile, could be teamed up to piece together a radio map of the sky equivalent to that provided by a single 'dish' of enormous (and quite impracticable) size. There was no love lost between Ryle and Hoyle; Ryle more than once announced results of his surveys that supposedly disproved Steady State, and Hoyle more than once rejected them.

Eventually the question was decided – provisionally, at least – by another route. In the late 1940s George Gamow (1904–68) and colleagues from George Washington University, Ralph A. Alpher (1921–) and Robert Herman (1914–), predicted as one of the consequences of what would later be termed the Big Bang, that throughout the universe there would today be residual radiation; this would be the same in all directions, and at a temperature of some $5°$ above absolute zero (5K). Gamow pictured the formation of the elements in the initial minutes of the explosion as coming from a primordial mixture of radiation and nuclear particles, which he called 'ylem', but this scheme ran into difficulties because of the lack of

Opposite: The Ryle radio telescope in Cambridge. It consists of four fixed and four rail-mounted dishes, each 13 metres (43 feet) in diameter, distributed along an east–west baseline of nearly 5 kilometres (3 miles). The use of these small telescopes in combination in a variety of positions over a period of days yields the same information as from one telescope of enormous size. Sir Martin Ryle's development of this method of 'aperture synthesis' won him a share of the 1974 Nobel Prize for physics.

The Very Large Array radio telescope near Socorro, New Mexico. It consists of 27 dish antennas, each 25 metres (82 feet) across. These are shared between three arms of a Y-shaped configuration, each arm being up to 21 kilometres (13 miles) long, and the dishes are used in conjunction by the method of aperture synthesis. The Y-shaped configuration gives a better overall performance across the sky than would be possible with an equivalent telescope restricted to an east–west axis.

a stable element of mass 5. Although the study was well-publicized, interest in these ideas declined because of the problems with the proposed scenario for element formation.

Unaware of Alpher and Herman's forecast, Robert H. Dicke at Princeton in 1964 proposed his own theory, which also predicted a background radiation a few degrees above absolute zero. Dicke's group set to work forthwith to build an instrument to detect this radiation, and so test the theory.

Meanwhile, oblivious of these theoretical ideas remote from their own concerns, two engineers at nearby Bell Telephone Laboratories, Arno Penzias (1933–) and Robert Wilson (1936–), were at work with a reflector originally designed to test communication by satellite. They took great pains to track down and eliminate all sources of interference, including two roosting pigeons, but they found that from every direction in the sky there came an unaccounted-for radiation corresponding to a temperature of about 3K. They consulted Dicke as to the possible cause, only to learn that he was actively searching for what they had found.

In the eyes of most cosmologists, the discovery in 1964–65 of this 'microwave background radiation' tipped the scales decisively against Steady State, which passed into the limbo of discarded theories. Indeed, it had always been motivated by views as to the form the universe ought to have, rather than the form it does have. Its begetters had pointed out that it was a theory that made clear predictions and so had the scientific merit of being vulnerable to disproof; and disproved it was. Yet historians remember earlier arguments that polarized the astronomical community, over whether the Earth or the Sun was at the centre of the universe, or whether nebulae were star clusters or clouds of luminosity – questions whose answers proved to be less straightforward that was expected; and they feel uncomfortable when that community is polarized yet again between opposing theories.

Both Steady State and Big Bang presupposed the recession of galaxies in all directions; yet even this interpretation of the evidence can awake unease. At root, the observational facts are that the light from small, faint galaxies, when turned into a spectrum, is found to be 'shifted' towards the red, and the more so, the fainter and smaller the galaxy. It is hard to doubt that 'small and faint' implies distant, but the redshift may be open to other interpretations. For example, it might be that the universe is static and that photons of light, in their long journeys across intergalactic space, somehow gradually lose energy and become of longer wavelength. However, all alternative theories are in some respect unsatisfactory, while the observed redshifts *behave* as if they represent an expansion, and for the present this continues to be the simplest and most satisfactory explanation of the observations.

There has indeed been something of a methodological dilemma: the theoretical cosmologist wanted the observing astronomer to produce 'facts' that would determine which was the correct model of the universe, but the very reduction of raw observations to produce such 'facts' required the assumption of a cosmological model which the observations were supposed to verify.

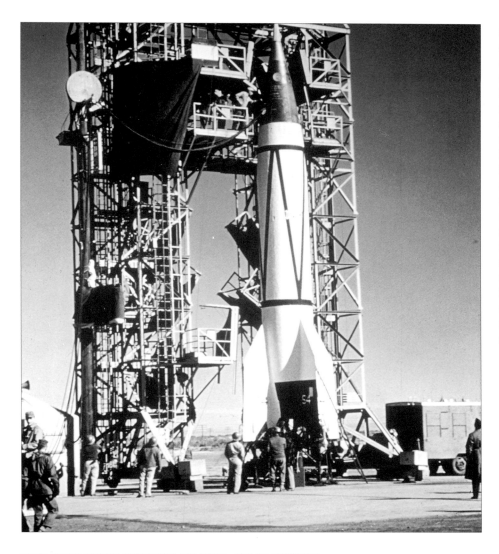

One of the German V-2 rockets captured by American forces at the end of World War II and used to give astronomers brief opportunities to study the sky from above the Earth's atmosphere. This one is in the gantry at the White Sands Missile Range, New Mexico, in 1948. V-2s were 14 metres (46 feet) tall and could climb as high as 150 kilometres (90 miles). Unfortunately the rocket would break up on impact, and so it was necessary either to separate off the parts that contained the scientific instruments so that these could descend more slowly, or to transmit the data by radio during the flight.

THE UNIVERSE HIDDEN BY THE ATMOSPHERE

As we have seen, visible light and radio waves are the forms of electromagnetic radiation that can penetrate the atmosphere. Indeed, radio waves find bad weather no obstacle, and radio telescopes can be employed, day and night, at low level sites with poor climates. But other regions of the spectrum – those that lie between the near infra-red and radio waves, and the short-wavelength ultra-violet, X-rays and gamma rays – are effectively blanketed from view by the atmosphere. Today, as a result of developments in spacecraft, the whole of the spectrum is available for study; new areas of investigation have been spawned, and the integration of astronomy and physics has become ever stronger. The different wavebands provide, in part, pictures of the universe at different temperatures.

The first attempts at observation above the atmosphere took the form of instruments carried in rockets. As early as 1946, captured German V-2 rockets were used

Space exploration of the Solar System

In 1959 the Soviet space probe Luna 3 provided the first view of the backside of the Moon; the most surprising discovery was that the Moon's far side lacked the large dark basaltic maria that characterize the near side facing the Earth. From this beginning, direct exploration of the solar system has blossomed, with spacecraft making close approaches to all the planets except Pluto, as well as to satellites, comets, and asteroids. These missions transmitted technical data of gravitational and magnetic fields, and also the now familiar 'postcards from space' that have revealed in detail many new worlds. Jupiter's satellite Io, with its active sulfurous volcanoes, and Uranus's satellite Miranda, with its fascinating combination of old, heavily cracked terrain and a young complex landscape with bright and dark scarps and ridges, are among the strange small worlds disclosed by unmanned instruments launched on lengthy journeys beyond the Moon.

The first high-definition close-ups of the Moon's surface came in July 1964, with the US lunar probe Ranger 7, which sent back thousands of televised images in preparation for the manned Apollo landings that followed five years later. American astronauts brought back hundreds of mineral specimens, which helped to establish the age and complex chemistry of the lunar origin.

In 1970 Venera 7 became the first spacecraft to send back signals from the surface of another planet, and the first actual pictures of Venus's landscape were produced by Soviet probes in 1975. Meanwhile, the American Mariner series documented the surfaces of Mercury, Venus, and Mars. The Viking probes, launched in 1975, gave magnificent views of the immense volcanic calderas and the gigantic Valles Marineris canyon system on Mars. The American Voyagers transmitted remarkable details of the weather on Jupiter and the ring and satellite systems of Saturn (in 1979–81), and then of Uranus and Neptune (in 1986 and 1989 respectively).

These explorations revitalized planetary studies, which had fallen into comparative disfavour earlier in the century in competition with the more glamorous investigations into cosmology and astrophysics.

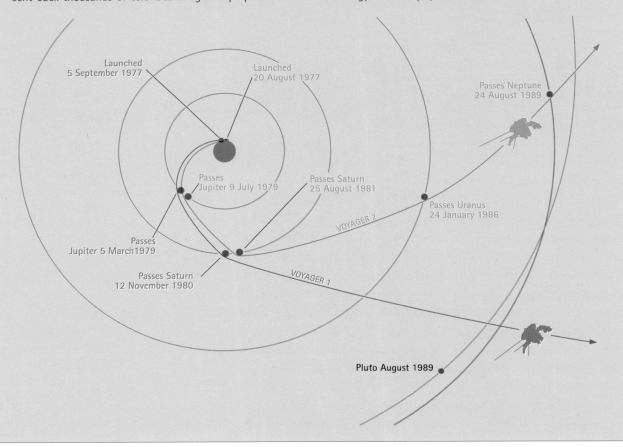

to obtain the first ultraviolet observations of the Sun, but within a few years the V-2s were superseded by rockets designed for space research. Since a rocket is above the atmosphere for only a matter of minutes, the data gathered by the early flights provided little more than crude snapshots that confirmed there were windows on the universe as yet unutilized.

This naturally whetted the astronomical appetite. Fortunately for would-be observers, the international rivalry in space that followed the launching in 1957 of the first Russian *sputnik* resulted in the provision of a whole series of astronomical satellites, orbiting the Earth and recording data in all the hitherto missing regions of the spectrum. US 'orbiting astronomical observatories' were launched in 1968 and 1972, and the International Ultraviolet Explorer (a joint European–UK–US venture) in 1978. The American series of Small Astronomical Satellites was inaugurated in 1970 with the launch from the coast of Kenya of SAS-1, which generated a catalogue of 161 X-ray sources; this was followed by SAS-2, launched in 1972, which carried equipment for measurement of gamma rays, and SAS-3, launched in May 1975 and carrying X-ray detectors. The Netherlands–UK–US Infrared Astronomical Satellite was launched in 1983, and in its eleven-month active life assembled a sensitive and complete mapping of the infrared sky. With the data from these and subsequent satellites, astronomers have in a few short decades enormously enlarged the horizons of their science, and strengthened its links with mainstream physics.

THE DISCOVERY OF EXOTIC OBJECTS

We have seen how already in the first half of the twentieth century, the scope of astronomy was enlarged until its range extended from complex subatomic processes to models of the universe involving General Relativity. At the same time it was increasingly realized that physical conditions in space were altogether more extreme than could be achieved in any laboratory, and that one could not rule out objects and processes that had seemed to belong rather to science fiction.

Yet the instincts of even the most forward-looking astronomers were conservative. Eighteenth-century speculators had discussed the characteristics of stars so dense that light would be prevented from leaving them by the strength of their gravitational attraction (see page 236); and according to Einstein's General Relativity, such bizarre objects (today's 'black holes') were theoretically possible as end-products of stellar evolution, provided the stars were massive enough for their inward gravitational attraction to overwhelm the repulsive forces at work. In 1935, Subrahmanyan Chandrasekhar correctly argued at a meeting of the Royal Astronomical Society that white dwarf stars more than half again as massive as the Sun could not withstand the intense tug of gravity and would collapse further; but Sir Arthur Eddington, even though he was himself a leading advocate of Relativity, poured scorn on the young man's proposals, finding it impossible to accept that Nature behaved in so extraordinary a fashion. Before the decade was out, J. Robert

This home-made array of poles and wires covering more than 4 acres was used at Cambridge to discover the first pulsar. Early radio telescopes at Cambridge took a number of forms, being intended for sky surveys rather than for the study of individual objects. It was therefore unnecessary for them to be fully steerable, as was the Jodrell Bank telescope (see pages 354–5), which could track an object.

Oppenheimer (1904–67) and Hartland Snyder (1913–62) focused attention on the process of gravitational collapse itself, and showed that General Relativity was needed to describe the dynamics of the process, thus laying the groundwork for what is called 'black hole physics'.

It was in the 1960s that observers with radio telescopes recognized two examples of the near-incredible types of objects that exist in the astronomer's universe. On three occasions in 1962 the Moon passed in front of one of the radio sources found during Ryle's third Cambridge survey, and observations of these 'occultations' allowed observers at Parkes radio telescope observatory in New South Wales to show that the object was a double radio source, one component of which they identified with a blue, star-like object of the thirteenth magnitude. This prompted the Dutch-born astronomer Maarten Schmidt (1929–) to use the Palomar 200-inch reflector to study the spectrum of the star.

The spectrum was peculiar indeed: superimposed on a continuum were bright emission lines that seemed to be of an unknown element, but which Schmidt eventually recognized as the series of lines of hydrogen investigated as long ago as 1885 by the Swiss physicist Johann Balmer. But these lines had an extraordinarily large redshift, which implied that the star-like object lay far outside our Galaxy – provided, that is, that the redshift was interpreted in the usual way.

There was acrimonious debate over the nature of this 'quasi-stellar radio source' or 'quasar', for if it was at the vast cosmological distance implied by its large red-shift, an enormous intrinsic luminosity was required. As time passed, quasars with greater and greater redshifts were discovered, including ones whose redshifts implied that they were receding with over 90 per cent of the speed of light, and that they were situated thousands of millions of light years away. To be visible at such distances, quasars had to be perhaps a million times more luminous than a complete galaxy; yet they had also to be small, as their luminosity could sometimes vary by a factor of 2 in a matter of hours. Only a supermassive black hole offered any hope of a physical explanation of the processes involved. Small wonder that some astronomers refused to believe the redshifts had been correctly interpreted.

The second extraordinary type of object was discovered four years later. Antony Hewish (1924–), a Cambridge radio astronomer, and his student, Jocelyn Bell, had constructed a crude but effective radio telescope consisting of poles and wires strung out across a four-acre site. Their purpose involved the study of the impact of the interplanetary plasma (material formed of ions and electrons moving freely) on the waves from distant radio sources. In the course of their investigations, they chanced upon a weak source that eventually proved to be giving out regularly-spaced pulses with a period of just over a single second. It seemed that such a very rapid transmission must have been generated on Earth, perhaps at a military installation – or if not, by 'Little Green Men' elsewhere in the universe – but it eventually became clear that the source was celestial and natural.

The 'pulsating radio star' or 'pulsar' proved to be a rotating neutron star, a tiny but immensely dense star of a type whose existence had been predicted (see page 305) as long ago as 1934 by Fritz Zwicky and Walter Baade, shortly after the discovery of the neutron itself by James Chadwick. It was rotating once a second, and each time it did so its beam of radiation swept the Earth, much as a lighthouse beam sweeps the surrounding water. Hundreds of pulsars have since been discovered, all rotating with astonishing speed: that at the heart of the Crab Nebula has a period of only 1/30th of a second.

The Arecibo Observatory in Puerto Rico, where a huge radio telescope dish 305 metres (930 feet) in diameter is sited in a natural hollow. It was with this instrument in 1974 that Russell A. Hulse and Joseph H. Taylor discovered the first binary pulsar. A number of pulsars are binary (two stars bound together by gravity, one of which is a radio pulsar), and in about half of these, the pulsar rotates a hundred or more times every second. In the pulsar discovered by Hulse and Taylor, two neutron stars move around each other in a little under eight hours, in very eccentric orbits. The resulting data on the two stars and their behaviour have been used in very precise tests of General Relativity.

Pulsars have proved to be of immense significance. These neutron stars can form only through the collapse of less dense stars, proving that such collapses do take place. In a neutron star, the stupendous pressure towards further gravitational collapse is resisted by the forces between the neutrons; but calculations show that in a star only a few times more massive, these forces would be insufficient, and the star would become a black hole.

THE VIOLENT UNIVERSE

The discoveries of the 1960s and 1970s gave astronomers an entirely different view of the pace of astronomical events. The notion that celestial evolution moved at an ever-stately march gave way to a cosmos filled with sudden happenings. From the split-second oscillations of the pulsars to the comparatively swift light variations of the quasars, cosmic events took place far more rapidly than the billion-years-evolutionary cycle of stars like the Sun. New stars were observed in formation, some of which would last only a few hundred thousand years, and by the 1980s astronomers realized that the cataclysmic final collapse of a supernova core could be reckoned in milliseconds. Even the staggering explosion of the Big Bang, with the formation of hydrogen and helium, was clocked at about three minutes.

Astrophysicists began to appreciate that much of the unexpected radiation outside the visible portion of the spectrum arose from non-thermal processes. That is, the radiation could not be associated with the random motion of hot atoms, but instead with highly ordered motions such as electrons moving in magnetic fields. For two or three decades astronomers were fond of calling their new vision 'the violent universe', but gradually the term fell out of use, while the expression 'high energy astronomy' gained increasing currency.

At the same time that astronomers were becoming familiar with such violent and rapid events as 'the first three minutes' of creation or the implosion-explosion of supernovae, they discovered that these diverse phenomena were linked in the explanation of how the elements have come to be. In the intense non-equilibrium interactions in the initial minutes of the Big Bang, photons of pure energy were transformed into protons and electrons, which soon combined into hydrogen and helium, but the absence of a stable mass 5 isotope prevented heavier elements from forming. In contrast, in the slow equilibrium processes in the cores of evolving stars, it was possible to build up carbon, oxygen, and some higher elements, while in the catastrophic collapse of supernovae many heavier elements could be formed and blown out into space; this cosmic debris then became the material for a subsequent generation of stars. This theoretical picture of element formation was conceived largely by Geoffrey and Margaret Burbidge (1925– , 1919–), Fred Hoyle, and William Fowler (1911–95) and independently by A. G. W. Cameron (1925–). One of the most profound insights of astronomy in the second half of the twentieth century, it has provided a solid basis for observational investigations of the evolving chemical composition of the heavens, and dramatic confirmation that our universe has a long history of slow evolutionary change.

IN CONCLUSION

This short final chapter has done no more than mention some of the instruments, observational techniques, and theories that have been developed in the second half of the twentieth century, and offer samples of the extraordinary objects that populate the universe of the contemporary astronomer.

The gulf that separates the modern astronomer from the general public is nothing new. Those at the frontiers of astronomy, with the learning required to understand the heavens, have always been an elite, set apart from the rest of society to whom this esoteric knowledge has filtered down in much-diluted forms. This was no doubt true in prehistoric times, and it was certainly true in China and in Mesoamerican cultures such as that of the Maya. In the Middle Ages, Ptolemy's *Almagest*, with its epicycles, deferents, and equants, was a closed book to all but a very few. Copernicus's *De revolutionibus* proved equally impenetrable, and Kepler's achievements went unappreciated by intellects of the calibre of Galileo and Descartes. The implications of Newton's *Principia* were left to a tiny band of brilliant mathematicians. With Einstein, the gulf became still wider: in 1920, the organizer of the 'Great Debate' (see page 329), C.G. Abbot, rejected relativity as a possible topic, saying 'I pray to God that the progress of science will send relativity to some region of space beyond the fourth dimension, from whence it may never return to plague us'. And since then, the technical and conceptual barriers facing the student of astronomy have grown steadily more daunting.

Perhaps only once in history has there been substantial harmony between the scholar's universe and that of the ordinary people. In the Latin Middle Ages, God was in his heaven with the angels and the blessed, and mankind was on *terra firma* at the centre of the cosmos, no place of honour but a position that fitted man's significance in the moral order. Yet even then, the epicycles of the astronomers caused dissension in the Faculties of Arts; and it would not be long before Copernicus's reform would lead the English poet, John Donne (1573–1631), to write:

> The New Philosophy calls all in doubt,
> The Element of fire is quite put out;
> The Sun is lost, and th'Earth, and no man's wit
> Can well direct him where to look for it…
> 'Tis all in pieces, all coherence gone;
> All just supply, and all Relation.

Earlier in this book we discussed Plato's challenge to his contemporaries, in effect inviting them to answer the question of whether the universe is an intelligible cosmos. Today, as down the centuries, the answer might be: 'Yes, but only just.'

Reference guide to
The Cambridge
Illustrated History of
Astronomy

Chronology

Glossary

Further Reading

Chronology

3500BC

Construction of Newgrange.

3000 Construction of Stonehenge I.   Egyptian religious calendar.
Egyptian administrative calendar.

2500

2000

Old Babylon: Venus records (place-value
1500   numerical system).

1000

Babylonian *Enuma* takes
definitive form.

Babylonians identify
Metonic cycle.

Eratosthenes, size of the Earth.

Hipparchus, models for the Sun and
the Moon, star catalogue, precession.

*Systematic
Babylonian records*   700

600

Greek cosmology

500

Greeks recognize sphericity of Earth.
Plato poses problem of planets.   400
Eudoxus, models of spheres.
Aristotle's world picture.   300
Aristarchus, distances of the Sun and the Moon.

Apollonius, eccentres, epicycles,   200
deferents.   *Greek
planetary models
based on
Babylonian records*   100BC

AD100

Ptolemy, *Almagest, Planetary Hypotheses, Tetrabiblos.*

200

Medieval Christendom   300

Martianus Capella.   Macrobius.   Calcidius.   400

Boethius.   500

Medieval Islam

632 Death of Muhammad.

House of Wisdom in Baghdad.   *Translations
from Greek*
al-Battani.

al-Sufi.
ibn Yunus.   ibn al-Haytham.

al-Zarqali.
   1120–25 Cairo Observatory.
Averroës.
   al-Bitruji.

1259 Maragha Observatory founded for al-Tusi.

al-Shatir.

1420 Samarkand Observatory built by Ulugh Beg.

1474 Peurbach, *New
Theories of the Planets.*

1575–80 Istanbul Observatory.

600

Bede.   700

800

900

Astrolabe enters West.   1000

1100

Translations from Greek and Arabic

1200

Anonymous: *Theory of the Planets.*   Sacrobosco: *Sphere.*

1300

1400

Invention of printing.
1496 *Epitome* of the *Almagest.*   1500
1543 Copernicus: *De revolutionibus.*

1570

1580

1572 'Tycho's Nova'.

1577 Tycho finds comet to be celestial.

1576–97 Tycho Brahe's observatory on Hven

1590

1600

1596 Kepler, *Cosmographic Mystery*.

1602 Tycho's star catalogue published.

1604 'Kepler's nova'.

1610

1609 Kepler, *New Astronomy*; first two laws. Galileo's first telescopic observations.

1610 Galileo, *Starry Messenger*.

1611 Kepler designs 'astronomical' telescope.

1613 Galileo, *Letters on Sunspots*.

1620

1619 Kepler's third law.

1630

1627 Kepler, *Prutenic Tables*.

1632 Galileo, *Two Great World Systems*.

1640

1640 Gascoigne's micrometer.

1644 Descartes, *Principles of Philosophy*.

1650

1660

1659 Huygens elucidates Saturn's rings.

1663 Gregory designs reflector.

1665 *Philosophical Transactions* begun.

1667 Paris Observatory founded. Mira's period identified.

1670

1672 Newton shows white light composite.

1675 Greenwich Observatory founded.

1677–78 Halley at St Helena

1680

1679 *Connoissance des Temps* founded.

1690

1687 Newton, *Principia*.

1690 Hevelius's star catalogue.

1695 Halley's Comet recognized as periodic.

1700

1710

1720

1721 Halley's papers on symmetric star system.

1725 Flamsteed's *British Catalogue*.

1730

1728 Newton, *System of the World*.

1729 Bradley's paper on aberration of light.

1740

1750

1750 Wright, *An Original Theory*.

1751–53 Lacaille at Cape of Good Hope .

1757 Michell argues most double stars binary.

1758 Achromatic lens described by Dollond.

1759 Return of Halley's Comet.

1760

1761 Trial begins of Harrison's H4 chronometer. Transit of Venus widely observed.

1767 *Nautical Almanac* founded. Transit of Venus widely observed.

1770

1780

1781 Messier's final catalogue of nebulae. W. Herschel discovers Uranus.

1783 Goodricke (and Pigott) suggest Algol eclipsing binary.

1785 W. Herschel's cross-section of Galaxy.

1783–1802 W. Herschel's sweeps for nebulae

1790

1789 W. Herschel's 40-foot reflector completed.

1796 Laplace, *Exposition*.

1799 Laplace, *Traité*, vol. 1.

1800 *Monatliche Correspondenz* founded.

1801 Piazzi discovers Ceres.

1800

1803 W. Herschel confirms binary stars.

1810

1814–15 Fraunhofer maps lines in solar spectrum.

1818 Bessel, *Fundamenta astronomiae*.

1820 Foundation of Royal Astronomical Society.   1820

1823 *Astronomische Nachrichten* begun.
                                      1824 9½-inch Dorpat refractor mounted.

                                                                                                                          1830

                                                                                    1834–38 J. Herschel
                                                                                    at Cape of Good Hope

1837 Struve announces parallax of Vega.
                                              1838 Bessel announces parallax of 61 Cygni.
1839 Pulkovo Observatory founded; 15-inch refractor.   1840

1842 Corona and prominences observed during solar eclipse.
                                                    1843 Schwabe announces sunspot cycle.
1844 Bessel argues Sirius and Procyon have unseen companions.
1845 'Leviathan of Parsonstown' reflector completed. Spiral structure of nebula recognized. First daguerreotype of Sun.
1846 Supposed resolution of Orion Nebula. Neptune (and satellite) discovered.
1847 Harvard 15-inch refractor completed. W. Struve's layered model of the Galaxy.
                                                            1849 Gould's *Astronomical Journal* founded.   1850

1852 Sabine announces sunspots linked with magnetic storms.

1856 Pogson's proposed magnitude scale published. First silver-on-glass astronomical mirrors.
1857 Clerk Maxwell shows Saturn's rings made of particles.   1858 De la Rue photographs sunspots by collodion process.
1859 Tempel finds nebulosity around Merope in Pleiades. Bunsen and Kirchhoff associate elements with
      spectral lines in laboratory. Publication of *Bonner Durchmusterung* begins.   1861–2 Kirchhoff's map of solar spectrum with   1860
1862 Ångstrom shows hydrogen present in solar atmosphere.   elements identified.
      Clark observes companion of Sirius. 1863 Astronomische Gesellschaft founded.
1864 Donati examines light of comet with spectroscope. Huggins shows a nebula to be formed of gas.
1866 Schiaparelli links August meteors with comet.
1868 Secchi describes four types of stellar spectra.
1870 Young observes reversing layer during solar eclipse.   1870
1872 Draper photographs spectrum of Vega.

1876 Dry gelatine plates used in photography.
                                      1877 Hall discovers two moons of Mars.
                                                                                                                          1880

1882 Gill's photograph of comet contains numerous stars.

1885 Nova S Andromedae encourages opposition to 'island universes'.

1887 Pickering initiates study of spectroscopic binaries. *Carte du Ciel* inaugurated.

1890 Lockyer, *Meteoritic Hypothesis*.   1890

1897 Yerkes 40-inch refractor completed.

1900 *Astronomischer Jahrsbericht* first published, with literature for 1899.   1900

1904 International Union for Co-operation in Solar Research founded.
                                                          1905 Hertzsprung suspects existence of a class
1906 Kapteyn's *Plan of Selected Areas*.   of 'giant' stars.
1908 Mt Wilson 60-inch reflector completed.

1910

1912 Leavitt's period–luminosity relation for Cepheids in Small Magellanic Cloud.

1913 Russell presents first 'H–R diagram' for field stars (stars not in clusters).

1914 Slipher announces large radial velocities of spiral nebulae. Shapley's theory of pulsating stars. Adams and Kohlschütter establish method of spectroscopic parallaxes.

1916 Van Maanen claims evidence of rotation in M 101.

1917 Mt Wilson 100-inch reflector completed.

1918 Shapley proposes 'Big Galaxy'. Publication begins of Henry Draper catalogue of stellar spectra.

1919 International Astronomical Union founded.

1920   1920 'Great Debate' between Shapley and Curtis. Saha publishes theory of ionization in stellar atmospheres.

1923 Hubble finds Cepheid variable in Andromeda Nebula.

1925 Hubble demonstrates Andromeda Nebula is independent galaxy. Payne, *Stellar Atmospheres*.

1926 Eddington, *The Internal Constitution of the Stars*.

1927 Oort analyses stellar motions to study structure of Galaxy.

1929 Hubble shows galaxies receding in conformity with 'law of redshift'.

1930   1930 Trumpler demonstrates existence of interstellar dust in plane of Galaxy. Discovery of Pluto.

1931 Chandrasekhar investigates structure of white dwarf stars.

1934 Baade and Zwicky publish theory of neutron stars; show novae fall into two classes.

1939 Bethe's detailed theory of nuclear source of stellar energy.

1940

1944 Baade announces discovery of stellar populations.

1946 V2 rockets make ultraviolet observations of Sun.

Early space research

1948 '48-inch' (1.25-metre) Schmidt telescope completed, Palomar Mountain (USA). 200-inch telescope completed, Palomar Mountain (USA).

1948 V2 rockets detect solar X-rays; Aerobee rockets introduced.

1950

1952 Baade announces revision of distance scale.

1932 Jansky constructs antenna, detects radio waves from Milky Way.

Radio astronomy's early years

1937 Reber constructs 9-metre dish steerable in elevation.

1939 Outbreak of World War II leads to intensive development of radar.

1942 Detection of radio waves from the Sun.

1945 End of World War II releases radar equipment and personnel for scientific work.

1949 First identification of radio sources with optical objects.

Ground-based optical telescopes since 1950   *Radio astronomy since 1950*

1951 Detection of 21-centimetre line.

1952 Optical identification of Cygnus A.

1955

1957 Jodrell Bank (UK) 250-foot (fully) steerable dish.

1959 120-inch telescope, Mt Hamilton (USA).

1960

1961 Parkes (Australia) 64-metre steerable dish.

1963 Arecibo (Puerto Rico) 305-metre dish in natural hollow. First quasars identified.

1964 Cambridge (UK) one-mile telescope. Detection of microwave background radiation.

1965 Jodrell Bank/Malvern (UK) VLBI, 127-kilometre baseline.   1965

1967 Canadian VLBI, 3,074-kilometre baseline.

1968 Announcement of discovery of first pulsar.

1970 Westerbork (Netherlands) 3-kilometre telescope.   1970

1972 Cambridge (UK) 5-kilometre telescope. Effelsberg (Federal Republic of Germany) 100-metre steerable dish.

1973 3.8-metre telescope, Kitt Peak (USA). 1.24-metre UK Schmidt telescope, Siding Spring (Australia).

1975 3.9-metre Anglo-Australian Telescope, Siding Spring (Australia).   1975 First observations with Very Large Array, Socorro (USA).   1975

1976 4-metre US telescope, Cerro Tololo (Chile). 6-metre telescope, Mt Pastukhov (USSR).

1977 3.6-metre European Southern Observatory Telescope, La Silla (Chile).

1978 3.8-metre UK Infrared Telescope, Mauna Kea, Hawaii.

1979 3.6-metre Canada–France–Hawaii Telescope, Mauna Kea, Hawaii. 3-metre US NASA Infrared Telescope Facility, Mauna Kea, Hawaii. Multiple Mirror Telescope, Mt Hopkins (USA).   1980 Very Large Array fully operational.   1980

1984 3.5-metre telescope, Calar Alto (Spain).

1985

1986 James Clerk Maxwell Telescope, 15-metre millimetre and submillimetre dish, Mauna Kea, Hawaii.

1987 4.2-metre UK telescope, Canary Islands.

1989 3.5-metre European Southern Observatory's New Technology Telescope, La Silla (Chile).   1989 Sweden–European Southern Observatory 15-metre submillimetre dish, La Silla, Chile.

1990 US Caltech Submillimeter Observatory, 10.4-metre submillimetre dish, Mauna Kea, Hawaii.   1990

1991 '10-metre Keck I' (9.8-metre) US Next Generation Telescope, Mauna Kea, Hawaii.

1993 US Very Long Baseline Array VLBI, baseline up to 8,000 kilometres.

VLBI: Very Long Baseline Interferometer. ESA: European Space Agency.

Space research since 1950

1955

1957 First USSR sputnik launched.

1959 USSR Luna 3 gives first images of back of Moon.

1960

1962 US Aerobee rocket detects X-ray star.

1964 US Ranger 7 gives images of Moon's surface.

1965

1968 US Orbiting Astronomical Observatory OAO-2 launched.

1969 First human on Moon.

1970 1970 USSR Venera 7 sends signals from Venus. US Small Astronomical Satellite SAS-1 launched.

1972 European Space Research Organization ultraviolet satellite TD-1 launched. US Orbiting Astronomical Observatory OAO-3 ('Copernicus') launched. US Small Astronomical Satellite SAS-2 launched.

1974 Astronomical Netherlands Satellite (Netherlands and US) launched. US Mariner 10 photographs Mercury, Venus.

1975 1975 USSR Venera 9 photographs surface of Venus. US Small Astronomical Satellite SAS-3 launched. ESA gamma ray satellite Cos-B launched.

1976 US Viking probes photograph surface of Mars.

1977 US Voyagers 1 and 2 launched.

1978 International Ultraviolet Explorer (ESA, UK, USA) launched. US High Energy Astrophysical Observatory HEAO-2 ('Einstein Observatory') launched.

1979 Voyagers 1 and 2 visit Jupiter.

1980 1980 Voyager 1 visits Saturn.

1981 Voyager 2 visits Saturn.

1983 Infrared Astronomical Satellite (Netherlands, UK, USA) launched. ESA X-ray satellite EXOSAT launched.

1985

1986 Voyager 2 visits Uranus. Giotto (ESA) and other spacecraft intercept Halley's Comet.

1989 Voyager 2 visits Neptune. US Galileo probe to Jupiter launched. ESA Hipparchos astrometric satellite launched.

1990 1990 US Hubble Space Telescope launched. Röntgenstrahlen Satellit (X-ray satellite) ROSAT (Germany, also UK and USA) launched.

1991 US Compton Gamma Ray Observatory launched.

1992 US Extreme Ultraviolet Explorer EUVE launched.

1993 Hubble Space Telescope optics repaired. Japan Asca X-ray satellite launched.

1995 Galileo visits Jupiter. ESA Solar and Heliospheric Observatory launched. ESA Infrared Space Observatory launched.

Glossary

A

aberration of light the small, constantly varying displacement in the observed position of a star, caused by the velocity of the Earth-based observer's orbit around the Sun.

absolute brightness see *luminosity.

absolute magnitude see *magnitude.

absorption band (line) a dark band (line) superimposed on a continuous spectrum, caused by the absorption of light from an incandescent source in passing through a gas of lower temperature.

achromatic lens a lens of two or more components, designed to eliminate as far as possible the effects of *chromatic aberration.

active optics techniques for making rapid corrections in the shape of a telescopic mirror or radio dish, in response to temporary changes from its designed shape.

alidade a sighting bar in an astrolabe or other instrument.

altazimuth the mounting of a telescope or other instrument so that it may be moved independently in *altitude or *azimuth about a horizontal and a vertical axis.

altitude the angle of a celestial body above the horizontal.

angular velocity the rate of change of an angle.

annual parallax the small, constantly varying displacement in the observed position of a star, caused by the displacement of the Earth-based observer from the centre of the solar system (see also *parallax).

aphelion the position in the orbit of a planet or comet where it is furthest from the Sun.

apogee the position in the orbit of the Moon where it is furthest from Earth.

apparent brightness see *brightness.

apparent magnitude see *magnitude.

arc minute one-sixtieth of a degree (1′).

arc second one-sixtieth of an arc minute; 1/3600th of a degree (1″).

armillary sphere a skeletal sphere with graduated rings representing circles on the celestial sphere, used for instruction or observation.

asteroid a minor planet; a small body in orbit about the Sun, most being in near-circular orbits in the gap between Mars and Jupiter.

astrolabe a medieval instrument for measuring *altitudes and for computations involving movements of celestial bodies (see pages 64–7).

astronomical unit the distance of the Earth from the Sun, usually taken as half the major axis of its elliptical orbit.

astrophysics the study of celestial bodies by analysis of their light.

atmospheric refraction the bending of the path of light from a celestial body by the Earth's atmosphere.

azimuth angle in the horizontal plane, usually measured eastwards from the north point.

B

backstaff a navigator's instrument for measuring the *altitude of the Sun.

Big Bang the moment in the past when, according to certain models, the universe began to expand from an initial condensed state.

binary star two stars held together by their mutual attraction, each in orbit about their common centre of gravity.

black hole a celestial body so massive that its gravitational pull prevents light from escaping from it.

brightness the *apparent* brightness of a celestial object as seen by an observer on the Earth; for *absolute* brightness, see *luminosity.

C

CCD see *charge-coupled device.

Cepheid a pulsating *variable star whose prototype is Delta Cephei.

charge-coupled device an electronic detector sensitive over a wide range of wavelengths, in recent years replacing the photographic plate in many uses.

chromatic aberration a defect in a refracting telescope in which not all wavelengths of light are brought to the same focus, causing a coloured fringe around the image of a star.

chromosphere part of the outer gaseous layers of the Sun, briefly visible during a total eclipse.

climate a climatic zone, a belt of the Earth's surface between two parallels of *latitude.

collapse, gravitational the process whereby a star, star system or *nebula collapses into itself as a result of the mutual gravitational pull of its constituents.

collodion process a photographic process involving a solution of guncotton etc. in a mixture of alcohol and ether.

colour excess a change in the *colour index of a star; the amount of reddening suffered by the starlight in passing through interstellar dust.

colour index of a star, the difference between its *photographic magnitude and *visual magnitude.

complement of an angle, the angle which, when added to the given angle, makes one right angle of 90°.

concave of a lens, one that diverges a beam of light, being thinner at the centre than the edge, in contrast to *convex.

concentric spheres in Greek astronomy, a geometrical model formed of a nest of spheres with a common centre, used to reproduce a planet's motion.

conjunction the time when two bodies of the solar system have the same celestial *longitude, especially the apparent near approach of two planets as one overtakes the other.

convex of a lens, one that converges a beam of light, being curved in the manner of a magnifying glass, in contrast to *concave.

corona the outermost parts of the Sun's atmosphere, visible in a total eclipse.

cosmos the universe, considered as organized and lawlike rather than at the mercy of chance.

cotangent the inverse of *tangent.

crepe ring one of the inner (and relatively faint) rings of Saturn.

cross-staff an instrument used in the Middle Ages for measuring the angle between two objects.

culmination the passage of a celestial body across the observer's *meridian.

D

decan in Egyptian astronomy, one of 36 stars or groups of stars used for telling the time at night (and possibly participating as divinities in rituals).

declination angle north or south of the celestial *equator, the counterpart of *latitude on Earth.

deferent in Greek astronomy, a circle that 'carries' on its circumference either

the Sun or the centre of another circle (*epicycle).

dip of a magnetic needle swinging in a vertical plane, the angle by which it departs from the horizontal.

discrete radio source a radio source located in a precisely defined direction.

divided-lens micrometer an accessory in a *reflector, for measuring small angles; the *objective of a *refractor when divided along a diameter for the same purpose.

Doppler Effect a change in the observed frequency of light (or other radiation) caused by the motion of the body towards or away from the observer.

double star two stars in the sky so close that at first they appear to be one star.

doublet lens a lens consisting of two elements, e.g. an *achromatic lens.

dwarf star one of the commonest stars, of low mass (like the Sun), as distinct from the rarer giant stars.

E

eccentric circle in Greek astronomy, a circle that is not centred on the Earth.

eccentricity of an ellipse, a measure of how much the foci depart from the centre.

eclipsing binary a close *binary star in which at least one star periodically eclipses its companion, usually also a *spectroscopic binary.

ecliptic the observed annual path of the Sun in a *great circle around the sky.

electromagnetic radiation a flow of energy (for example, light and radio waves) produced when electrically charged particles are accelerated.

electron a stable elementary particle that is a constituent of all atoms.

elementary in Aristotelian cosmology, the region below the Moon, occupied by the four elements (earth, water, air, and fire).

elongation the angle separating a planet from the Sun; the angle planet–Earth–Sun.

emission band (line) a band (line) in the *spectrum of a glowing gas under low pressure.

ephemeris a table of (daily) positions of a heavenly body.

epicycle in Greek astronomy, a small circle whose centre moves on a *deferent and which carries a planet or another small circle.

equant point in Greek astronomy, the seat of uniform angular motion.

equator, celestial the circle in which the heavenly sphere is cut by the plane of the Earth's equator.

equatorial a form of mounting for a telescope or other instrument in which one axis is directed to the celestial North or South Pole.

equinox one of the two occasions in the year (in March and September) when the Sun crosses the celestial *equator.

extragalactic nebula a term used early in the twentieth century for *galaxy.

F

faculae active regions in the upper part of the *photosphere of the Sun that appear unusually bright.

fixed star in Greek astronomy, a star that always maintains its position relative to the other stars, in contrast to a planet; a star, in the modern sense of the term.

focal length in a telescope, the distance between the *objective lens or mirror and the point where the light comes to a focus.

focus of an ellipse, two points *A* and *B* on the major axis such that *AP* + *BP* is a constant for all points *P* on the ellipse.

focus, empty in an (elliptical) planetary orbit, the focus not occupied by the Sun.

Fraunhofer lines narrow, dark *absorption lines cutting across the continuous *spectrum of the Sun or a star.

Fraunhofer spectrum the spectrum of the Sun when seen in detail.

fundamental particles the simplest forms of matter.

G

galaxy a large aggregation of stars, star clusters, and other objects, analogous to our Galaxy or Milky Way system.

gamma rays electromagnetic radiation with the shortest wavelengths (less than about 10^{-12} metres).

Giant Branch a region of an *H–R diagram populated by large and highly luminous stars.

globular cluster a spherical system of hundreds of thousands of stars.

graduated marked with divisions (of an arc or circle in an instrument for measuring angles).

great circle in the sky, a circle centred on the observer, dividing the sky into two halves.

H

H–R diagram a graph correlating the *luminosities and *spectral types of stars, developed independently by Hertzsprung and Russell.

heliacal rising the first reappearance of a star, star cluster or planet in the dawn sky, after some weeks lost in the glare of the Sun.

heliacal setting the last appearance of a star, star cluster or planet in the

evening sky, before some weeks lost in the glare of the Sun.

heliometer a telescope with a divided object-glass micrometer for the accurate measurement of angles.

Hertzsprung–Russell diagram see *H–R diagram.

hippopede the planetary path in the form of a figure-of-eight, generated by two of a nest of *concentric spheres.

hour circle on an astrolabe, a circle to indicate time by the division of day and of night each into twelve equal parts.

HST Hubble Space Telescope.

I

impetus the theory developed in the fourteenth century to provide a cause for the motion of projectiles.

inclination the angle between the *ecliptic and the plane of a planet's orbit.

inequality in the study of lunar and planetary orbits, a departure from a simple pattern.

inertia in Newtonian mechanics, the tendency of a body at rest to stay at rest, or of a body in movement to continue to move in the same direction and with the same speed.

infrared radiation radiation at wavelengths between those of visible light and radio waves.

intercalary in a calendar, the days, weeks or months added when necessary to keep in step with the natural cycle.

intrinsic brightness see *luminosity.

inverse-square law Newton's law of gravitational attraction, in which the force diminishes with the square of the distance.

ionization a process whereby an atom or molecule loses one or more electrons.

K

kinetic energy the energy of a body by virtue of its motion, equal to the work it could do in coming to rest.

L

latitude degrees north (or south) of the Earth's equator; on the sky, the angular distance of an object north or south of the *ecliptic.

light year the distance travelled by light in one year (nearly 10^{13}km).

light-gathering power W. Herschel's term for the size of a telescopic mirror or *objective lens.

longitude on Earth, degrees east or west of a standard location (today, Greenwich); on the sky, the angle of an object measured eastwards along the *ecliptic from the spring *equinox point.

luminosity the total energy emitted by a star summed over all the wavelengths; used loosely to mean absolute *magnitude (absolute brightness, intrinsic brightness).

M

macrocosm the cosmos (the universe as a whole), in contrast to a microcosm or individual living body.

Magellanic Clouds two irregular *galaxies, visible to the naked eye in the southern sky, and our nearest extragalactic neighbours.

magnification the amount by which a telescope magnifies.

magnitude a scale for measuring the brightness of celestial objects in which a star of magnitude 1.0 is 100 times brighter than a star of magnitude 6.0; *apparent* magnitude, magnitude as observed from Earth; *absolute* magnitude, magnitude as the same star would appear if at a standard distance of 10 *parsecs.

Main Sequence the region in an *H–R

diagram occupied by the majority of the stars.

mean a precise measure of 'average'.

mechanical philosophy the doctrine that explanations of natural phenomena must be in terms of impacts between bodies.

meridian the circle in the sky passing through the celestial North and South Poles and the observer's zenith.

meteor a fragment of fragile interplanetary material, observed as a 'shooting star'.

meteorite a fragment of solid interplanetary material large enough to penetrate our atmosphere and reach the surface of the Earth.

Meteoritic Hypothesis Lockyer's hypothesis that stars originated out of solid particles rather than gas.

Metonic cycle a calendric cycle based on the near equality between 19 years and 235 lunar (*synodic) months.

microcosm a living body, seen as organized analogously to the cosmos or macrocosm.

micrometer a telescope accessory for measuring small angles.

microwave background radiation a weak radio signal, the same from all directions, interpreted as a relic of the *Big Bang.

mural quadrant a *quadrant mounted on a north–south wall and used to measure the altitudes of celestial bodies at their *culmination.

muwaqqit a timekeeper in a mosque.

N

nebula a celestial body observed as an ill-defined patch of light.

neutron a particle present in the nuclei of all atoms except hydrogen.

neutron star a highly dense star that has undergone gravitational collapse, as a result of which its core is composed primarily of neutrons.

Next Generation Telescope a modern telescope whose design features make use of the most recent technology.

NGT a *Next Generation Telescope.

nocturnal a sixteenth-century (and later) instrument for telling the time at night, especially at sea.

nova a newly appeared star (*nova stella*).

nutation the 'nodding' of the Earth's axis (period 18.6 years) caused mainly by the pull of the Moon on the non-spherical Earth.

O

objective (lens or 'object glass') the principal lens of a refracting telescope.

obliquity of the ecliptic the angle between the *ecliptic and the celestial *equator.

Olbers's Paradox the problem of explaining the observed darkness of the night sky, which contrasts with what would be expected in an infinite universe with stars (or galaxies) regularly distributed.

open cluster a loosely defined cluster of stars (such as the Pleiades), found in the disc of the Galaxy.

opposition the location of the Moon or one of the outer planets, when opposite to the Sun in the sky.

P

parabola the curve formed by the intersection of a circular cone and a plane parallel to its axis.

parallax the change in the observed position of an astronomical object caused by a change in the observer's position (see also *annual parallax).

parameter a measured quantity used in calculations.

parsec a unit of distance used in stellar astronomy, the distance of a star with an *annual parallax of 1 *arc second; 1 parsec = 3.26 *light years.

perigee the position in the orbit of the Moon where it is nearest the Earth.

perihelion the position in the orbit of a planet or comet where it is nearest the Sun.

period–luminosity relationship a correlation between the periods and mean *luminosities of *Cepheids, useful as a distance indicator.

periodic variation a regular change, especially in the orbit of the Moon or a planet or in the *luminosity of a variable star.

perturbations an irregularity in a planetary orbit caused by the attractive pull of another planet.

photographic magnitude a measure of *magnitude as recorded on a photograph.

photosphere the visible surface layer of the Sun, from which its white light is emitted.

place-value the writing of numbers in which significance is given to the positions of the component symbols (as in the arabic numerals we use today).

planet in Greek astronomy, a 'wandering' star (Sun, Moon, Mercury, etc.); since Copernicus, a major satellite of the Sun.

planetary nebula a celestial body with the light of a nebula and the disc of a planet (W. Herschel).

polar circles the Arctic and Antarctic Circles.

precession the motion of the Earth's axis about the pole of the *ecliptic in a period of 25,800 years, caused by the pull of Sun and Moon on the non-spherical Earth.

projection in the *astrolabe, the association of a point in the sky with a point on the astrolabe plate by means of lines drawn from the celestial South Pole.

prominences streamers of glowing gas visible in the outer layers of the Sun.

Q

qibla in Islamic practice, the direction towards Mecca.

quadrant an instrument with a graduated arc of 90°, usually of brass, used for measuring angles.

quadrature the Moon near its quarters when the angle Sun–Earth–Moon is a right angle, and similarly for exterior planets when they are at a *longitude 90° east or west of the Sun.

quadrivium in medieval studies, the four mathematical subjects (arithmetic, music, geometry, astronomy).

quantum mechanics, quantum theory modern atomic theory developed from the hypothesis that radiation from electrons is emitted in discrete packets or *quanta*.

quasar a 'quasi-stellar object', the optical counterpart of a strong, though extremely distant, radio source.

quintessence in Aristotelian cosmology, the (fifth) element of which the heavens are composed.

R

radial velocity velocity towards or away from the observer ('line of sight velocity').

radio source a source of radio waves from outside the solar system.

radio star a star with an unusually strong emission at radio wavelengths (originally: any *discrete radio source).

radius vector in Kepler's second law, the line from the Sun to a planet.

redshift the displacement to longer wavelengths of features in a spectrum, particularly of distant *galaxies, thought to result from the *Doppler Effect.

reduction the correction in the observed position of a heavenly body to eliminate effects due to instrumental errors, *aberration, *nutation, etc.

reflector a telescope in which a mirror is used to collect the rays of light.

refractor a telescope in which a lens (*objective) is used to collect the rays of light.

retrogression the interval during which a planet reverses its normal west-to-east motion across the sky.

right ascension the counterpart on the sky of terrestrial longitude, measured eastwards from the intersection of the *ecliptic and the celestial *equator in Aries.

RR Lyrae star a type of pulsating star similar to the regular *variable star so named.

S

secular variation a change in the orbit of the Moon or planet that accumulates indefinitely.

selective absorption the scattering of starlight by interstellar dust, with consequent changes in the colour of the star.

sextant an instrument with a graduated arc of 60°, used for measuring angles; a portable navigational instrument in which mirrors effectively double the 30° arc.

sidereal time time measured by the stars rather than by the Sun (the sidereal day is about 4 minutes shorter than the twenty-four hours of the solar day).

sine of an angle in a right-angled triangle, the length of the opposite side divided by the hypotenuse.

solar apex the position in the sky towards which the solar system is travelling (in the constellation Hercules).

solstices the dates in June and December when the Sun reaches its maximum distance from the celestial *equator and so passes overhead inhabitants in one of the *tropics.

spectral type the classification of a star by the features in its *spectrum.

spectrograph, spectroscope an instrument for studying the *spectrum of light.

spectroheliogram a photograph of the Sun in the light of one line in the spectrum (for example, the red hydrogen line), revealing *faculae and *prominences.

spectroscopic binary a *binary star whose nature is revealed by the study of its *spectrum.

spectroscopic parallax the distance of a star inferred from its *spectral type and apparent *magnitude.

spectrum the spreading of a beam of light or other radiation according to wavelength, as in the colours of the rainbow.

spherical aberration the blurring in a telescopic image formed by a spherical mirror or lens, caused by the failure of the light rays to converge to a single point.

spiral nebula a nebula of spiral form, but used sometimes as a general name for *galaxies.

Steady State a cosmological model in which the universe appears broadly the same, everywhere and at all times.

sublunary in Aristotelian cosmology, in the terrestrial region ('below the Moon').

supernova the cataclysmic explosion of a star at the end of its evolution.

synodic month the interval of time between successive *conjunctions of the Sun and Moon.

T

tangent of an angle in a right-angled triangle, the side opposite the angle divided by the side adjacent.

transit (i) of a celestial body, when it crosses the *meridian; (ii) the passage of one of the inner planets across the face of the Sun.

trigon three signs of the zodiac forming an equilateral triangle.

trivium in medieval studies, the three literary subjects (grammar, rhetoric, logic).

tropics (i) the circles in the sky marking the extreme positions north and south of the celestial *equator reached by the Sun at the *solstices; (ii) the corresponding circles on Earth.

U

ultraviolet radiation radiation at wavelengths 'beyond the violet' end of the visible *spectrum, in the range from about 10^{-8} to 10^{-7} metres.

V

variable star a star whose apparent *brightness varies significantly, either periodically or irregularly.

visual magnitude a measure of *magnitude as seen with the human eye, as distinguished from *photographic magnitude.

volvelle in Ptolemaic astronomy, a component of a cardboard replica of a planetary model, used to calculate positions or to demonstrate the Ptolemaic system.

vortex in the cosmology of Descartes, a whirlpool of matter surrounding the Earth, the Sun, etc.

W

white dwarf a small, dim star of high density, at the end of its evolution.

X

X-rays radiation at shorter wavelenths than the ultraviolet, in the range from about 10^{-11} to 10^{-8} metres.

Z

zenith the point in the sky directly overhead the observer.

zij in Islamic astronomy, a table for calculating planetary positions.

zodiac the belt around the sky in which the Sun, Moon, and principal planets are always found, conventionally divided since Antiquity into twelve equal regions or 'signs', each 30° in extent and named after a contiguous constellation.

An *asterisk before a word indicates another entry.

Further Reading

General reading

For a similar range of topics to that covered in the present work, see *The Fontana/Elsevier History of Astronomy and Cosmology* by John North (Fontana, London, and Elsevier, New York, 1995). Though minimally illustrated, this paperback is elegantly written by a leading historian of astronomy. It includes an extended bibliographical essay covering both popular and specialist writings.

One other survey of the history of astronomy, by a leading astronomer, is still of value, though somewhat dated: *A History of Astronomy*, by A. Pannekoek (publ. in Dutch in 1951; English transl., George Allen & Unwin, London, 1961; reprinted in paperback, Dover Publications, New York, 1990).

For accounts of the lives and work of all the major figures in history of astronomy (except those still living at the time of publication), see the *Dictionary of Scientific Biography*, ed. by C.C. Gillespie (16 vols, Charles Scribner's Sons, New York, 1970–80, plus later additions). In many cases the entry in the *DSB* is quite simply the best discussion available, and the work is available in most major libraries.

An in-depth study of the history of astronomy down to 1950 is contained in *The General History of Astronomy*, under the general editorship of Michael Hoskin (4 vols in 7 parts, Cambridge University Press, Cambridge, 1984–, in progress); the relevant volumes and parts are listed below in the reading lists for individual chapters.

Many history of science journals contain occasional articles in history of astronomy, but the one journal dedicated to the subject is *Journal for the History of Astronomy*, ed. by Michael Hoskin (quarterly plus annual *Archaeoastronomy* supplement, Science History Publications, 16 Rutherford Road, Cambridge CB2 2HH, UK).

An anthology of interesting essays on topics ranging over the whole period of this book is *The Great Copernicus Chase, and Other Adventures in Astronomical History*, by Owen Gingerich (Sky Publishing Corporation, Cambridge, Mass., and Cambridge University Press, Cambridge, 1992). Another anthology, this time mainly on topics in astronomy since the invention of the telescope, is *The Astronomical Scrapbook: Skywatchers, Pioneers, and Seekers in Astronomy*, by Joseph Ashbrook (Sky Publishing Corporation, Cambridge, Mass., and Cambridge University Press, Cambridge, 1984).

Chapter 1: Astronomy before History

Aveni, Anthony F., *Ancient Astronomies* (St Rémy Press, Montreal, and Smithsonian Books, Washington DC, 1993).

Aveni, Anthony F., *Empires of Time: Calendars, Clocks and Cultures* (Basic Books, New York, 1989)

Aveni, Anthony F., *Skywatchers of Ancient Mexico* (University of Texas Press, Austin and London, 1980)

Aveni, Anthony F. (ed.), *The Sky in Mayan Literature* (Oxford University Press, Oxford and New York, 1992)

Aveni, Anthony F., and Gary Urton (eds), *Ethnoastronomy and Archaeoastronomy in the American Tropics* (New York Academy of Sciences, New York, 1982)

Burl, Aubrey, *Prehistoric Astronomy and Ritual* (Shire Publications, Princes Risborough, Aylesbury, HP17 9AJ, UK, 1983)

Hodson, F. R. (ed.), *The Place of Astronomy in the Ancient World* (Oxford University Press, for the British Academy, London, 1974)

Ruggles, C. L. N., *Astronomy in Prehistoric Britain and Ireland* (Yale University Press, New Haven, 1997)

Ruggles, C. L. N. (ed.), *Archaeoastronomy in the 1990s* (Group D Publications, 81 Park Road, Loughborough, LE11 2HD, UK, 1993)

The *Archaeoastronomy* supplement to *Journal for the History of Astronomy* is devoted to the subject matter of this chapter.

Chapter 2: Astronomy in Antiquity

Clagett, Marshall, *Ancient Egyptian Science*, vol. 2: *Calendars, Clocks, and Astronomy* (American Philosophical Society, Philadelphia, 1995)

Crowe, Michael J., *Theories of the World from Antiquity to the Copernican Revolution* (Dover Publications, New York, 1990)

Heath, Thomas L., *Greek Astronomy* (Dent, London, 1932; reprinted by Dover Publications, New York, 1991)

Hodson, F. R. (ed.), *The Place of Astronomy in the Ancient World* (Oxford University Press, for the British Academy, London, 1974)

Neugebauer, Otto, *The Exact Sciences in Antiquity* (2nd edn, Yale University Press, Providence, RI, 1957; reissued by Dover Publications, New York, 1969)

Pedersen, Olaf, [and Mogens Pihl], *Early Physics and Astronomy* (Macdonald, London, and American Elsevier, New York, 1974; 2nd edn, Cambridge University Press, Cambridge, 1993)

Thurston, Hugh, *Early Astronomy* (Springer-Verlag, New York, 1993)

Van der Waerden, B. L., *Science Awakening II: The Birth of Astronomy* (Noordhoff International, Leiden)

Van Helden, Albert, *Measuring the Universe: Cosmic Dimensions from Aristarchus to Halley* (University of Chicago Press, Chicago and London, 1985)

See also *The General History of Astronomy*, vol. 1: *Antiquity to the Renaissance*, Part A: *Astronomy in Antiquity* (Cambridge University Press, Cambridge, in preparation); and the five extended essays dealing with astronomy in Egypt, Mesopotamia and India, in vol. xv of *Dictionary of Scientific Biography*, ed. by C. C. Gillispie (Charles Scriber's Sons, New York, 1978)

Chapter 3: Islamic Astronomy

King, David A., *Astronomy in the Service of Islam* (Variorum, Aldershot, 1993)

King, David A., and George Saliba, *From Deferent to Equant* (New York Academy of Sciences, New York, 1987)

Saliba, George, *A History of Arabic Astronomy: Planetary Theory during the Golden Age of Islam* (New York University Press, New York and London, 1994)

See also *The General History of Astronomy*, vol. 1: *Antiquity to the Renaissance*, Part B: *Astronomy in the Middle Ages and the Renaissance* (Cambridge University Press, Cambridge, in preparation)

Chapter 4: Medieval Latin Astronomy

Crowe, Michael J., *Theories of the World from Antiquity to the Copernican Revolution* (Dover Publications, New York, 1990)

Gingerich, Owen, *The Eye of Heaven: Ptolemy,*

Copernicus, Kepler (American Institute of Physics, New York, 1993)

Grant, Edward, *Planets, Stars and Orbs: Medieval Cosmology, 1200–1687* (Cambridge University Press, Cambridge, 1993)

Lindberg, David C. (ed.), *Science in the Middle Ages* (The University of Chicago Press, Chicago and London, 1978)

Pedersen, Olaf, [and Mogens Pihl], *Early Physics and Astronomy* (Macdonald, London, and American Elsevier, New York, 1974; 2nd edn, Cambridge University Press, Cambridge, 1993)

Rosen, Edward, *Three Copernican Treatises* (Columbia University Press, New York, and Oxford University Press, London, 1939; reprinted Dover Publications, New York, 1959, Octagon Books, New York, 1971)

Van Helden, Albert, *Measuring the Universe: Cosmic Dimensions from Aristarchus to Halley* (University of Chicago Press, Chicago and London, 1985)

Westman, Robert S. (ed.), *The Copernican Achievement* (University of California Press, Berkeley and Los Angeles, 1975)

See also *The General History of Astronomy*, vol. 1: *Antiquity to the Renaissance*, Part B: *Astronomy in the Middle Ages and the Renaissance* (Cambridge University Press, Cambridge, in preparation)

Chapter 5: From Geometry to Physics: Astronomy Transformed

Aiton, Eric J., *The Vortex Theory of Planetary Motions* (Macdonald, London, and American Elsevier, New York, 1972)

Caspar, Max, *Kepler* (2nd English edn, Dover Publications, New York, 1993)

Drake, Stillman, *Discoveries and Opinions of Galileo* (Doubleday, Garden City, NY, 1957)

Drake, Stillman, *Galileo at Work: His Scientific Biography* (University of Chicago Press, Chicago and London, 1978)

Galileo, *Dialogue Concerning the Two Chief World Systems*, transl. by Stillman Drake (University of California Press, Berkeley and Los Angeles, 1953; rev. edn, 1967)

Galileo, *Sidereus Nuncius, or The Sidereal Messenger*, transl. by Albert Van Helden (University of Chicago Press, Chicago, 1989)

Jardine, Nicholas, *The Birth of History and Philosophy of Science* (Cambridge University Press, Cambridge, 1984)

Sharratt, Michael, *Galileo, Decisive Innovator* (Blackwell, Oxford, 1994; now published by Cambridge University Press)

Stephenson, Bruce, *Kepler's Physical Astronomy* (Spring Verlag, New York, 1987; paperback, Princeton University Press, Princeton, NJ, 1994)

Stephenson, Bruce, *The Music of the Heavens: Kepler's Harmonic Astronomy* (Princeton University Press, Princeton, NJ, 1994)

Thoren, Victor E., *The Lord of Uraniborg: A Biography of Tycho Brahe* (Cambridge University Press, Cambridge, 1990)

Van Helden, Albert, *Measuring the Universe: Cosmic Dimensions from Aristarchus to Halley* (University of Chicago Press, Chicago and London, 1985)

Westfall, Richard S., *Essays on the Trial of Galileo* (Vatican Observatory, 1989; distributed by The University of Notre Dame Press, Notre Dame, Indiana)

Wilson, Curtis, *Astronomy from Kepler to Newton: Historical Studies* (Variorum, Aldershot, 1989)

See also *The General History of Astronomy*, vol. 2: *Planetary Astronomy from the Renaissance to the Rise of Astrophysics*, Part A: *Tycho Brahe to Newton*, ed. by R. Taton and C. Wilson (Cambridge University Press, Cambridge, 1989)

Chapter 6: Newton and Newtonianism

Armitage, Angus, *Edmond Halley* (Nelson, London, 1966)

Bennett, J.A., *The Divided Circle: A History of Instruments for Astronomy, Navigation and Surveying* (Phaidon Christie's, Oxford, 1987)

Christianson, Gale E., *In the Presence of the Creator: Isaac Newton and his Times* (Free Press, New York, 1984)

Coyne, G.V., M. Heller and J. Zycinski (eds), *Newton and the New Direction in Science* (Vatican Observatory, 1988; distributed by Libreria Editrice Vaticana, Vatican City)

Grant, Robert, *History of Physical Astronomy* (Henry G. Bohn, London, 1852)

Grosser, Morton, *The Discovery of Neptune* (Harvard University Press, Cambridge, Mass., 1962)

King, Henry C., *The History of the Telescope* (Charles Griffin, London, 1955)

Koyré, Alexandre, *Newtonian Studies* (Harvard University Press, Cambridge, Mass., 1965)

Ronan, Colin A., *Edmond Halley: Genius in Eclipse* (Doubleday, Garden City, NY, 1969; Macdonald, London, 1970)

Westfall, Richard S., *Never at Rest: A Biography of Isaac Newton* (Cambridge University Press, Cambridge, 1980)

See also *The General History of Astronomy*, vol. 2: *Planetary Astronomy from the Renaissance to the Rise of Astrophysics*, Part A: *Tycho Brahe to Newton*, and Part B: *The Eighteenth and Nineteenth Centuries*, ed. by R. Taton and C. Wilson (Cambridge University Press, Cambridge, 1989, 1995)

Chapter 7: The Astronomy of the Universe of Stars

Armitage, Angus, *William Herschel* (Thomas Nelson, London, 1962)

Buttmann, Günther, *The Shadow of the Telescope: A Biography of John Herschel* (Lutterworth Press, Guildford and London, 1974)

Crowe, Michael J., *Modern Theories of the Universe from Herschel to Hubble* (Dover Publications, New York, 1994)

Ferris, Timothy, *Coming of Age in the Milky Way* (The Bodley Head, London, 1988)

Hoskin, Michael, *Stellar Astronomy: Historical Studies* (Science History Publications, Cambridge, 1982)

Hoskin, Michael, *William Herschel and the Construction of the Heavens* (Oldbourne, London, and American Elsevier, New York, 1963)

King, Henry C., *The History of the Telescope* (Charles Griffin, London, 1955)

Lubbock, Constance A. (ed.), *The Herschel Chronicle: The Life-story of William Herschel and his Sister Caroline Herschel* (Cambridge University Press, Cambridge, 1933)

Whitney, Charles A., *The Discovery of Our Galaxy* (Angus & Robertson, London, 1971)

See also *The General History of Astronomy*, vol. 3: *Stellar Astronomy, Instrumentation and Institutions from the Renaissance to the Mid-nineteenth Century* (Cambridge University Press, Cambridge, in preparation)

Chapter 8: The Message of Starlight: The Rise of Astrophysics

Berendzen, Richard, Richard Hart and Daniel Seeley, *Man Discovers the Galaxies* (Neale Watson Academic Publications, New York, 1976)

Christianson, Gale E., *Edwin Hubble: Mariner of the Nebulae* (Farrar, Straus & Giroux, New York, 1995)

Clerke, Agnes C., *A Popular History of Astronomy During the Nineteenth Century* (Adam & Charles Black, Edinburgh and London, 1st edn 1885, 2nd edn 1887, 3rd edn 1893, 4th edn 1902)

Clerke, Agnes C., *The System of the Stars* (Longmans, Green, London, 1890; 2nd edn, Adam & Charles Black, London, 1905)

Crowe, Michael J., *Modern Theories of the Universe from Herschel to Hubble* (Dover Publications, New York, 1994)

Ferris, Timothy, *Coming of Age in the Milky Way* (The Bodley Head, London, 1988)

Hearnshaw, J.B., *The Analysis of Starlight: One Hundred and Fifty Years of Astronomical Spectroscopy* (Cambridge University Press, Cambridge, 1986)

Hermann, Dieter B., *The History of Astronomy from Herschel to Hertzsprung* (Cambridge University Press, Cambridge, 1984)

Hubble, Edwin, *The Realm of the Nebulae* (Yale University Press, New Haven, Conn., 1936; reprinted by Dover Publications, New York, 1958)

Krisciunas, Kevin, *Astronomical Centers of the World* (Cambridge University Press, Cambridge, 1988)

Lang, Kenneth R., and Owen Gingerich (eds), *A Source Book in Astronomy and Astrophysics, 1900–1975* (Harvard University Press, Cambridge, Mass., and London, 1979)

Leverington, David, *A History of Astronomy from 1890 to the Present* (Springer, London, 1995)

North, J.D., *The Measure of the Universe: A History of Modern Cosmology* (Oxford University Press, Oxford, 1965; Dover Publications, New York, 1990)

Paul, Erich Robert, *The Milky Way Galaxy and Statistical Cosmology, 1890–1924* (Cambridge University Press, Cambridge, 1993)

Osterbrock, Donald E., *James E. Keeler: Pioneer American Astrophysicist* (Cambridge University Press, Cambridge, 1984)

Osterbrock, Donald E., John R. Gustafson and W. J. Shiloh Unruh, *Eye on the Sky: Lick Observatory's First Century* (University of California Press, Berkeley and Los Angeles, 1988)

Smith, Robert, *The Expanding Universe: Astronomy's 'Great Debate' 1900–1931* (Cambridge University Press, Cambridge, 1982)

Struve, Otto, and Velta Zebergs, *Astronomy of the 20th Century* (Macmillan, New York and London, 1962)

See also *The General History of Astronomy*, vol. 4: *Astrophysics and Twentieth-century Astronomy to 1950*, ed. by Owen Gingerich (Cambridge University Press, Cambridge; Part A, 1984, Part B, in preparation)

Chapter 9: Astronomy's Widening Horizons

Edge, David O., and Michael J. Mulkay, *Astronomy Transformed: The Emergence of Radio Astronomy in Britain* (Wiley-Interscience, New York, 1976)

Hey, J.S., *The Evolution of Radio Astronomy* (Elek Science, London, 1973)

Hirsch, Richard F., *Glimpsing an Invisible Universe: The Emergence of X-ray Astronomy* (Cambridge University Press, Cambridge, 1983)

Lang, Kenneth R., and Owen Gingerich (eds), *A Source Book in Astronomy and Astrophysics, 1900–1975* (Harvard University Press, Cambridge, Mass., and London, 1979)

Lightman, Alan, and Roberta Brawer, *Origins: The Lives and Worlds of Modern Cosmologists* (Harvard University Press, Cambridge, Mass., 1990)

Needel, Allan (ed.), *The First 25 Years in Space* (Smithsonian Institution Press, Washington, 1989)

Smith, R.W., *The Space Telescope: A Study of NASA, Science, Technology and Politics* (Cambridge University Press, Cambridge, 1989)

Sullivan, W.T. III (ed.), *The Early Years of Radio Astronomy* (Cambridge University Press, Cambridge, 1984)

Verschuur, Gerrit L., *The Invisible Universe: The Story of Radio Astronomy* (The English Universities Press, London, and Springer-Verlag, New York, 1974)

Picture acknowledgements

Every effort has been made to obtain permission to use copyright material; the publishers apologise for any errors or omissions and would welcome these being brought to their attention.

ii-iii, 18–19, Luke Dodd/Science Photo Library; iii (inset), 142, 148–149, 165, 205, from J Hevelius: *Machinae Coelestis Pars Prior*, 1673, by permission of the Syndics of Cambridge University Library; 3, Werner Forman Archive/Smithsonian Institution; 5, Cambridge University Collection of Air Photographs, copyright reserved; 7, 37, 239, 241, 242, 244, 245, 254b, 262t, 288b, 289r, 291, 292, 311l, 315r, 320, 331l,b, 332, 341, 342, Private Collection; 8, 268, 313, courtesy of Professor Owen Gingerich; 10–11, photograph by Chris Jennings; 12, South American Pictures; 13, courtesy of Professor A.F. Aveni; 15, Sächsische Landesbibliothek, Dresden; 16–17, Werner Forman Archive; 20, from A.M Stephen: *Hopi Journal*, Part II, Map 4, 1936, by permission of the Syndics of Cambridge University Library; 22, 23, 27, copyright British Museum; 25, 31 (from P. Apianus: *Cosmographia*), 60, 77 (from *Urania Ptolomaeus*, 1538), 88 (from *Astronomicum Caesareum*, fo. FII), 89 (from *Regiomontanus: Kalendar*), 92, 109 (from J A Magini: *Novae Ephemerides*, 1582), 121 (from Johannes Kepler: *Tabulae Rudolphinae*), 169r, collection of Professor Owen Gingerich; 26, Yale Babylonian Collection (YBC 7289); 30, Biblioteca Estense Universitaria, Modena; 32, 177, from Francesco Barozzi: *Cosmographia in Quatvor Libros*, 1585; 33, from William Cunningham: *The Cosmographical Glasse*, 1559; 42, Biblioteca Nazionale Marciana, Venezia; 45, 95, 143t, copyright Royal Observatory, Edinburgh; 46, from Peurbach: *Novae Theoricae Planetarum* (1460) 1482, by permission of the Syndics of Cambridge University Library; 48, British Library, London (Oriental and India Office Collections); 49, from John Thomson: *Illustrations of China and its Peoples*, 1873, by permission of the Syndics of Cambridge University Library; 51, 54, Istanbul University Library; 57, Sonia Halliday/Istanbul University Library; 59, Forschungs- und Landesbibliothek Gotha/Foto-Atelier Louis Held/Renno; 63 (MS Marsh 139, fo. 16v), 71 (MS Canon. Class. Lat. 257, fo. 1v), The Bodleian Library, University of Oxford; 64l, 80, 202, Museum of the History of Science, Oxford/photo by Paul Freestone; 64r, Flammarion-Giraudon/Bibliothèque de l'Arsenal, Paris; 65, The Warden and Fellows of Merton College, Oxford/photo by Thomas-Photos, Oxford; 69, Royal Library, Copenhagen (MS GKS 277, 2°, fo. 143r); 72, Michael Holford; 73, The Wellcome Institute; 74, Biblioteca Estense, Modena (MS Lat 209 f 9v); 75, from Hartmann Schedel: *Liber Chronicarum*, 1493; 76, from Gregor Reisch: *Margarita Philosophica*, 1504, by permission of the Syndics of Cambridge University Library; 78, from Petrus de Alliaco: *Concordancia Astronomie cum Theologia*, 1490; 79t, The Metropolitan Museum of Art, Robert Lehman Collection, 1975 (1975.1.31). Copyright 1981 by the Metropolitan Museum of Art; 79b (Harl Ms 4940, fo. 28), 82l (Ms Cotton Nero D. vii, fo. 20r), 131 (Maps C.6.c.2), 145 (C.112.c.3, from William Gilbert: *De Magnete*), by permission of the British Library, London; 81, from J Werner & P Apianus: 'Introductio Geographia' in *Doctissimas in Verneri Anno-*

tationes, 1533; 82–3, copyright Bibliothèque Royale Albert 1er, Brussels (Ms IV.111, fo. 13v); 84, courtesy of the University of St Andrews, School of Physics and Astronomy; 85t, from Petrus Apianus: *Cosmographia, per Gemmam Phrysum...Restituta*, 1539; 85b, 143b, 174, 219, courtesy of the Whipple Museum of the History of Science, University of Cambridge; 87, Edimedia/Bibliothèque Nationale; 91l, Bayer. Staatsbibliothek, Munich; 91r, Muzeum Okregowe w Toruniu (District Museum, Torun); 96, from Copernicus: *De Revolutionibus*, by permission of the Syndics of Cambridge University Library; 98, Gavnø Godskontor (Gavnø Castle), Denmark; 100t, from Tycho Brahe: *De Stella Nova*, 1573; 100b (from J Bayer: *Uranometria*, 1603), 129 (from C Scheiner: *Rosa Ursina*, 1626–30), 187 (MS Herschel W.5/5, no. 3), 188 (from Herschel's *Collected Works*, Vol. 1, 1912), 209, 234, 237, 265 (Schwabe Notebook 2, 1826, p.274), 288t (RAS Photos 2, No C4/13), courtesy of the Royal Astronomical Society Library; 102 (from Regiomontanus: *De Cometae Magnitudine... Problemata XVI*, 1531), Houghton Library, Harvard University/photo courtesy of Professor Owen Gingerich; 103, from the Department of Prints and Drawings of the Zentralbibliothek, Zurich; 104l, 105 (courtesy of the Whipple Museum of the History of Science, University of Cambridge), 106, 112, from Tycho Brahe: *Astronomiae Instauratae Mechanica*, 1598; 107 (from J Blaeu: *Gran Atlas ou Cosmographie Blaviane*, 1667), The Houghton Library, Harvard University; 110, British Library, London/Bridgeman Art Library, London; 114, Österreichische Nationalbibliothek, Wien/COD.2554, fo. 1v; 115, Fondation St-Thomas, Strasbourg; 116, 200, from Johann Kepler: *Mysterium Cosmographicum*, 1596; 117, from Johann Kepler: *Epitome Astronomiae Copernicanae*, 1618–21; 120, Crawford Collection, Royal Observatory, Edinburgh/photo courtesy of Professor Owen Gingerich; 122, 147, 171, 178–9, 181, 207, 213, National Maritime Museum, London; 127, Biblioteca Nazionale Centrale, Florence; 123, courtesy of the Istituto e Museo di Storia della Scienza, Florence/photo by Franca Principe; 124, 125, from Galileo Galilei: *Starry Messenger*; 126, from Galileo Galilei: *Dialogue on the Two Great World Systems*, 1632; 128, from Galileo Galilei: *Letters on Sunspots*; 132, from Giambattista Riccioli: *Almagestum Novum*, 1651; 135, Louvre, Paris/Giraudon/Bridgeman Art Library, London; 137, from René Descartes: *Principles of Philosophy*, 1644; 141, from Thomas Digges: *Perfit Description of the Caelestiall Orbes*, 1576; 150t, Mary Evans Picture Library; 150b, from Christiaan Huygens: *Systema Saturnium*, 1659; 151, 231, courtesy of the National Portrait Gallery, London; 153, 206, 248, by permission of the President and Council of the Royal Society; 155, Bibliothèque Nationale/Laurois Giraudon; 159, courtesy of the Trustees of the Portsmouth Estates/photo by Jeremy Whitaker; 163, from B le B de Fontenelle: *Oeuvres Diverses*, Vol. 1, 1728, by permission of the Syndics of Cambridge University Library; 166–7, courtesy of Roger Gaskell Rare Books; 168 (from C Wolf: *Histoire de l'Observatoire de Paris*, 1902), 168–9, 177, 257, courtesy of Observatoire de Paris; 170l, 183, 191, 218, Science Museum/Science & Society Picture Library; 170r (from W H Smyth: *Speculum Hartwellianum*, 1860), 208, 216,

217, 258, 262b, 263, 266, 267, 270r, 271b, 272–3, 274, 275, 277, 278, 279, 280, 281, 283, 286, 287, 295, 296, 298b, 299, 303, 305, 307, 308, 309, 311r, 315l, 319, 321, 323, 324, 325, 327, 353, courtesy of Institute of Astronomy, Cambridge (IOA); 172, from Halley: *Tabula*, 1705; 173, from *Mémoires de l'Académie Royale des Sciences*, 1760; 180, the Master and Fellows, Magdalene College, Cambridge; 184, 193, courtesy of Académie des Sciences, Paris; 186l, from David Gregory: *Astronomiae Elementa*, 1702; 186, from Christiaan Huygens: *Cosmotheoros*, 1698; 192, courtesy and copyright of the Astrophysikalisches Institute Potsdam; 194, *L'Illustration*/Sygma; 198, 199, from Johann Kepler: *De Stella Nova*, 1606; 212, from Robert Hooke: *An Attempt to Prove the Motion of the Earth*, 1674; 214, courtesy of Osservatorio Astronomico di Palermo; 222 (Add MS 3965, ff. 280r, 74r), by permission of the Syndics of Cambridge University Library; 226, reproduced by permission of Durham University Library; 227, from Thomas Wright: *An Original Theory of the Universe*, 1750; 232 (MS Gunther 36, ff. 125v), 233, 243, 252, 253, Museum of the History of Science, Oxford; 238, from a volume (c. 1771) of *Mémoires de l'Académie Royale des Sciences*; 240, from *Philosophical Transactions*, vol. 75, 1785; 246, photo courtesy of William Herschel Society, Bath, with kind permission from Observatorio Astronomico de Madrid; 247, with kind permission from Observatorio Astronomico de Madrid; 249, The Royal Borough of Kensington & Chelsea Libraries and Arts Service; 250t, from G F Chambers: *A Handbook of Descriptive Astronomy*, 1877; 250b, from John Herschel: *Results of Astronomical Observations*, 1849; 251, from N.-L. de Lacaille: *Coelum Australe*, by permission of the Syndics of Cambridge University Library; 254t, 310 (photo by Professor Owen Gingerich), courtesy of Lord Rosse; 260, Mt Wilson and Las Companas Observatories, courtesy American Institute of Physics (AIP), Emilio Segrè Visual Archives; 270l (courtesy IOA), 289l (courtesy IOA), 312, 318 (courtesy IOA), Yerkes Observatory Photo; 276, The Library, Mt Stromlo Observatory of the Australian National University (courtesy of Dr David Dewhirst); 284, 285, Lowell Observatory Photograph; 294, 344–5, Mary Lea Shane Archives of the Lick Observatory; 298t, courtesy of the Harvard College Observatory; 300, courtesy AIP, Emilio Segrè Visual Archives, Margaret Russell Edmonson Collection; 306, reproduced by permission of The Huntington Library, San Marino, California; 328 (courtesy Professor Owen Gingerich), 331r, 340, The Observatories of the Carnegie Institution of Washington; 330, 336, Palomar Observatory/California Institute of Technology; 345, European Space Agency/Science Photo Library; 347, Royal Observatory, Edinburgh/Science Photo Library; 349, Gary Ladd/Science Photo Library; 350, Roger Ressmeyer, Starlight/Science Photo Library; 351, NASA; 352, courtesy NRAO/AUI, photo by Bell Telephone Laboratories; 356, courtesy of Bruce Elsmore; 357, NRAO/AUI/Science Photo Library; 359, US Naval Reserve Laboratory; 362, courtesy of the Mullard Radio Astronomy Observatory, Cambridge; 363, Dr Seth Shostak/Science Photo Library; 354–5, Jodrell Bank/Science Photo Library

Index